More praise for *Einstein's Mistakes*

"A theoretical physicist by training, Mr. Ohanian doesn't write like one. . . . On the whole *Einstein's Mistakes* is original and fresh. . . . To see Einstein's wanderings not as the strides of a god-like genius but as the steps and missteps of a man—fallible and imperfect—does not diminish our respect for him but rather enhances it."
—Darrin M. McMahon, *Wall Street Journal*

"Ohanian explores lesser known aspects of Einstein's scientific endeavors, tracing how some of the discoveries came about, their link to works of those before him and the mistakes that he made in the process." —*Nature*

"Ohanian, a former editor of the *American Journal of Physics*, proceeds through the years, rarely missing a paper, speech, interview, or controversy. . . . Ohanian's book delves . . . deeply into physics and into Einstein's thought processes. . . . A sophisticated overview of modern physics, including more of Einstein's missteps than readers usually encounter." —*Kirkus Reviews*

"Ohanian is to be commended for trying to provide a general audience with an unflinching, comprehensive look at Einstein's science. His explanations of the physics are a cut above those of most popular books on Einstein. . . . Ohanian, to his credit, points to various problems that Einstein scholars have passed over in silence." —Michel Janssen, *Physics Today*

"With his usual clarity, Ohanian brings to life an important era in the development of physics. He underscores the human strengths and frailties of the protagonists and demonstrates how bumpy the road to scientific discovery really is. An insightful and engaging read for experts and non-experts alike."
—Kevork Spartalian, Department of Physics, University of Vermont

"[A] clear picture of how the scientific and cultural worlds of early 20th-century Europe influenced Einstein's theories about physical phenomena. The reader plunges deeply into Einstein's life and discovers the fundamental forces that shaped his achievements . . . for both scientist and non-scientist alike."   —John Boccio, Department of Physics and Astronomy, Swarthmore College

"*Einstein's Mistakes* is a superlative intellectual biography of Einstein and an excellent introduction to the complex, complicated and often counterintuitive discipline of modern physics. . . . Anyone interested in the history of science should read this fascinating and rewarding book."
—Mark Greener, *Fortean Times*

"Hans Ohanian writes that Albert often let his intuition overrule flawed proofs and shaky math. Maybe you'll feel better about your own flubs."
—*Discover*

"[Ohanian] writes fluently, with a good eye for a telling anecdote."
—Graham Farmelo, *Times Higher Education Supplement*

"A compelling portrait of a titan who stumbled his way into immortality."
—Bryce Christensen, *Booklist*, starred review

"*Einstein's Mistakes* comes from a physicist who offers a biography of Einstein by way of analyzing his failures and mistakes: as such it provides an involving survey which considers the history of physics and Einstein's mistakes as well as those of other leading scientists over the decades. An involving, moving survey."   —*Midwest Book Review*

"This is a wonderful book, entertaining, informative, full of interesting material. I think everyone will enjoy reading it."
—David Goodstein, California Institute of Technology

"This is a unique book. Einstein's scientific mistakes are so masterfully presented and thoroughly analyzed that reading the book will enrich the knowledge of the specialist and the general reader alike about fundamental aspects of Einstein's real and attempted contributions to twentieth-century physics."
—Vesselin Petkov, Concordia University, Montreal, author of *Relativity and the Nature of Spacetime*

"Ohanian makes one big mistake himself: The title of the book. This book tells us a lot more than *Einstein's Mistakes*. It is a wonderful and insightful description of how science, and all creative work, really takes place. To paraphrase Kepler (quoted by Ohanian), it reveals the wondrous and twisted roads that lead us to knowledge."

—Stephen Krashen, Professor Emeritus, University of Southern California, author of *The Power of Reading*

"A wonderfully interactive and unusual biography of Einstein: Each of Einstein's 'mistakes' challenges us to try and follow the workings of his incredible mind."

—Wolfgang Rindler, Professor of Physics, University of Texas at Dallas, author of *Essential Relativity: Special, General, and Cosmological*

Einstein's Mistakes

THE HUMAN FAILINGS OF GENIUS

Hans C. Ohanian

W. W. NORTON & COMPANY | NEW YORK | LONDON

TO MY OLDEST FRIEND

For information about permission to reproduce selections from this book, write to
Permissions, W. W. Norton & Company, Inc., 500 Fifth Avenue, New York, NY 10110

For information about special discounts for bulk purchases, please contact
W. W. Norton Special Sales at specialsales@wwnorton.com or 800-233-4830

Manufacturing by RR Donnelley, Bloomsburg

Library of Congress Cataloging-in-Publication Data

Ohanian, Hans C.
Einstein's mistakes : the human failings of genius / Hans C. Ohanian.
p. cm.
Includes bibliographical references and index.
ISBN 978-0-393-06293-9 (hardcover)
1. Physics—History. 2. Physics—Philosophy. 3. Science—History.
4. Einstein, Albert, 1879–1955. I. Title.
QC7.O33 2008
530.09—dc22
2008013155

ISBN 978-0-393-33768-6 pbk.

W. W. Norton & Company, Inc., 500 Fifth Avenue, New York, N.Y. 10110
www.wwnorton.com

W. W. Norton & Company Ltd., Castle House, 75/76 Wells Street, London W1T 3QT

1 2 3 4 5 6 7 8 9 0

ERRORS ARE THE PORTALS OF DISCOVERY.

—James Joyce, *Ulysses*

Contents

Preface

This book is a forensic biography that dissects Einstein's scientific mistakes. It deals only tangentially with the mistakes in his personal life, which have already been sufficiently exposed by some of his biographers. And it does not deal at all with the mistakes that misguided souls imagine they perceive in his theories of special and general relativity. It focuses, instead, on the missteps that Einstein made in his search for his theories and the misconceptions he had about some of the subtleties of his own creations. These mistakes in his scientific work are less well known than the mistakes in his personal life, but they are ultimately more important. As Einstein himself said, "What is essential in the life of a man of my kind lies in *what* he thinks and *how* he thinks, not in what he does or suffers."

The several years I spent exploring Einstein's mistakes were an exciting and enjoyable experience for me. Not, I hope, because of *Schadenfreude* (roughly translated as "gloating," but gloating is a behavior, whereas *Schadenfreude*, literally "joy of harm," is an emotion; it is an almost uniquely German word, with an exact equivalent only in Chinese). But, rather, because these mistakes made Einstein appear so much more human. They occasionally brought him down from the Olympian heights of his great discoveries to my own level, where I could imagine talking to him as a colleague, and maybe bluntly say, in the give-and-take of a friendly discussion among colleagues, "Albert, now *that* is really stupid!"

For my research on Einstein's mistakes, I relied on primary sources, that is, his own books, papers, lectures, and letters (many of them printed in *The Collected Papers of Albert Einstein*). However, for the biographical and historical background, I relied more heavily on secondary sources, such as the excellent and well-balanced biography by Albrecht Fölsing.

The translations from German to English are my own, because I found that existing translations are often unreliable. In casual remarks and in letters, Einstein often used colorful idiomatic expressions, and translators not thoroughly familiar with the German vernacular have taken pratfalls over these (*traduttori, traditori*, say the Italians). Two examples will illustrate the point. Einstein's famous pronouncement "Raffiniert ist der Herrgott aber boshaft ist Er nicht" is commonly translated as "The Lord is subtle, but not malicious," and Abraham Pais even used this translation as the title for his biography of Einstein, '*Subtle is the Lord . . .*'. As a German speaker, Pais should have known better. The German word *raffiniert* has a rather negative connotation; its correct translation is "cunning" or "crafty," and thus, "The Lord is cunning, but not malicious." Another amusing example is the comment that Einstein made about Marie Curie, describing her as a *Häringseele*, which is usually mistranslated as having the "soul of a herring." Despite its etymology, *Häringseele* does not mean there is something fishlike about a person; it is merely an epithet for a lean, gaunt, or meager person. Hence Einstein's comment, "Frau Curie ist sehr intelligent, aber eine *Häringseele*, das heisst arm an jeglicher Art Freude und Schmerz" translates as "Frau Curie is very intellient, but meager in emotion, that is to say, impoverished in any kind of joy or pain." Poor Marie—in France she was pilloried for her affair with Paul Langevin, and in the English-speaking world she was filleted as a fish.

I thank several friends and colleagues who read the manuscript and provided helpful comments. Also, I thank Angela von der Lippe, my editor at W. W. Norton, who tried her best to keep me from making a fool of myself. If she did not succeed, it is entirely my own fault.

Chronology of Einstein's Mistakes

1905   Mistake in clock synchronization procedure on which Einstein based special relativity.

1905   Failure to consider Michelson-Morley experiment.

1905   Mistake in "transverse mass" of high-speed particles.

1905   Multiple mistakes in the mathematics and physics used in calculation of viscosity of liquids, from which Einstein deduced size of molecules.

1905   Mistakes in the relationship between thermal radiation and quanta of light.

1905   Mistake in the first proof of $E = mc^2$.

1906–1907   Mistakes in the second, third, and fourth proofs of $E = mc^2$.

1907   Mistake in the synchronization procedure for accelerated clocks.

1907–1915   Mistakes in the Principle of Equivalence of gravitation and acceleration.

1911   Mistake in the first calculation of the bending of light.

1913   Mistake in the first attempt at a theory of general relativity.

1914   Mistake in the fifth proof of $E = mc^2$.

1915   Mistake in the Einstein-deHaas experiment.

1915   Mistakes in several attempts at theories of general relativity.

1916   Mistake in the interpretation of Mach's principle.

1917   Mistake in the introduction of the cosmological constant (the "biggest blunder").

1919   Mistakes in two attempts to modify general relativity.

1925–1955   Mistakes and more mistakes in the attempts to formulate a unified theory.

1927   Mistakes in discussions with Bohr on quantum uncertainties.

1933   Mistakes in interpretation of quantum mechanics ("Does God play dice?").

1934   Mistake in the sixth proof of $E = mc^2$.

1939   Mistake in the interpretation of the Schwarzschild singularity and gravitational collapse (the "black hole").

1946   Mistake in the seventh proof of $E = mc^2$.

"I will resign the game"

O n Tuesday, June 24, 1969, Albert Einstein drove Donald Crowhurst into madness. Crowhurst was a participant in the single-handed around-the-globe sailboat race organized by the *Sunday Times* of London, and he was alone in his trimaran *Teignmouth Electron*, becalmed in mid-Atlantic, in the Sargasso Sea, about 700 miles southwest of the Azores. There were no eyewitnesses to his descent into madness, but we know what happened because he meticulously recorded the events in his logbook. On that fateful day, he began to read Einstein's book *Relativity, The Special and the General Theory*. The book was first published in 1917, just before Einstein became a celebrity, and it explains his theories in more or less nontechnical terms. It sold a great many copies, went through fifteen editions, was translated into a dozen languages, and is still in print. Most of these copies remained mostly unread, but—like the currently fashionable books by Stephen Hawking—they presumably added some cachet to the private libraries of intellectuals and their fellow travelers.

Crowhurst had stowed the book, and a few others, aboard his yacht to while away long boring days with no wind. He was an electrical engineer, and his training in mathematics and physics was certainly adequate for coping with this book. After reading a dozen pages, he came across a paragraph that, quite literally, blew his mind. Einstein proposed to test the simultaneity of two events—say, two lightning strokes—at two widely separated points A and B by observing the arrival of light signals from these events at the midpoint between A and B. Discussing the travel times of the light signals from the points A and B to the midpoint M, Einstein claimed:

> That light requires the same time to traverse the path $A \rightarrow M$ as for the path $B \rightarrow M$ is in reality neither a *supposition nor a hypothesis* about the physical nature of light, but a *stipulation* which I can make of my own free will in order to arrive at a definition of simultaneity.[1]

Crowhurst simply could not believe this. He knew that—because of the motion of the Earth—it was by no means self-evident that the speed of light relative to the Earth would be the same in all directions, and he thought it outrageous that Einstein would pretend to settle this question by stipulation. In his logbook he wrote:

> I said aloud with some irritation: "You can't do THAT!" I thought, "the swindler." Then I looked at a photograph of the author in later years. The essence of the man rebuked me. I re-read the passage and re-read it, trying to get to the mind of the man who wrote it. The mathematician in me could distinguish nothing to mitigate the offending principles. But the poet in me could eventually read between the lines, and he read: "Nevertheless I have just *done* it, let us examine the consequences."[2]

But Crowhurst's attempts to understand the consequences led him down a dizzying spiral into madness. For the next seven days, he wrote some 25,000 words at a furious rate, almost nonstop, taking only an occasional short break, perhaps for a nap or a bite to eat. He wrote like a man possessed, frantically filling his logbook with a string of nonsensical pseudoscientific sentences, such as:

I introduce this idea $\sqrt{-1}$ because [it] leads directly to the dark tunnel of the space-time continuum, and once technology emerges from this tunnel the "world" will end (I believe about the year 2000, as often prophesied) in the sense that we will have access to the means of "extra physical" existence, making the need for physical existence superfluous . . .

And yet, and yet—*if* creative abstraction is to act as a vehicle for the new entity, and to leave its hitherto stable state it lies within the power of creative abstraction to produce the phenomenon!!!!!!!!!!!!!! We can bring it about by creative abstraction![3]

He thought that he had discovered the means to impose his free will on the material world, and that he was thereby solving all the problems of mathematics, physics, and engineering:

Mathematicians and engineers used to the techniques of system analysis will skim through my complete work in less than an hour. At the end of that time problems that have beset humanity for thousands of years will have been solved for them.

Toward the end, he became obsessed with time, and he labeled each paragraph with the hour, minute, and second it was written, although these precise times were quite meaningless. In his writing frenzy he had forgotten to wind his chronometer; it had stopped, and he had restarted it on the basis of a rough guess of local time according to the position of the Moon in the sky. Finally, on July 1, at 11 hours, 17 minutes, 0 seconds, he wrote:

11 17 00 It is the time for your
move to begin
I have not need to prolong
the game
It has been a good game that
must be ended at the
I will play this game when
I choose I will resign the
game 11 20 40 There is
no reason for harmful

Crowhurst stopped in midsentence, at the bottom of the page. He ripped the chronometer from its bulkhead mount, and then—presumably at 11 hours, 20 minutes, 40 seconds—he went on deck and threw the chronometer and himself into the sea. His trimaran was found six days later by a passing cargo ship, adrift and abandoned, with his logbook still lying on his desk, opened to the last page of his writing. His body was never found.

ALTHOUGH EINSTEIN'S BOOK was the proximate cause of Crowhurst's descent into madness, the stress of the preceding months was undoubtedly a contributing factor. Early in the race, Crowhurst had recognized that his boat was unsuitable to do battle with the violent weather south of Cape Horn, and he had decided to win the race by fraud: he zigzagged back and forth in the South Atlantic, out of sight of the normal shipping lanes, whereas in his radio transmissions he pretended to sail past the Cape of Good Hope, across the Pacific, and around Cape Horn. This was dishonest, but it was clear evidence that Crowhurst was much more sane than the other competitors who actually did take their boats around the globe along this perilous route. One of the other competitors, the Frenchman Bernard Moitessier, who was in the lead after rounding Cape Horn, insanely chose to sail around the route a second time and head for the Pacific Islands, "where there is plenty of sun and more peace than in Europe,"[4] thereby losing the race, but winning eternal *gloire* in the eyes of the French nation.

Crowhurst kept two logbooks: one with a fake description of his pretended voyage, and one with a description of his real track and his private thoughts. From the latter, we know that his fraud troubled him, and that he went through several crises of conscience, almost confessing his guilty secret in his radio transmissions. In this tormented state of mind he was ready to be pushed over the edge by any additional mental overload—and Einstein's book gave him that push.

Crowhurst was driven into madness by Einstein's assertion that the equality of travel times for light signals in opposite directions is a matter of free will, but when Crowhurst at first exclaimed that this was not permissible and that it was a swindle, he came close to the truth. The equality of these travel times is not a stipulation made by an act of free will but a hypothesis, which is experimentally testable and is found to be valid because the laws of nature make it valid. In treating this as a stipulation,

Einstein made a conceptual mistake—he adopted the right equality, but for the wrong reason. In theoretical physics, the end does not justify the means. The theoretical physicist is not only expected to obtain the right results, but he is also expected to give us a coherent understanding on how to reach these results. Physics is at least as much about understanding as it is about doing calculations, a point that beginning physics students often fail to appreciate. Einstein's misconception about the real meaning of the equality of speeds of light signals in opposite directions was a big mistake.

And, as we will see, he made other mistakes. Einstein was a pioneer exploring unknown territory, and such pioneers rarely find the best and shortest path to their goal. Einstein's exploratory paths were often tortuous, with many erratic twists and turns. Physicists who followed Einstein into the new territories were equipped with more patience and better mathematical tools, and they built firmer and straighter roads than Einstein's meandering paths. A century of such road-building has revealed that almost all of Einstein's seminal works contain mistakes. Sometimes small mistakes—mere lapses of attention—sometimes fundamental failures to understand the subtleties of his own creations, and sometimes fatal mistakes that undermined the logic of his arguments.

But the examination of these mistakes also reveals that they often were highly fruitful. Einstein was able to develop his theories despite his mistakes, and sometimes because of them. He used his mistakes as stepping stones to reach his grand discoveries. We can say of Einstein what Arthur Koestler said of Johannes Kepler, the seventeenth-century astronomer who discovered the laws of planetary motion by tottering along a meandering path on which he stepped from mistake to mistake until he reached his goal: "The measure of Kepler's genius is the intensity of his contradictions, and the use he made of them."[5] Koestler called Kepler a sleepwalker—he intuitively knew where he wanted to go and managed to get there, without quite knowing how. Like Kepler, Einstein was a sleepwalker.

Einstein's
Mistakes

I

"A lovely time in Berne"

Einstein recollecting his life in Berne in a letter to his friend Conrad Habicht. There, in 1905, he completed five influential papers on physics, including his celebrated paper on special relativity.

A lbert Einstein published his original paper on the theory of special relativity in 1905. He was then living in Switzerland, employed at the Federal Patent Office in Berne as a patent clerk, 3rd Class. He had recently been granted Swiss citizenship, and he was the proud owner of a Swiss passport. But Einstein did not send his theory of relativity nor any of his other great works to a Swiss journal. He knew that to get his works worldwide attention, he had to publish in one of the large German journals, and so he sent all his early works to *Annalen der Physik*, the preeminent German physics journal published in Leipzig, the city then at the center of the European book trade.

Germany was the world's leading producer of books, sausage, sauerkraut, coal, canons, armor plate, church organs, hardware, glassware, toys, pencils, aspirin, dyes, cologne, ladies's stockings, playing cards, candles, and long operas. It was also the world's leader in research in physics. Just as painters converged on Paris to learn from the French masters, physicists converged on Berlin and Munich and Heidelberg to study the latest developments in theoretical and experimental physics. And they all published in *Annalen*, the largest physics journal in Germany. *Annalen* had a long and international list of subscribers, and even some of the then-small and unknown colleges and universities in the United States subscribed to it. If you visit an American university library founded during the 1800s, you are likely to find musty volumes of *Annalen* somewhere in a basement or an annex.

Einstein was Swiss by choice and patent clerk by accident. He was born in 1879 in the small city of Ulm, in the Swabian region of southern Germany. His father and mother came from Jewish merchant families of nearby towns, but they regarded themselves as Germans first and Jews second. Their own very German and modern first names, Hermann and Pauline, and the names they gave to their children, Albert and Maja, reveal their rejection of Jewish customs. They were not religious, did not attend services at the synagogue, and they happily ate pork roasts and sausages whenever they felt like it. His father took pride in not practicing Jewish rites in his home, which he regarded as relics of an ancient superstition. Instead of the birkat ha-mazon, he was wont to recite the poetry of Schiller or of Heine.

According to a tradition going back to medieval times, the Ulmers are mathematicians. How this tradition began is unclear—no native of Ulm is to be found in the ranks of world-class mathematicians. It is true, however, that Johannes Kepler, the famous sixteenth-century astronomer and mathematician who discovered the laws of planetary motion, was born in the same corner of Swabia, in the nearby town of Weil-der-Stadt, and maybe the Ulmers thought that Kepler's halo extended as far as Ulm.

If the air of Ulm carries some mathematical miasma, it did little good to Einstein. He always remained a rather mediocre mathematician. In his autobiography (or "obituary," as he liked to call it) written in his late years in Princeton, he confessed, "I had excellent teachers . . . , so I really could have gotten a profound mathematical education . . . That I neglected mathematics to a certain extent had its cause not only in that my interest in natural sciences was stronger than in mathematics, but also in the following strange experience. I saw that mathematics was split up into many specialties, each of which could absorb the short lifespan granted to us. Thus I saw myself in the position of Buridan's ass, which was unable to decide on a particular bundle of hay . . ."[1] Einstein preferred to leave any difficult mathematical labors to others, such as his mathematician friend Marcel Grossmann, during his years in Zurich and the several other mathematical assistants he later employed during his years in Berlin and Princeton, whose job it was to grind out mathematical details that Einstein found too troublesome. He called them his *Rechenpferde*, or his "calculating horses," a reference to Clever Hans the horse, a sensational equine celebrity of those days that seemed to be able to answer questions on arithmetic by tapping its hoof.

ALTHOUGH EINSTEIN DID NOT INHERIT much mathematical talent from Ulm, he did inherit several of the character traits traditionally associated with Swabians: the stubbornness and the roguish sense of humor of the Swabian peasant; an inclination to mysticism; and a habit of *grubeln*, or protracted, agonized brooding over whatever was on his mind. Such a habit of brooding is a definite asset for a theoretical physicist or mathematician, and in his late years in Princeton, when confronted with a problem, Einstein would often say, "I will a little tink,"[2] and he would light his pipe and sink into intense *grübeln*.

Mysticism may seem to be a strange trait in a scientist, but many of Einstein's pronouncements sound like the reveries of a mystic: "The most beautiful thing we can experience is the mysterious. It is the fundamental emotion that stands at the cradle of true art and true science. He who does not know it and can no longer wonder, no longer feels amazement, is as good as dead, a snuffed-out candle . . . ,"[3] and "Everyone who is seriously involved in the pursuit of science becomes convinced that a spirit is manifest in the laws of the universe, one that is vastly superior to that of man . . . ,"[4] and "I want to know how God created this world. I am not interested in this or that phenomenon, in the spectrum of this or that element. I want to know his thoughts. The rest are details,"[5] and again, "When I assess a theory, I ask myself, if I was God, would I have arranged the universe in this way?"[6]

For Einstein God was merely a figure of speech. When Einstein spoke of God, he merely meant that the universe is under the sway of absolute, pervasive, and permanent laws; and his desire to glimpse the mind of God was merely his quaint way of expressing his quest for these laws. This, of course, is the quest of all scientists. But Einstein's approach to this quest was different in that he made his profound discoveries in the manner of a mystic—he relied on inspirations, on irrational intuitive insights, and not on methodical, dispassionate logical analysis of observational and experimental facts. Such facts played little role in his thinking. Instead, he preferred to contemplate intangibles: the beauty and simplicity, the internal consistency and necessity he perceived in his theories. Einstein's mindset was attuned to the words that Beethoven used as the theme for his last string quartet: "Must it be? Yes, it must be!"

On occasion, Einstein not only ignored the observational and experimental facts, but he even denied them. Asked what would be his reaction to observational evidence against the bending of light predicted by his theory of general relativity, he answered, "Then I would feel sorry for the good Lord. The theory is correct anyway."[7] In this occasional denial of facts and in the stubborn reliance on his own inspirations and revelations, Einstein was a kind of St. Francis of physics, or St. Francis Einstein of the Daffodils, as William Carlos Williams called him in a poem.[8] Some photographers sensed this mystical spirituality in Einstein, and they succeeded in capturing this feeling in their portraits of him.

Einstein was notoriously stubborn. Nothing illustrates this better than an anecdote told by Ernst Straus, one of Einstein's mathematical assistants during his later years at Princeton. He and Einstein had finished writing a report, and they were looking for a paper clip. Opening a lot of drawers, they finally found one, but it was bent out of shape and unusable. Opening some more drawers, they found a box full of brand-new paper clips. Einstein took one of those and started to reshape it into a tool to straighten the bent clip. Straus, in astonishment, asked Einstein what he was doing, and he answered, "Once I am set on a goal, it becomes difficult to deflect me"—and he admitted to Straus that this would make a good anecdote.[9]

Einstein considered his stubbornness an important asset. He said, "All I have is the stubbornness of a mule; no, that's not quite all, I also have a nose."[10] Straus explains that this stubbornness was very important to Einstein, because he felt that the task of a scientist is to find the most important question, and then to pursue it relentlessly. He thought that scientific greatness was primarily a matter of character, the determination not to compromise or to accept incomplete answers.

From his parents, Einstein inherited the Jewish intellectual tradition, with its deep respect for learning and its penchant for rabbinical disputation and abstruse dogmatism, in the manner of Baruch Spinoza, the sixteenth-century Dutch-Jewish philosopher. From them he also inher-

Portrait from 1947, showing Einstein in a contemplative mood. *(Photo by Philippe Halsman, © Halsman Estate)*

ited the soft, mellifluous Swabian dialect and intonation, with an overabundance òf the southern German diminutives ending in "*erl*"—such as
Weiberl, Kinderl, Bilderl, Fensterl—that infested not only his speech, but also
his letters to friends and relatives. When Einstein learned English late in
life, he retained the Swabian intonation—if you want to hear what a Swabian sounds like when he tries to speak English, listen to a recording of a
speech by Einstein. In Shaw's *Pygmalion*, Professor Higgins identified the
birthplace of a bystander after listening to just a few words; the professor
could no doubt have identified Einstein's birthplace just as quickly.

And from his mother, an accomplished piano player, Albert inherited
a love for music and some measure of musical talent. He began playing
the violin as a child, and he continued playing throughout his life. After
he became a world-famous physicist, he occasionally gave public concerts,
for the benefit of Zionist causes. His favorite composers were Bach and
Mozart. His playing was more remarkable for its passion than for its art-
istry—in the opinion of one professional musician, Einstein's bowing was
like that of a woodsman sawing a log.[11] But Einstein would brook no criticism of his playing, and occasionally he displayed a hot temper when he
thought his playing was not appreciated.[12]

THE EINSTEINS MOVED to Munich soon after the birth of
Albert, or Albertle, as he was called within the family. There his father
and his uncle founded Einstein & Cie, for the manufacture of electric
machinery, especially electric generators and illumination equipment. A
revolution in electric technology was sweeping across Europe, and Einstein senior expected that his factory would profit from the development
of this new technology. The family lived in a house next to the factory.
Little Albertle grew up in Munich, attending a nearby elementary school
and then several years of high school, or the German "Gymnasium."

Einstein recalled that as a child he was much intrigued by a compass:
"A wonder of this sort I experienced as a child of 4 or 5, when my father
showed me a compass. That this needle behaved in such a fixed manner
did not fit at all with the kind of happening that could find à place in our
unconscious conceptual framework (action tied to direct 'contact'). I still
remember—or think I do—that this experience made a deep and permanent impression on me."[13] Between the ages of twelve and sixteen he
taught himself basic geometry and mathematics, including the principles

of differential and integral calculus. He also read a series of books on the natural sciences, which inspired him to become a physicist.[14]

It is said of Swabians that they acquire wisdom only at an age of forty, the "Swabian age," but Albertle acquired his wisdom early. He was a precocious child, and acted the part. He was a good student, better than most in his class, but his teachers found him too full of questions, too full of himself, and too disrespectful. Albertle thought his teachers contemptible; he described his teachers in elementary school as drill masters and army sergeants, and those in high school as lieutenants.

In reaction to the thoroughly unreligious attitude of his parents, the adolescent Albertle briefly became obsessed with the religion of his forefathers. He zealously studied the Scriptures and even composed a hymn to G_d, which he would inflict on his family at all hours. But he soon recovered from this religious folly: ". . . [it] reached an abrupt end at the age of twelve. Through the reading of popular scientific books I soon reached the conviction that much in the stories in the Bible could not be true . . . Suspicion of every kind of authority grew out of this experience . . . an attitude that has never left me."[15]

EINSTEIN & CIE at first enjoyed some success, but it then came into competition with larger, more powerful corporations that pushed it out of business. This was in part because the larger corporations enjoyed better financing, but in part because Einstein senior backed the wrong horse. At the end of the nineteenth-century the advantages of AC versus DC transmission of electric power on long-distance transmission lines were hotly disputed. Einstein senior placed his bets on DC, whereas, by a little calculation, his son could have demonstrated to him that transmission of AC power, at high voltage, is much more efficient. Because of this advantage, AC gradually displaced DC, and Einstein senior was one of the losers in a typical case of the "shakeout" that invariably follows the expansion of capitalism into a new technology. Einstein senior moved his factory and his family to Pavia, in Italy, where he expected less competition. However, he decided to leave Albert behind in Munich, so he could finish his high school education there.

Albert wanted to escape from the rigid discipline of his German high school. Even many years later, in his autobiography, he complained, "It is a grave mistake to believe that the joy of seeing and searching can be pro-

moted by coercion and a sense of duty. I think that one could eliminate the
voraciousness of even a healthy beast of prey, if it were possible, by means
of a whip, to force it to devour continuously when it is not hungry."[16] He
found the military drills especially objectionable, and he dreaded the prospect of military service, which was imposed on all young German males at
age seventeen. When one of his teachers scolded him for disrespect and

In Pavia, the Einstein family
lived on the second floor of
this small-*palazzo*, in the same
rooms that almost a hundred
years earlier had been occupied by the Italian romantic
poet Hugo Foscolo.

The Einstein factory on the
outskirts of Pavia, now con-
verted into an apartment
house.

told him that it would be better for the class if Albert were to remove himself from the school, Albert took the teacher at his word, packed his bags, and escaped to Italy. There he convinced his parents to let him finish high school in Switzerland, in Aarau, near Zurich.

Albert's Swiss school was a progressive, nondenominational school, which even taught Darwin's theory of evolution, still somewhat of a novelty in those days. (As regards Darwin, the Swiss high schools of more than a hundred years ago were more enlightened than many American high schools of today, which pander to vociferous Christian sects that confuse religion with science and infest classrooms with delusional pseudoscientific theories, such as creationism and intelligent design.) In old age, Albert remembered his school fondly: "By its liberal spirit and by the austere earnestness of its teachers . . . this school made an unforgettable impression on me; by comparison with six years of schooling in an authoritarian German Gymnasium . . . I became acutely aware how much an education directed toward freedom of action and responsibility is superior to an education resting on drill, imposed authority, and ambition."[17]

But Albert's appraisal was skewed by the special dispensations that he enjoyed as a foreigner in Switzerland. Toward the end of the nineteenth-century, militarism was as rampant in Switzerland as it was in Germany. Deeply impressed by the speed and efficiency with which the Prussians had defeated the Austrians and the French in the wars of 1866 and 1870, the Swiss had adopted Prussian drill and discipline for the military training of the students in their high schools. (Even today, military training is still compulsory, and all young Swiss males are required to keep an SIG 550 assault rifle and fifty rounds of ammo at home, to protect the cuckoo clocks or whatever.) As a foreigner, Albert was excused from these compulsory military exercises, and he could spend idyllic afternoons flirting with his landlord's daughter while his Swiss classmates had to perform long hours of tiresome military drill, Prussian fashion, supplemented by cross-country marches, combat exercises, and firing practice with live ammunition.

AFTER RECEIVING HIS high school diploma in 1896, Albert enrolled as a physics student at the Federal Polytechnic Institute in Zurich (now usually called the ETH, the acronym of its unpronounce-

The main building of the Zurich Polytechnic Institute, renamed ETH in 1911, but still called the "Poly" by its students today.

The house in which Einstein lodged as a student. With Swiss precision, a tablet affixed to the house gives a list of the relevant dates: "Here resided 1896–1900 the great physicist and lover of peace ALBERT EINSTEIN, 1879–1955, a 1901 citizen of Zurich."

able German name of *Eidgenössische Technische Hochschule*). Meanwhile the Einstein factory had fallen into bankruptcy, and it became impossible for his father to defray the costs of Albert's university education. A wealthy uncle and aunt intervened and paid for Albert's years as student in Zurich. The aunt probably did not know that Albert ungraciously referred to her as the "horror aunt."[18] With his monthly allowance, Albert could afford room and board in a house near the Polytechnic, and he could afford to join his fellow students in the cafés on Bahnhofstrasse, the ritziest street in Zurich, where he spent most afternoons. He was also fond of the luxurious Grand Café Metropol (now Grand Café), on the bank of the Limmat River, where once a week he would meet with his closest friend and classmate, Marcel Grossmann. With some other friends he often went sailing on Lake Zurich, and sailing was to remain one of his favorite pastimes in later years.

Einstein was a rather erratic student, skipping many classes and barely completing the minimum of work for graduation, while devoting most of his time to independent study of more advanced topics in physics: "I played the truant a great deal and, at home, studied the masters of theoretical physics with religious zeal."[19] For exams, he relied on notes taken by Grossman. Einstein's grades were mediocre, and he made a poor impression on his professors by his frequent absences from class and by his sophomoric and disrespectful attitude. His physics professor scolded him: "You are a very clever boy, Einstein, an extremely clever boy, but you have one great fault; you'll never let yourself be told anything."[20] Einstein retaliated by addressing the professor as Herr Weber instead of the customary Herr Professor Weber, a childish show of disrespect he had also employed in school in Munich.[21] This insult must have irritated Herr Professor Weber all the more because he had initially been favorably impressed by Einstein and had invited him to attend his lectures even before enrollment at the Polytechnic. On one occasion, by failing to follow the prescribed procedures, Einstein triggered the explosion of some laboratory equipment and injured his hand, an accident that confirmed his reputation as a rash know-it-all.

During his student years, Einstein resigned his German citizenship and applied for Swiss citizenship. The citizenship was granted after investigation of his character by a police detective, who reported in Einstein's favor, saying that he was "a very zealous, diligent, and extremely respectable man,"[22] and that, besides, he was abstinent.

Grand Café Metropol, now called Grand Café, one of Einstein's favorite haunts in his student days in Zurich.

If the detective had dug a bit deeper, the report might have been less favorable. Einstein was abstinent from drink, but not from sex. He had debauched the only female student in his class, Mileva Marič, a bright young woman from Serbia. Like Albert, Mileva had developed the ambition to become a physicist at an early age, and she had enrolled at the Polytechnic because it was one of the few universities in which women were admitted to the study of physics. Distracted by the torrid, clandestine affair, she failed her exams; and distraught by the pregnancy that followed hard upon, she failed in a second, and final, try at her exams. This ended her career in physics. Einstein sent her home to her parents in Serbia, where she gave birth to a daughter. Albert wrote to Mileva, making a show of his fatherly love and interest: "I love her so much and I don't even know her yet! Couldn't she be photographed once you are totally healthy again? Will she soon be able to turn her eyes to something?"[23] But he never made any effort to see this beloved daughter. Soon after, the infatuated Mileva returned to Switzerland leaving the daughter behind, and the infant child disappeared from the historical record without a trace—maybe she died at a young age, maybe she grew up in some corner of Serbia, and maybe her descendants still live there today. The secret of Albert's illegitimate daughter was well kept by his family; most of the letters referring to the daughter were destroyed, and the secret was not revealed until thirty years after Albert's death, when some remaining letters between Albert and Mileva (or "Jonzerl" and "Schnoxl," as they called each other) were published.

AFTER HIS GRADUATION in 1900, Einstein hoped for some academic post, perhaps as assistant to one of his professors. But in view of his poor record as student, none of the professors at the Polytechnic had any interest in employing Einstein. He sent dozens of postcards to professors throughout Germany, Austria, and Italy offering his services, but got no takers. "I will soon have honored all physicists from the North Sea to the southern extreme of Italy with my offers,"[24] he reported in a letter to Mileva. The financial aid from his rich uncle ended with graduation, and he spent a miserable two years trying to make a living as a part-time private tutor.

For a while he tutored in Schaffhausen, near the German border. Then he moved to Berne and earned a meager and intermittent income by giving private lessons in physics and mathematics. He was saved from starvation by his friend Grossmann, who helped him to secure the position of patent clerk at the federal patent office.

Einstein was relieved to land a job with a steady income. In contrast to the hardships of the preceding years, the modest salary of this job looked like riches to him. "I am doing well," he wrote in a letter to a friend, "I am an honorable federal ink pisser with a regular salary. Besides, I ride my old mathematico-physical hobbyhorse and saw on my violin."[25] Einstein found that his work at the patent office was not excessively demanding, and that at the end of a normal working day, he still had the time and the appetite for scientific research. On a slow day he could even do some of his research at his desk at the patent office, hiding his notes whenever his supervisor entered the room.

Much later, he ruminated on the advantages of this job. He reckoned that a practical profession "is a blessing for a man of my kind. The academic career places a young man under a kind of compulsion to produce scientific papers in impressive quantities—a temptation to superficiality . . . Most practical professions are of such a kind that a man of ordinary

The windows on the third floor of this house at Kramgasse 49 in Berne open into the living room of the apartment in which Einstein and Mileva lived in 1905. It is now a museum. Einstein later said, "The special relativity theory came into being at Kramgasse 49 in Berne and the beginnings of general relativity theory likewise in Berne." *(Courtesy of Bern Tourism, Bern, Switzerland)*

talents is capable of achieving what is required of him . . . If he has deeper scientific interests, he can burrow into his favorite problem in parallel to his professional duties. He does not have to be depressed by the fear that his endeavors will remain fruitless."[26]

Conversations with two friends he made at Berne, Conrad Habicht and Maurice Solovine, provided a valuable stimulus for his research. With them, Einstein formed a three-man discussion group, which they called the "Olympia Academy." Years later, he recalled the Olympia Academy with pleasure: "It surely was a lovely time in Berne, when we conducted our merry Academy, which surely was less childish than those highly respected ones, which I later was to come to know closely."[27] They met frequently for readings and discussion of various books on physics and on philosophy, some of which had a deep influence on Einstein's ideas about relativity.

After Einstein began working in the patent office, he married Mileva.

They lived in several rented apartments in Berne, including an apartment on the second floor of Kramgasse 49, near the center of the city and within easy walking distance from work. Here his first son was born in 1904, and here also were born several of his great ideas, including relativity, in 1905.

EINSTEIN'S FIRST PUBLICATION in *Annalen* was a rather insignificant paper on capillarity in 1901, which in a slightly modified form he also used for a thesis he presented to the University of Berne. Then he published several papers on thermodynamics, none of which were of lasting importance. Today, all of Einstein's early papers make rather boring reading, as do most other papers published in the volumes of *Annalen* of those years.

But in 1905, his miracle year, Einstein reached a mental flash point that triggered a frenetic burst of creativity. He worked like a man possessed, displaying fits of furious mental activity, which often left him in a state of desperate confusion, but finally yielded sharp insights. His intense mental efforts left him totally exhausted, and when he finished his work he collapsed into his bed for two weeks.[28]

Between March and September of that year, he completed no less than five papers that were to exert a permanent and profound influence on physics: first the description of light as a stream of quanta of light (later called *photons*) and the explanation of the photoelectric effect; then a determination of the size of molecules; then the theory of Brownian motion and how this provided conclusive evidence for the existence of atoms; then the theory of relativity; and finally the relation between energy and mass. The first of these was to lead to the award of the Nobel Prize; the second and third were to convince many skeptics that atoms and molecules were fact, not fiction;[29] the fourth was to cast Einstein in the role of inventor of relativity; and the fifth was to become famous for the formula $E = mc^2$ and its infamous connection to nuclear energy and the atomic bomb.

All these papers were published in volumes 17 and 18 of *Annalen*. These volumes have now become collector's items; a copy fetches over $10,000 at auction, and libraries that do not keep the volumes under lock and key are liable to lose them to thieves. If you examine volumes 17 and 18, you will find that they contain nothing much else that is remarkable—the other works contained in them have been forgotten, and so have most of their authors, many of them highly respected Herr Doktor Professors at then-celebrated German universities, all overshadowed by the inventions of a patent clerk.

But if 1905 was the year of miracles, it was also the year of mistakes. As we will see, 1905 produced a larger crop of Einstein's mistakes than any other year in his long and illustrious career. Four out of the five famous papers he produced during that year were infested with flaws.

IN ESSENCE, RELATIVITY ASSERTS that uniform, unaccelerated motion is always relative. When a body is in uniform motion, this motion is meaningful only in relation to another body. You may have had a strong intuitive impression of this relativity of motion while sitting in a train, waiting for its departure. (Examples of this sort often make outsiders groan that relativists seem obsessed with trains and elevators; if you are sick of trains, you can replace the train waiting at the station by an airliner waiting at the departure gate.) Distracted by a book or a conversation, you suddenly perceive that another train is slowly sliding past your window and you have a moment of confusion: Did you fail to notice the moment of departure? Is your train rolling or is the other train? You can tell only that there is relative motion, and even if you look down at the ground, and determine which train is moving, you are merely establishing this motion relative to the ground.

This relativity of motion underlies a famous anecdote (probably apocryphal) about Einstein: He was traveling by train from Switzerland to Baden-Baden, and he asked the conductor, "When does Baden-Baden stop at this train?" Apocryphal or not, Einstein's question has been adopted as the title for a piece of sculpture consisting of a railroad dolly carrying a large clock, on display at the Museum of Modern Art in New York.

For the precise description of position and motion, physicists select a reference point and they measure distances relative to this reference point. For instance, the reference point might be the control tower of an airport, and relative to this reference point the position of, say, an aircraft can then be described by the east-west distance, the north-south distance, and the vertical distance. These distances are called the *space coordinates*, and, because we need three coordinates to pinpoint the position of the aircraft, we say that space is three dimensional. However, to describe events at the aircraft—for example, the collision of a seagull with the aircraft—we also need to specify the *time coordinate*, or the time at which the event happened. If we know the space coordinates and the time coordinate of an event, we know the where and the when of the event—and these are always the first and most basic questions that an investigator has to ask when collecting information about the event.

Physicists like to imagine that the space around the reference point is filled up with a lattice of metersticks and also an array of clocks. The metersticks measure the space coordinates of any event and the clocks measure the time coordinate. Such an imaginary lattice of metersticks in conjunction with an array of clocks is called a *reference frame*.

The three space coordinates in conjunction with the time coordinate are called the spacetime coordinates. Since spacetime comprises four coordinates, we say that it is four dimensional. This notion of four-dimensional spacetime was introduced by the mathematician Hermann Minkowski, one of Einstein's professors at the Polytechnic Institute in Zurich. In elementary school and in high school, Einstein was a good student, consistently receiving the highest notes in mathematics. But when he enrolled as a student at the Polytechnic, Einstein adopted a know-it-all attitude, and he paid little attention to Minkowski's lectures and skipped many. Minkowski described him as a "lazy dog," and, years later, upon the publication of the theory of relativity, he commented, "I really would not have believed him capable of it."[30]

After recovering from his astonishment at the unsuspected abilities of this lazy dog, Minkowski proceeded to reformulate the core of Einstein's theory in elegant mathematical language; he introduced the notion of the four-dimensional spacetime and expressed the ideas of relativity in terms of the geometry of this four-dimensional spacetime. Einstein at first objected to Minkowski's geometrical reformulation of relativity as "superfluous erudition,"[31] but he and other physicists soon recognized the advantages of Minkowski's geometrical methods, and these methods were later to play a crucial role in the development of general relativity.

Different reference frames can have different motions in relation to one another. For instance, the reference frame of an aircraft might have a speed of 900 klicks per hour relative to the reference frame of the ground (*klick* is US Army slang for "kilometer," a usage commendable for its brevity; 900 klicks per hour is about 600 miles per hour). And if the aircraft hits some turbulence, its reference frame will bounce up and down and maybe tilt sideways. Such irregularities in the motion make it awkward to use this reference frame for the description of physical phenomena, and physicists prefer to use reference frames that move uniformly, without any acceleration.

A reference frame with uniform, unaccelerated motion along a straight line is called an *inertial reference frame*. A reference frame attached to the ground is approximately an inertial reference frame, and so is any reference frame with uniform motion relative to the ground. (Actually, the reference frame attached to the ground is not exactly inertial, because the Earth rotates around its axis and also orbits around the Sun. These circular motions involve a change in the direction of the motion, which means the motion is not uniform—it is accelerated motion. However, the accelerations associated with the circular motion of the Earth about its axis and the orbital motion of the Earth around the Sun are small, and they can often be neglected, so the reference frame attached to the ground can be regarded as nearly inertial. Astronomers, who are obsessively exact, prefer a reference frame attached to the Sun, which gives a more accurate implementation of an inertial reference frame.)

AT THE END of the nineteenth-century, physicists were confronted with a puzzle regarding the speed of light. This puzzle had originated about a hundred years earlier, about 1805. It had then been recognized that light was a wave, with oscillations of very short wavelength and very high frequency (only much later was it discovered that the oscillations in a light wave consist of electric and magnetic disturbances, or electric and magnetic fields, as physicists like to call them).

The nineteenth-century physicists thought that the propagation of a light wave was analogous to the propagation of a sound wave. Sound requires a medium for its propagation—it can propagate in air, water, wood, etc. Likewise, light was thought to require a medium for its propagation, a medium permeating our immediate environment and also all of interplanetary and interstellar space, so as to permit the propagation of light everywhere. This medium was called the *ether*, a name taken from the ancient term for the stuff medieval astronomers had imagined to fill the celestial regions beyond the Moon (it is sometimes called the *luminiferous ether*, to distinguish it from the chemical compound $(C_2H_5)_2O$ used for anesthesia, also called ether). The ether was thought to be very tenuous, so it would not impede the motion of the planets, but strongly elastic, so its vibrations would yield the high speed of a light wave, 300,000 klicks per second, or about 200,000 miles per second.

The ether was assumed to be at rest relative to the Sun, and thus the speed of light was assumed to be 300,000 klicks per second with respect to the Sun, or with respect to the Solar System. Since the Earth moves at a speed of 30 klicks per second around the Sun, nineteenth-century physicists expected that in the reference frame of the Earth, a light signal traveling in the same direction as the Earth should have a reduced speed (the Earth is running away from this signal, and the signal therefore takes longer to pass by the Earth), whereas a light signal traveling in the opposite direction should have increased speed (the Earth is running head-on into this signal, and the signal therefore rushes by the Earth). It was expected that the speed of light obeys the intuitively obvious addition and subtraction rules for speeds, so relative to the reference frame of the Earth, the light signal traveling in the same direction as the Earth should have a speed of 300,000 − 30 klicks per second, and the light signal traveling in the opposite direction should have a speed of 300,000 + 30 klicks per second.

And here was the puzzle confronting the physicists of the late nineteenth century: several experiments had sought to detect this dependence of the speed of light on the speed of the Earth, and they all failed to detect any effect whatsoever—the speed of light seemed completely unaffected by the motion of the Earth.

THE MOST INGENIOUS among these experiments was contrived in 1881 by the American physicist Abraham Michelson during a sabbatical in Berlin. Michelson was born in Poland to Jewish parents who later emigrated to the United States and settled in Nevada. He began the study of physics while a midshipman at the US Naval Academy at Annapolis, a position he had gained by sheer chutzpah. When told that the quota for midshipman appointments was filled, he had traveled to Washington, DC, and parked himself in the anteroom of President Ulysses S. Grant's office at the White House. At the end of the day, Grant had agreed to talk to him, and—impressed by the persistence and initiative of this midshipman candidate—he had given him an overquota appointment.

At Annapolis, Michelson soon discovered that he was more interested in physics than in the navy, and he designed an improved apparatus for

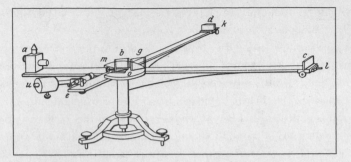

An illustration reproduced from Michelson's paper in the *American Journal of Science*, showing the interferometer he used in his experiment in Potsdam in 1881. One arm of the interferometer is parallel to the direction of motion of the Earth, the other perpendicular. The lamp (*a*) on the left of the instrument sends a light beam toward the angled, half-silvered mirror (*b*) at the center. Here the light beam splits into one part that races back and forth along the arm *bc* and another part that races back and forth along the arm *bd*.

the measurement of the speed of light by timing the round-trip of a light signal over a long "racetrack," with a rotating mirror at the far end. He improved this experiment throughout his life, and he obtained a final result of 299,796 klicks per second for the speed of light, with an uncertainty of ± 4 klicks per second (this was right on target: the modern value is 299,792 klicks per second, which is, indeed, within 4 klicks per second of Michelson's value).

On the strength of his reputation for high-precision measurements in physics, Michelson obtained a leave of absence from the navy, to go to Berlin for postgraduate study in optics. In those days, the Germans were leaders in optics and in almost all other branches of physics, and any American physicist who could afford it went to study in Germany. While Michelson was investigating interference effects with light under the mentorship of Hermann von Helmholtz, it occurred to him that these effects could be exploited for an extremely sensitive experiment comparing the speeds of light in two different directions. He designed an instrument for this purpose, had it built by one of the best instrument makers in Berlin, and set it up in a laboratory in the basement of the nearby astrophysical observatory in Potsdam. Michelson noted that the instrument was so extraordinarily sensitive that the vibration produced by stamping on the pavement 100 meters from the observatory would spoil its operation.

The instrument has two perpendicular arms of exactly equal lengths that serve as "racetracks" for light beams in these two perpendicular directions. Michelson aimed a light beam at the center of the instrument, where a half-silvered (that is, semitransparent) mirror divides this light beam into equal pieces, one racing along one arm, one along the other. These light beams reflect at mirrors at the ends of the arms and race back toward the center, where they again encounter the half-silvered mirror and recombine into a single light beam that emerges from the front of the instrument. If the round-trip speeds along the two arms are exactly equal, the light waves in the two returning light beams arrive at the center in step, and they interfere constructively, so the wave crest of each light wave coincides with the wave crest of the other. This results in a strong emerging light beam. But if the round-trip speeds are different, the light waves arrive slightly out of step and interfere less constructively, resulting in a weaker emerging light beam. Perhaps they even interfere destructively, with total cancellation of the light beam, if the crests of one wave coincide with the troughs of the other wave. Thus, just a small difference in speeds can result in a drastic change in the intensity of the emerging recombined light beam.

Michelson's invention of this instrument, now called a Michelson interferometer, was nothing short of brilliant. Michelson immediately recognized that it could be used not only for a sensitive comparison of the speeds of light along the arms, but also for a sensitive comparison of the lengths of the arms—if one of the arms is shortened slightly (or the mirror at the end is moved slightly), the recombined light beam will change from, say, constructive interference to destructive interference. Thus, a change of length of just a fraction of one wavelength of light can be easily detected. Apart from its use to detect differences in the speed of light, the Michelson interferometer was widely used to measure small differences of length, and for this very practical application of the instrument, Michelson was awarded the Nobel Prize in 1907, the first award to an American physicist.

The Michelson interferometer continues to be used for precise length comparisons even today. For instance, the LIGO detector for gravitational waves near Baton Rouge, Louisiana, is a Michelson interferometer with arms consisting of evacuated pipes 4 klicks (or 2.5 miles) long in which laser light beams are made to travel back and forth several hundred times before they are recombined. This Michelson interferometer can detect a change of length as small as a billionth of a billionth of a meter, which is the change of length that might be produced by the effect of a gravitational wave arriving from a distant supernova explosion.

THE RESULTS OF MICHELSON'S 1881 experiment in Potsdam were negative. Michelson oriented one arm of his interferometer parallel to the direction of the motion of the Earth, and the other perpendicular. The round-trip time for a beam of light along the parallel arm should then have been longer than for the perpendicular arm. But Michelson found no difference at all. The sensitivity of his instrument was adequate to detect a speed of the Earth relative to the ether of 5 klicks per second, and the negative result of the experiment established that the speed, if any, was smaller than that. With the ether at rest in the Solar System, the speed of the Earth relative to the ether should have been 30 klicks per second, the orbital speed of the Earth. Thus, the negative results of Michelson's experiment were in flat contradiction to the naïve ether theory.

Michelson was a firm believer in the ether theory of light. To explain his negative results, he resorted to a modification of the ether theory. He argued that maybe the Earth entrains the ether in its immediate vicinity and drags it along, in which case the Earth would also "drag" along the light waves, and we would then not see any effect of the motion of the Earth on the speed of light. But this explanation was in conflict with the behavior of starlight in astronomical telescopes. Astronomers had observed that the motion of the Earth around the Sun causes starlight to enter their telescopes at a slant, just as streaks of rain enter the cockpit of your convertible at a slant when you are driving through the rain with an open roof; and this slanting, or "aberration," of starlight would not occur if the ether near the telescope was being dragged along by the Earth.

Michelson returned to the United States, quit the navy, and repeated a more precise version of his interferometric experiment in a collaboration with Edward Morley at the Case School in Cleveland; thereafter the experiment became known as the Michelson-Morley experiment. This more precise version still did not detect any effect of the motion of the Earth on the speed of light. But Michelson saw this as further evidence of ether drag, and he persisted in his belief in the ether theory to the end of his life. Even

The apparatus with which Georg Joos repeated the Michelson-Morley experiment on the speed of light in 1930. It is on display at Deutsches Museum, Munich.

after relativity had been widely accepted and the idea of the ether had been rejected by most physicists, he lamented, "But without a medium, how can the propagation of light waves be explained? . . . How explain the constancy of propagation, the fundamental assumption . . . if there be no medium?"[32]

The Michelson-Morley experiment became one of the favorite tests of relativity, and it was repeated many more times, by other experimenters. The most accurate repetition was performed by the German physicist Georg Joos who, in 1930, set up a Michelson interferometer in a large evacuated tank in the cellars of the Zeiss factory at Jena. By then, the dark shadows of the Nazi dictatorship were falling over Germany and over German science, and Joos hardly mentioned the Jew Einstein at all in his description of the experiment. He has been faulted for that, but it could be argued that he was merely giving Einstein tit for tat: in all his own writings, Einstein hardly ever mentioned the Michelson-Morley experiment, so perhaps it served him right that the experimenters didn't mention him.[33]

Oddly, it seems that Einstein may have completed his work on relativity in 1905 without at all being aware of the Michelson-Morley experiment. Historians still dispute what Einstein knew about this experiment and when he knew it. Einstein's own statements about this issue are contradictory. In an interview, some years before his death, he claimed that he was not aware of the Michelson-Morley experiment until after 1905, and he said, "Otherwise, I would have mentioned it in my paper." In a second interview, he said, "This is not so easy. I am not sure when I first heard of the Michelson experiment. I was not conscious that it had influenced me directly during the seven years that relativity had been in my life." And he added that he must have been aware of the result before 1905, because he had "simply assumed this result of Michelson to be true."[34]

For Einstein, the Michelson-Morley experiment was of little importance, because he arrived at relativity not by brooding about experiments and observational facts, but by brooding about theoretical aspects of electricity and magnetism, especially the equations that the Scottish physicist James Clerk Maxwell had formulated in the 1860s to describe electric and magnetic fields. During one of Einstein's visits to Cambridge in the 1920s, one of his hosts said to him, "You have done great things, but you stand on Newton's shoulders," and Einstein replied, "No, I stand on Maxwell's shoulders."[35]

EINSTEIN'S GRAND 1905 PUBLICATION on relativity is divided into four parts, like the parts of a classic French (or Swiss) four-course dinner: there is an appetizer, an entrée, a main course, and a dessert (but no wine—Einstein was abstinent). The appetizer serves up the fundamental principles on which Einstein proposed to build his theory; the entrée presents some immediate consequences of these principles; the main course consists of ponderous mathematical deductions of various results concerning electric and magnetic fields of moving bodies (heavy going here); and the dessert consists of a few tidbits to delight experimenters. Quite appropriately, the title of Einstein's paper is derived from the main course—he called it "On the Electrodynamics of Moving Bodies."

For the foundation of the theory, Einstein laid down two principles: the Principle of Relativity and the Principle of the Constancy of the Speed of Light. The first principle simply says that the laws of physics in every inertial reference frame are the same as those in every other inertial reference frame. This means that there is no way to detect the motion of a reference frame, except by comparing the motion of this reference frame with that of another—thus, motion of a reference frame is always relative motion.

The second principle says that the speed of light is an absolute constant: in every reference frame it is the same in all directions, and in every reference frame it is the same as in every other reference frame.

The first principle is in accord with our everyday intuition, as illustrated by the example of the passenger sitting in the train at the station, although it goes much beyond everyday experience in that Einstein assumes that the Principle of Relativity is valid for all speeds, no matter how large. In contrast to Newton, who regarded relativity as a consequence of his laws of mechanics, Einstein regarded the principle of relativity as a fundamental law of physics, which represents a generalization based on experience. Thus, Einstein's view on relativity reverts to that of Galileo, who also regarded relativity as representing direct experience. The modern view on the Principle of Relativity is almost the opposite to that of Newton's. Whereas Newton used the laws of physics to deduce relativity, particle physicists today use the Principle of Relativity as an aid in deducing the laws of physics—they use the Principle of Relativity as a "sieve" for discriminating between acceptable laws and unacceptable laws.

The second principle is very much against everyday intuition. We expect that the speed of light should behave like the speed of sound. When we pursue a sound signal at high speed, its speed relative to us is reduced, and if we ride in a supersonic jet, we can catch up with the sound signal, overtake it, and leave it behind. Likewise, we expect that when we pursue a light signal, its speed should be reduced, and if we could achieve superluminal speed, we would expect to catch up with the light signal, overtake it, and leave it behind.

Einstein demanded that all the laws of physics should be brought into compliance with his two principles, and he investigated what modifications had to be made in the laws of physics to bring them into compliance. As a first step toward this goal, Einstein determined how the time coordinate and the space coordinates measured in one reference frame are related to the corresponding coordinates measured in another reference frame moving at some speed relative to the first. These transformation equations for the coordinates form the basic mathematical machinery for the study of relativity.

Today, these transformation equations are called the Lorentz transformation equations, because, unbeknownst to Einstein, they had already been discovered a year earlier by Hendrik Antoon Lorentz, an eminent

Dutch physicist, then the leading authority on the theory of electricity and magnetism. But Lorentz did not recognize what treasure lay hidden in these transformation equations; he had found them while studying the relativity of electric and magnetic fields, and he failed to recognize that the equation had much wider implications.

FROM THE TRANSFORMATION EQUATIONS Einstein was able to extract several amazing results: the time dilation of moving clocks, the length contraction of moving metersticks or other rigid bodies, the relativity of simultaneity (what is simultaneous in one reference frame is not necessarily simultaneous in another reference frame), the combination rule for speeds (which differs from the intuitive addition or subtraction rule), the increase of mass with speed, and, later, the relation between energy and mass, $E = mc^2$. He also deduced how the electric and magnetic fields generated by electric charges and electric currents increase and decrease when the charges and currents are in motion, and how the electric and magnetic forces are altered when the charges and currents are in motion. These results on the transformation of electric and magnetic fields gave Einstein's paper its name, but, like the Lorentz transformation equations, they had actually already been obtained by Lorentz a year earlier.

Among Einstein's results, the weirdest were the time dilation and the length contraction. Einstein deduced that when a clock is in motion at high speed relative to a reference frame, its rate is slower than that of an identical clock at rest in the reference frame. This is completely contrary to our everyday experience with watches and chronometers: we can carry a watch with us in a high-speed airplane trip around the world, and when we return home, we find that our watch hasn't lost anything by comparison with a similar watch left at home.

The reason why the time dilation does not show up under these conditions is that at airplane speeds the time dilation is very, very small. For a trip around the world at the speed of a typical airliner, 900 klicks per hour, the time dilation is only 0.3 microsecond (that is, a millionth of a second), much too small to register on a watch.[36] Even for a trip around the world in the International Space Station, at 28,000 klicks per hour, the time dilation remains quite small, only 3 microseconds for each trip around the Earth. If you want a large time dilation, you need a speed comparable with the speed of light, maybe 90 percent of the speed of light or 99 percent. At 90 percent of the speed of light, time is dilated by a factor of 2.3; and at 99 percent of the speed of light, it is dilated by a factor of 7.1. For a speed closer to the speed of light, the time dilation becomes even larger, as large as you please. (If you have an electronic calculator, you can amuse yourself by calculating time dilations; the time dilation factor equals 1 divided by the square root of $1 - V^2/c^2$, where V is the speed of the clock and c is the speed of light, 300,000 klicks per second.)

Einstein thought that the best bet for detecting the time dilation would be by exploiting the rotation of the Earth. Because of this rotation, a point on the Earth's equator moves at a speed of 1,700 klicks per hour relative to a point on the pole, and a clock at rest on the equator should therefore suffer a time dilation relative to a clock on the pole. However, Einstein was mistaken in this, because for clocks placed at sea level on the surface of the Earth, the time dilation from rotation is always exactly canceled by an opposite gravitational time dilation arising from general relativity. This is one mistake we should not hold against Einstein: in 1905, the gravitational time dilation was not yet known—Einstein discovered this dilation several years later.

The time dilation applies not only to clocks but also to all physical, chemical, and biological processes. A Polish physicist facetiously proposed a relativistic refrigerator: to preserve a chicken carcass, simply put it into a rocket and make it move at high speed. The chemical alterations of the meat will be slowed down (as they are when the carcass is cooled in a regular refrigerator), and the meat will stay fresh.

For an astronaut moving at high speed relative to the Earth, the time of the astronaut's "body clock" will be dilated, and the astronaut will age at a slower rate than her teammates on Earth. If motion at a speed *larger* than the speed of light were possible, time could not only be slowed down, but it could be reversed, so the traveler returns to her starting point *before* she started the trip. A limerick that physics professors are wont to recite draws attention to this:

> *There was a young lady named Bright*
> *Who traveled much faster than light.*
> *She set out one day in a relative way·*
> *And returned the previous night.*

However, it is not possible to accelerate a body to a speed as large or larger than the speed of light—to achieve the speed of light would require an infinite amount of energy, which is not attainable. With their gigantic accelerators physicists can make particles move at a speed close to the speed of light, but they cannot achieve a speed equal to (or larger) than the speed of light.

The time dilation is mutual: when the astronaut moves at high speed relative to the Earth's reference frame, her time is dilated relative to Earth time; conversely, because the Earth moves at high speed relative to the astronaut's reference frame, Earth time is dilated relative to the astronaut's time. This mutual time dilation seems counterintuitive: how can the astronaut's clocks run slow relative to the Earth clocks, and, concurrently, the Earth clocks run slow relative to the astronaut's clocks? But careful analysis shows there is no logical inconsistency in this provided we keep the clocks moving (if we stop the relative motion of the reference frames, then which clocks are late will depend on which reference frame we elect to accelerate to achieve the stopping of the motion).

IN EINSTEIN'S DAYS, no sufficiently accurate clocks were available for a direct test of the time dilation. But around 1960 the Hewlett-Packard Company developed highly accurate portable atomic clocks, about the size of a suitcase. In atomic clocks, the "ticking" is controlled by the oscillations of a sample of molecules or atoms instead of the oscillations of a pendulum, as in a grandfather clock. These new portable atomic clocks were routinely used to compare the synchronizations of clocks installed in metrology laboratories at different locations. The portable clocks were transported from one location to another on commercial aircraft, accompanied by couriers. The clocks needed a reserved seat of their own, and this led to some amusing incidents. On one occasion, by a mix-up at the travel office, the courier presented to the astonished aircraft crew a ticket and a boarding pass made out to "Mr Clock." And the couriers quickly learned that it was expedient to call the clocks *electronic* rather than *atomic*, to allay the fears of passengers and crew.[37]

Scientists took advantage of such portable atomic clocks to confirm the relativistic time dilation resulting from the motion of the aircraft. In a test performed in 1972, Pan Am, TWA, and American Airlines made available a seat for Mr Clock on their regular commercial flights, and the clock was sent on a long trip, circling the world twice, first eastward, and then westward. The captains of the aircraft supplied flight data, so the scientists could determine the speed of the aircraft and calculate the expected time dilation according to Einstein's formula. At the end of the trip, the reading of the clock was compared with the reading of another atomic clock that had stayed on the ground, at the US Naval Observatory in Washington. This test of the time dilation confirmed Einstein's deduction—at the end of the trip, the traveling clock was late by a fraction of a microsecond, in agreement with expectations.

Even before 1960, a rudimentary test of time dilation had been performed by physicists studying cosmic rays. Most of the cosmic rays that strike the surface of the Earth are high-speed muons, particles very similar to electrons but of larger mass. These muons are produced in the upper atmosphere by the impact of other, more energetic, "primary" cosmic rays arriving from our own and other galaxies. When one of these energetic cosmic rays strikes the atoms in the upper atmosphere, it creates a shower of muons, some of which travel downward toward the surface of the Earth.

Muons are highly unstable particles; they last only a couple of millionths of a second, decaying quickly into other particles. Because of their short lifetime, all the muons should decay before completing the trip to the surface of the Earth, and we should not observe any muons arriving at the surface. But time dilation prolongs the lifetime of the muons—at the typical, high speed of the muons, the time dilation in the Earth's reference frame is a factor of about 9, and because of this, the muons survive long enough to reach the surface. Thus, the detection of cosmic-ray muons coming through the atmosphere provided the first observational evidence of time dilation.

THE OTHER WEIRD RESULT deduced by Einstein was the length contraction. The length of any kind of rigid body—a meterstick, an airplane, a meteorite, a planet—moving at high speed through a reference frame is shortened along the direction of motion. Like the time dilation, the length contraction is minuscule at everyday speeds. For an airliner flying at 900 klicks per hour, the length contraction is about a billionth of a centimeter, less than the diameter of a single atom. But at high speed it can be much larger. The length contraction factor is simply the inverse of the time dilation factor; for instance at 90 percent of the speed

of light the time-dilation factor is 2.3, and the length-contraction factor is
1/(2.3), or 0.44.

The length contraction was proposed years before Einstein by Lorentz
and, independently, by the Irish physicist George FitzGerald, who were
attempting to find an explanation for the failure of the Michelson-
Morley experiment. Lorentz and FitzGerald hit upon the clever idea that
the negative result of the Michelson-Morley experiment could be explained
by a length contraction of the arm of the interferometer parallel to the
motion of the Earth; this would shorten the round-trip travel time for the
light beam along that arm, and make it equal to the travel time for the other
arm. FitzGerald was acutely aware that some mechanical cause had to be
found for this contraction, that is, the contraction had to result from some
compressional effect in the material of the arm, and he conjectured that
some modification of the internal electric forces in the material might be
the culprit.[38]

Lorentz investigated this conjecture theoretically and found that,
indeed, the electric forces are affected by the motion in just the right way
to bring about a contraction. However, although Lorentz correctly identified the basic physical cause for the contraction, he could perform only a
rather rudimentary, tentative calculation of this contraction, because the
lengths of pieces of solid materials, such as the brass arms in Michelson's
interferometer, cannot be calculated from the atomic structure by means
of Newtonian mechanics, but only by quantum mechanics—and that had
not yet been invented.

The contraction became known as the FitzGerald-Lorentz contraction,
and it is celebrated in another limerick:

There was a young fellow named Fisk,
Whose fencing was extremely brisk.
So fast was his action,
The FitzGerald contraction,
Reduced his rapier to a disk.

To reduce a rapier to the shape of a disk requires a speed almost equal
to the speed of light. The required speed is roughly equal to the speed
of the high-energy protons produced at the Fermilab accelerator near
Chicago—their speed is so high, that in 1 second a light ray would gain

only 150 meters on them. In the reference frame of such a proton, a 1-meter rapier lying on the ground next to the accelerator would be contracted to a 1-millimeter disk.

Like the time dilation, the length contraction is mutual. In the reference frame of the proton, the rapier is contracted. But in the reference frame of the rapier, or the reference frame of the Earth, the proton is contracted—instead of a small round ball, the proton is a thin disk, or a miniature pancake. Like the mutual time dilation, this mutual length contraction is counterintuitive, but it leads to no inconsistencies.

The length contraction is closely related to another weird effect discovered by Einstein: the relativity of synchronization. This means that the synchronizations of the clocks in different reference frames are different. For example, although the clocks on two bell towers in Berne and Zurich begin to ring noon at the same instant as reckoned in the reference of the Earth, they begin to ring at different instants as reckoned in a reference frame moving at high speed in the direction from Berne to Zurich—the Berne clock will be late in this reference frame.

This difference in synchronization implies the length contraction. To measure the length of a rapier moving at high speed (and pointing forward) in the direction from Berne to Zurich, the experimenters on Earth have to determine where the point and the pommel are at one instant of time. But when they do these measurements at one instant as reckoned in the reference frame of the Earth, they will be doing them at different instants as reckoned in the reference frame of the rapier—the measurement at the pommel will be late. This means that when they measure the position of the pommel, it will have moved forward, nearer to the position they measured for the point, and, evidently, this will result in a shortened result for the length of the rapier. Thus, the contraction emerges from the difference in synchronization of the clocks.

IN 1905, NEITHER the time dilation nor the relativity of synchronization was within experimental reach—the clocks of those days were not accurate enough to detect these effects. But in the final part of his paper, as dessert, Einstein offered one result that was within reach of experiment. He showed that to bring the equation of motion of a particle into compliance with relativity requires a modification of the mass of the particle at high speed. The mass of a high-speed particle must be larger than the mass of a slow particle—we might call this a mass dilation.

Einstein found one formula for the mass dilation of a particle accelerating along a straight line (like a car accelerating along a straight road), and another formula for the mass dilation of a particle accelerating centripetally while moving along a circle (like a car rounding a curve). The former mass is called the *longitudinal* mass, and the latter mass the *transverse* mass. Of these, the longitudinal mass is always larger than the transverse mass. For instance, the protons racing around the main accelerator ring at Fermilab, with a speed just 150 meters per second below the speed of light, have a transverse mass 1000 times as large as their normal mass and a longitudinal mass 30,000 times as large as their normal mass.

In 1905, the fastest available particles were electrons emitted by radioactive substances and electrons accelerated in cathode-ray tubes (vaguely similar to the TV tubes of today). Experimenters had measured the deflections of such electrons by electric and magnetic forces, and even before Einstein's publication of his paper on relativity they had become convinced that the masses of these electrons increased with speed. These experiments continued in the years after Einstein's paper, mainly in Germany and in France.

The early experiments seemed to be in agreement with calculations by Max Abraham, a physicist at Göttingen and later professor at Milan, who had modeled the electron as a rigid sphere filled with electric charge and used Newton's mechanics and Maxwell's equations to discover how the electron's own electric fields enhance its inertia, or its mass, at high speed. Other physicists had proposed various other theoretical models of the electron, incorporating length contraction, à la FitzGerald-Lorentz. In 1905, the best available experimental data on the increase of mass with speed seemed to be in better agreement with some of these other theoretical models than with Einstein's prediction.

But Einstein stubbornly refused to believe this. He declared, ". . . in my opinion, these theories should be ascribed a rather small probability because their basic postulates concerning the mass of a moving electron are not made plausible by theoretical systems that encompass wider complexes of phenomena."[39] His stubbornness paid off. Shortly after, experiments by Alfred Bucherer at the University of Bonn gave a good fit to Einstein's formulas, and Bucherer wrote to Einstein, "By careful experiments I have demonstrated the validity of the Principle of Relativity beyond all doubt."[40] This ringing announcement was somewhat premature; Bucherer's measurements were not as accurate as he thought they were, but the announcement certainly helped toward the acceptance of Einstein's theory. Within five years of Einstein's launch of his theory, Max Planck, then the preeminent German theoretical physicist, would declare, "In its breath and profundity, the revolution in the realm of the physical world view brought about by [Einstein's Principle of Relativity] can be compared only to that caused by the introduction of the Copernican world system."[41]

2

"And yet it moves"

According to legend, these words were spoken by Galileo after his
conviction for heresy for defending the proposition that
the Earth moves around the Sun.

Einstein is famous for formulating the theories of special and of general relativity, but Einstein did not discover relativity. The man who discovered relativity was Galileo Galilei, born in Pisa in 1564 and convicted of heresy in 1633 for maintaining that the Earth moves around the Sun, in contradiction to Church dogma that proclaimed that the Sun moves around the Earth.

Galileo was convicted by a tribunal of the Holy Office of the Inquisition in Rome and sentenced as a heretic on June 22, 1633, a day that lives in infamy. On that day, garbed in the white shirt of a penitent, he was carried on the back of a mule from his prison at the Palace of the Holy Office near piazza San Pietro to the Dominican Convent of Santa Maria Sopra Minerva, to hear his sentence.

The Convent of Santa Maria Sopra Minerva and the adjacent church of the same name, in piazza Minerva, behind the Pantheon, are built upon the ruins of an ancient Roman temple dedicated to Minerva, the goddess of wisdom. But the irony of sentencing a scientist upon the ruins

This narrow house, to the left of the church of Santa Maria Sopra Minerva, is the entrance to the Dominican convent. Behind this modest entrance lies a large complex of buildings, including a cloister and two courtyards, most of which now belongs to the Italian Parliament.

of a temple of wisdom was merely coincidental. The convent was the traditional place for sentencing heretics because it was the lair of the Dominicans, the bloodhounds of the Church, always on the prowl in search of heretical depravity. For the greater profit of God, they confiscated all the wealth and property of any heretic they convicted, which probably contributed to their zeal and most certainly contributed to the influence of their order.

The Italians called the Dominicans the *Domini canes,* "the dogs of God," and perhaps the leaders of the order had a sense of humor about this designation—the coat of arms of the Dominicans features a black-and-white dog, in chromatic concordance with the black-and-white robes worn by the members of the order. But there was no sense of humor when it came to heretics. Not long before, Giordano Bruno, a defrocked monk, his body broken by torture, had been carried to hear his sentence in the same way as Galileo, mounted on a mule. He had been sentenced as an unrepentant heretic, for asserting that all stars are Suns, with Earthlike planets and, perhaps, alien inhabitants. Then he had been remounted on his mule and taken to be burned at the stake in Campo dei Fiori, now a quaint flower and vegetable market, where you can find the spot of his execution marked with a memorial.

The Great Hall of the Dominican convent provided a suitable setting for passing judgment on heretics—its ceiling is covered with a splendid fresco showing in exquisite detail the butchery of the Albigensian heretics in the thirteenth century, which marked the beginning of the Inquisition.[1] On the day of Galileo's sentencing, the Great Hall was packed with dogs of God, prelates, clerks, and scribes. Galileo was led into the hall, and he was made to kneel on the marble floor before the seven inquisitors-general presiding over this session of the tribunal, cardinals all of them, arrayed in their blood-red robes.

The sentence was read to him:

> We say, pronounce, sentence, and declare that you, the said Galileo, by reason of the matters adduced in your trial, and by you confessed as above, have rendered yourself in the judgment of this Holy Office vehemently suspected of heresy, namely, of having believed and held the doctrine—which is false and contrary to the sacred and divine Scriptures—that the Sun is the center of the world and does not move from east to west and that the Earth moves and is not the center of the world . . .
>
> We condemn you to the formal prison of this Holy Office during our pleasure, and by way of salutary penance we enjoin that for three years to come you repeat once a week the seven Penitential Psalms . . .
>
> And so we say, pronounce, sentence, declare, ordain, and reserve in this in any other better way and form which we can and may rightfully employ.[2]

Still kneeling in front of the inquisitors-general, Galileo then read the formula of abjuration:

> I, Galileo, son of the late Vincenzio Galilei, Florentine, aged seventy years, arraigned personally before this tribunal and kneeling before you, Most Eminent and Reverend Lord Cardinals and Inquisitors-General against heretical depravity . . . swear that I have always believed, do believe, and by God's help will in the future believe all that is held, preached, and taught by the Holy Catholic and Apostolic Church.
>
> . . . with sincere heart and unfeigned faith I abjure, curse and

detest the aforesaid errors and heresies . . . and swear that in future
I will never again say or assert, verbally or in writing, anything that
might furnish occasion for a similar suspicion regarding me . . . So
help me God and these His Holy Gospels, which I touch with my
hands.[3]

According to legend, he muttered the words "and yet it moves" at the
end of his abjuration. But this is surely a fabrication—the many eyes of
the dogs of God were upon him, and any heretic who failed to give a convincing show of sincere repentance could expect no mercy. He may have
thought the words, but he must have kept them to himself.

INTELLECTUALLY, EINSTEIN WAS the son of Galileo Gali-
lei, and of Isaac Newton, and of James Clerk Maxwell. Einstein's theories
of relativity rest on foundations first laid by Galileo—the special theory
rests on Galileo's discovery of the relativity of motion, and the general theory rests on Galileo's discovery of the equal rates of acceleration of freely
falling bodies. Furthermore, both the special and the general theories
incorporate and extend Newton's laws, and both incorporate Maxwell's
equations. It is a wise father who knows all his sons; and, intellectually, it is
a wise son who knows all his fathers.

Einstein acknowledged his debt to Newton and to Maxwell, but he was
not fully aware of the extent of Galileo's fatherhood. In an introduction
he wrote for Galileo's celebrated *Dialogue Concerning the Two Chief World
Systems*, he faults Galileo for failing to produce a general mathematical
proof of relativity (impossible, because a general theory of mechanics
was lacking).[4] He fails to see that Galileo never intended to give a general
mathematical proof. Galileo regarded relativity as an empirical, observational fact, that is, a law of nature, and Einstein's own formulation of
the Principle of Relativity three hundred years later imitated Galileo's
in treating this principle as a law of nature and not as a mathematical
deduction from anything else.

In the language of physics, the dispute between Galileo and the Church
was a dispute over relativity: Galileo held that relativity was valid, and the
Church held that it was not. Thus, both Galileo and Einstein may be said
to owe their fame to relativity. However, "relativity" was not in Galileo's
vocabulary—it did not enter the language of physics until three hundred
years later. Instead of speaking of "relativity," Galileo spoke of how we can-

not distinguish between a reference frame at rest and a reference frame in motion by any phenomena that occur in these reference frames.

Galileo discovered the first indications of relativity in his groundbreaking studies of mechanics. He laid the first firm foundations for the science of mechanics, and he initiated the mathematical study of motion—he formulated what we might call the "geometry" of motion. And in his study of motion he came to recognize that motion is relative: the motion of a "movable" (by which he meant any kind of body in motion) can be described just as well relative to the ground or relative to some other "movable" that is itself moving relative to the ground. Thus, the motion of an arrow can be described relative to the ground or relative to a rider on a galloping horse, and in both descriptions the fundamental laws of motion·are exactly the same. This is the essence of relativity, and it led Galileo to think relativity might be applicable to all kinds of phenomena. For Galileo, relativity was merely a weapon that he developed to parry the attacks by the zealots addicted to the immobile Earth. He did not fully understand the implications of relativity, and he could never have guessed the pervasive role that it was to play in the twentieth century, after it reached perfection at the hands of Einstein.

THE GALILEIS WERE impoverished descendants of a leading Florentine family, and they liked to regard themselves as "patricians," not quite the same as noblemen, but near enough to lay claim to their own coat of arms, with a little ladder as their emblem. When Galileo was born, the family lived in Pisa, and his father was eking out a living by teaching music. It was a musical age—Palestrina, Orlando di Lasso, Gabrieli, Monteverdi, Frescobaldi, and Lully were contemporaries of Galileo's. Music and dancing were as much part of the education of a gentleman as arms, horsemanship, and hunting, although some musical instruments were deemed inappropriate; thus in Castiglione's *Book of the Courtier*, a sixteenth-century manual on how to be cool and win the hearts of women, gentlemen were warned against performing on wind instruments, because they deform the face.[5]

The young Galileo developed a talent for the invention and construction of mechanical toys, a talent that was also developed by Newton and by Einstein. At age ten, Galileo's family moved to Florence and he spent some years as a novice in a nearby monastery. There he revealed much aptitude and skill in literary studies, which formed the foundation for the lucid and elegant writing style that he later exhibited in his letters and books. Galileo felt an inclination toward monkery, but his father decided that a career in medicine would be more profitable—then, as now, doctors earned good money. At age seventeen, Galileo was sent back to Pisa to matriculate in the university as a student of medicine.

At the university, Galileo became acquainted and infatuated with mathematics. Over the objections of his father, he decided to become a mathematician, but lack of money forced him to withdraw from the university without taking a degree. Back in Florence, he invented a hydrostatic balance, the *bilancetta*, for measurement of the density of bodies by the method of Archimedes, the Greek natural philosopher of the third century BC who is remembered for jumping out of his bathtub and streaking through the streets of Syracuse with yells of "Eureka!" Galileo's balance was widely adopted by jewelers and goldsmiths to test the density of alloys. Galileo wrote an essay on the *bilancetta* and a book on the center of gravity of solids that brought him fame throughout Italy—he was described as "the Archimedes of his time." And fame brought employment—Galileo was appointed to the position of mathematical lecturer at the University of Pisa. There he began his original, incisive studies and experiments on the motion of falling bodies and projectiles, exposing the misconceptions held by his predecessors and contemporaries.

MOST OF THESE MISCONCEPTIONS derived from Aristotle, another Greek philosopher of the third century BC, tutor of Alexander the Great. His writings were extensive and encyclopedic, covering much of the knowledge of the time. His style may best be described as early Polonius, and his misconceptions about the animal kingdom are often good for a laugh (a twitch in the right testicle of a horse is a symptom of a disease that will lead to shedding of the hooves; but not to worry—they will grow right back). The scholars of the Middle Ages confused quantity with quality, and Aristotle's writings became standard fare at all medieval universities.

In science, Aristotle emphasized the direct observation of nature and he taught that theory must follow fact. Unfortunately he did not practice what he preached. He asserted that the speed of a falling body is directly proportional to its weight, but he never bothered to drop a large stone and a smaller stone simultaneously from his outstretched hands to the floor, which would have shown him that they hit the floor at the same instant. Free fall is of obvious practical importance when shaking apples off a tree or tossing the contents of a chamber pot out a window into the street below (as was customary in medieval towns). But the quantitative details of free fall are irrelevant for such a harvesting of apples or disposal of sewage, which perhaps explains why the details were not investigated.

With consummate showmanship, Galileo gave a dramatic demonstration of Aristotle's error about free fall by dropping balls of lead of different weights from a great height. As reported in a biography written by his student Viviani shortly after Galileo's death, ". . . these all moved with equal speeds, he showing this by repeated experiments made from the height from the Leaning Tower in the presence of other professors and all the students."[6]

In one of his own books, Galileo offers some more details: "But I, . . . who have made the test, can assure you that a cannon ball that weighs one hundred pounds (or two hundred, or even more), does not anticipate by even one span [a handbreadth] the arrival on the ground of a musket ball of no more than half [an ounce], both coming from a height of two hundred braccia [100 meters]."[7] Here Galileo exaggerates, by braccia and by pounds. Renaissance Italians were fond of building towers to advertise their *virtú* (their power, not their virtue, although the distinction was often lost on them), and Pisa in those days sported many towers, but none higher than the Leaning Tower (57 meters). Furthermore, the heaviest cannon balls in use at the time weighed no more than 100 pounds.

The professors of Pisa, who had been raised on Aristotle and taught from his treatises, were not grateful to Galileo for giving them the lie by dropping cannon balls at their feet. Galileo had a talent for making enemies—as Koestler said, he provoked "the cold, unrelenting hostility which genius plus arrogance minus humility creates among mediocrities."[8] He was a quarrelsome, proud, and stubborn man, easily offended, vain and possessive about his scientific discoveries, always eager for public attention. He had a choleric temperament, and in surviving books from his private library we find marginal annotations of "buffoon," "piece of asinity," "evil poltroon," "ungrateful villain."[9] He was fond of provoking arguments, and he delighted in covering his opponents with ridicule and skewering them with his keen sarcasm. His blunt exposure of the errors of

Aristotle made him many enemies among the mediocre professors of Pisa. They compelled him to leave the university, in spite of—or because of—his superior intellectual *virtú*.

GALILEO SOON GAINED a much better position as professor of mathematics at the somewhat more progressive University of Padua, where he stayed for eighteen years. Padua was then under the rule of Venice, and the Venetians gave Galileo his due, awarding him a generous salary. And Venice could afford it—although its power had started to decline, Venice was still the richest of the Italian city states.

The years at Padua were the most productive and creative of Galileo's life. His investigations on mechanics were completed during that time, although not published until his final years after his conviction for heresy. His lectures were a great success, attended by scholars from all parts of Europe. Besides the lectures for his own students, he also gave public lectures, which sometimes attracted crowds of a couple of thousand listeners. Galileo was an inveterate popularizer and impresario, always eager to attract the attention of the widest audience, and to achieve this he took to writing his books in Italian, instead of the Latin that was the language for scientific publications preferred by all his colleagues.

While at Padua Galileo invented, or reinvented, the telescope. He read reports that a Dutch optician had built a device that made distant objects appear closer, and he quickly figured out how to do this with two lenses. Galileo proceeded to build hundreds of telescopes, some with magnifications as large as 32×. A few of his telescopes are still in existence, and some are on display at the Museum of Science in Florence.[10]

After perfecting his telescopes, Galileo aimed them at the heavens and initiated a new era of spectacular discoveries in astronomy. In quick succession, he discovered mountains on the Moon, spots on the Sun, stars in the clouds of the Milky Way, the four large moons of Jupiter, the moonlike phases of Venus, and the rings of Saturn. These discoveries brought Galileo renown throughout Europe. His wide fame also brought more tangible rewards. With a keen instinct for self-advancement, Galileo named the moons of Jupiter the Medicean stars, in honor of Cosimo II, Grand Duke of Tuscany, of the Medici family. In the long run, the name did not stick—today astronomers call the four largest moons of Jupiter the Galilean satellites. But in the short run, this flattery paid off—Galileo was promptly appointed chief philosopher and mathematician to the Grand Duke, "Filosofo e Matematico primario del Serenissimo Gran Duca di Toscana." Galileo moved to Florence, with a large salary, with no teaching duties whatsoever, and with plenty of time for research and writing . . . and quarrels.

IT WAS DURING his Florentine period that Galileo took up the defense of the Copernican system. Until the sixteenth century, the generally accepted model of the universe was the Ptolemaic model, contrived by the Greek-Egyptian astronomer Ptolemy in the second century AD. In this system, the Earth is fixed at the center of the universe and the Sun and the planets circle around it. This system was consistent with Aristotle's teaching that the Earth cannot be in motion.

But early in the sixteenth century the Polish astronomer Nicolas Copernicus proposed a different system of the universe, with the Sun fixed at the center and all the planets, including the Earth, orbiting around the Sun. At first, the Copernican system was thought to have an advantage of simplicity over the Ptolemaic system, but this simplicity proved elusive. In both systems the planets were supposed to move along circular orbits, but for an accurate description of planetary motions it proved necessary to introduce numerous epicycles: each planet moves along a small circle, and the center of each such circle in turn moves along a larger circle around the fixed center of the universe. Copernicus had hoped to do away with these awkward epicycles, but this proved impossible, and he found that the epicycles required in his system were even more numerous than those in the Ptolemaic system.

The system that Copernicus described in his book *Revolutions of the Celestial Bodies* was so complicated that hardly anybody could read or understand it (historians of science have called it the book that nobody read). Nevertheless, the Copernican system had one crucial advantage: the motion of the planets in this system is periodic—whenever the planet completes one circuit around the Sun, it repeats this circuit according to exactly the same schedule again and again, forever. In contrast, in the Ptolemaic system, each planet completes its circuits around the Earth according to a variable schedule—as seen from the Earth, the planet sometimes completes the circuit sooner, sometimes later, so each circuit is different.

The heliocentric configuration and the periodicity of the planetary motions were the two essential clues that led the German astronomer Johannes Kepler, Imperial mathematician in Vienna and Galileo's contemporary, to a new understanding of the planetary orbits. Kepler had access to (and stole) the new, more accurate, observational data obtained by the Danish astronomer Tycho Brahe, and after lengthy calculations with these new data, he achieved a drastic simplification, establishing that the planets move along elliptical orbits around the Sun, not circles or combinations of circles. He also discovered the Law of Periods, relating the periods of the planetary motions to the sizes of the orbits.

Galileo became an ardent and vociferous partisan of the Copernican system, and this set him on a collision course with the Inquisition. The Inquisition was charged with the investigation and suppression of Catholically incorrect thoughts, or "heretical depravity." The field operatives of the Inquisition—who acted as investigators, prosecutors, and judges all in one—were Dominicans, an order that also supplied the Church with its chief theologians and scores of preachers. With his defense of the Copernican system, Galileo was baiting these dogs of God. The Church had not—and never did—formally declared the Copernican system as heretical. But it was viewed as "rash" and contrary to Scripture, and Galileo knew he was treading on dangerous ground.

Galileo also knew that the Catholic hierarchy had actually encouraged Copernicus to publish his book and that the Catholic Church had a long tradition of acceptance of various subtle interpretations of Scripture. Cardinal Robert Bellarmine, the most influential theologian of Galileo's day, later promoted to sainthood, expressed the eminently reasonable and enlightened view that "If there were a real proof that the Sun is in the center of the universe, . . . and that the Sun does not go around the Earth but the Earth around the Sun, then we would have to proceed with great circumspection in explaining passages of Scripture which appear to teach the contrary, and rather admit that we did not understand them than declare an opinion to be false which is proved to be true. But, as for myself, I shall not believe that there are such proofs until they are shown to me."[11] Thus, in contrast to the Bible-thumping Christian fundamentalist zealots of then and now, Bellarmine was willing to let science overrule the Bible, *if* there was real proof. And Galileo thought he had real, conclusive proof of the motion of the Earth in his theory of the tides, his pet theory, which he considered the culminating achievement of his life and to which he later devoted his *Dialogue Concerning the Two Chief World Systems*, the book that was to lead to his arrest and conviction for heresy.

CHI LINGUA HA, A ROMA VA (He who has a tongue, goes to Rome), says an Italian proverb. And in 1615, emboldened by the favorable reception accorded to his telescopic discoveries, Galileo decided to go to Rome, confident that by the power of his eloquence, the weight of his scientific arguments, the respect enjoined by his accomplishments, and the favor of friends in high places he would overcome the resistance to the Copernican ideas. In Rome he was well received and was granted several audiences with influential cardinals, to whom he argued his case. They listened with polite interest, but his arguments fell on deaf ears, and his scientific proof based on his theory of the tides failed to convince. Exactly what was wrong with his theory of tides eluded the cardinals and most everyone else in his time. But that tides had something to do with the Moon was general knowledge, and that must have made his theory rather questionable in the eyes of his contemporaries.

Galileo stubbornly and unwisely insisted. He was warned to "write freely, but keep outside of the sacristy."[12] The Tuscan ambassador, who was under instructions from the Grand Duke to assist Galileo, reported that "Cardinal del Monte and myself, and also several cardinals from the Holy Office, had tried to persuade him to be quiet and not go on irritating this issue. If he wanted to hold his Copernican opinion, he was told, let him hold it quietly and not spend so much effort in trying to have others share it. Everyone fears that his coming here may be very prejudicial and that, instead of justifying himself and succeeding, he may end up with an affront."[13]

The publicity that Galileo brought to the Copernican ideas attracted the attention of the Holy Office, and several theological experts, or qualifiers, were instructed to examine these ideas. Within a few days, they gave their expert opinion: on the basis of a dozen verses from the Bible, they declared the proposition that the Sun is at rest at the center of the universe as "foolish and absurd, philosophically and formally heretical, inasmuch as it expressly contradicts the doctrine of the Holy Scripture in many passages . . ."[14] The Holy Office instructed Cardinal Bellarmine to summon Galileo to an audience and read him the riot act. Accordingly, in the presence of notaries and a pack of dogs of God in their black and white robes, Bellarmine admonished Galileo to abandon his Copernican ideas as erroneous, and not to hold or defend them. And this Galileo promised to do—although he muttered about "the ignorance, malice and impiety of my opponents who had won the day."[15] Round one in the fight over the Copernican system ended in a draw—the Dominicans had marked a line in the sand that Galileo was ordered not to cross, but the dogs of God had merely been permitted to bark, not to bite. Galileo returned to Florence, and he remained unusually quiet for a few years.

IN 1623, MAFFEO BARBERINI was elevated to the papacy, taking the name Urban VIII. The new pope was a perplexing character, a Counter-Reformation pope with a Renaissance disposition, intelligent, sophisticated, skeptical, proud, and vainglorious. He was no zealot—he often gave the impression that he was following the precepts of Machiavelli rather than those of the Bible. It is said that upon the death of Cardinal Richelieu, he commented, "If there is a God, Cardinal Richelieu will have much to answer for; if not, he has done very well."[16] He brought nepotism to new heights, appointing two of his nephews as cardinals and enriching the Barberini family to an extent that amazed even the Romans, well-versed in corruption as they were. He completed many ambitious construction projects in Rome, such as the splendid Palazzo Barberini, the fountain of the Triton in the nearby piazza Barberini, and the baldachin in St. Peter's. He was an accomplished poet, composing hymns and rhymed transcriptions of biblical verses—and also a nicely crafted poem in praise of Galileo.

While still a cardinal, Barberini had been a strong supporter of Galileo's, and upon his elevation to the papacy, he eagerly invited Galileo to Rome. Galileo's visit was delayed by illness, but when he finally arrived in Rome, Urban showed him much favor and granted him several long and cordial audiences, during which he and Galileo argued against and for the Copernican System. No account of these audiences has come down to us, but from contemporary sources we know Urban had a mulish disposition, impermeable to persuasion. He loved to contradict, constantly giving decisions opposite to whatever was proposed to him, so much so that the Venetian ambassador resorted to the trick of proposing to him the opposite of whatever he really wanted.[17] Urban was exceedingly garrulous, and he expected to be admired by his listeners and addressed with unctuous reverence, especially by listeners who disagreed with him.

We can then readily imagine that Galileo's earnest arguments for the Copernican system and his theory of the tides were hotly and volubly contested by the pope with interminable objections, wise and otherwise. And Urban had the last word by dint of a cute piece of theological sophistry that had taken his fancy: Even if Galileo's theory gives a complete explanation of the tides, this does not mean that it is true, because God in His infinite power and wisdom could have produced the tides in many ways that are unthinkable to the human mind. Therefore a scientist can never claim to know the true mechanism of the tides or any other phenomenon. At best the scientist can only hope to "save the phenomenon," that is, he can say that it is *as if* the phenomenon is based on some proposed mechanism. But the scientist must always treat the proposed mechanism as fiction, as an unprovable hypothesis, because to offer a physical proof is to challenge God's omnipotence, and "it would be excessive boldness for anyone to limit and restrict the Divine power and wisdom to some particular fancy of his own."[18]

Galileo assumed that Urban's favor gave him license to discuss the Copernican system openly, and he began to write the draft of a book with the title *Dialogue on the Tides*. Urban encouraged him to proceed, but did not approve of the emphasis on the tides and demanded that the title be changed to *Dialogue Concerning the Two Chief Systems of the World*. Following a tradition that goes back to the dialogues of Plato, the book is written in the form of a conversation among three fictional persons: Salviati, an adherent of the Copernican system; Simplicio, an adherent of the Ptolemaic system; and Sagredo, their host, performing the function of an umpire. Galileo named Salviati and Sagredo for two deceased friends—the former a Tuscan gentleman, the latter a Venetian patrician—and Simplicio for a medieval philosopher who had written commentaries on Aristotle.

THE *DIALOGUE* OCCUPIES an odd position among the classics of the history of science, and not only because it was prohibited by the Catholic Church for two hundred years. It is an outstanding piece of expository writing, and its early chapters display an acute and clear grasp of the relativity of motion. But the later chapters, dealing with the Copernican system and the so-called proofs of the motion of the Earth are packed with glaring mistakes. These mistakes are so obvious that it seems incredible that Galileo was not aware of them. His blindness to the defects of his "proofs" of the motion of the Earth cannot be explained by normal ignorance or negligence, but only by some abnormal state of obsessive delusion—he could not bring himself to accept that his arguments made no sense and did not fit the facts. He desperately craved a proof of the motion of the Earth, and in his desperation he grasped at straws. We may even have to face the odious possibility that Galileo knew very well that his arguments were false, and that he engaged in deliberate fraud. If he committed fraud, he did it out of a deep conviction that he ultimately was right. He felt fanatically certain that the Earth moves, and he would stop at nothing to win the argument—for him, the end justified the means.

The *Dialogue* is divided into four chapters, each representing one day of discussions among the protagonists. The first two chapters are devoted to a careful exposition of why the translational and rotational motions of the Earth, proceeding with an immense speed, produce no discernible effects on its inhabitants. The third chapter discusses the Copernican system, and the fourth chapter presents Galileo's novel theory of the ocean tides.

The first two chapters display Galileo's elegant expository writing style at its best, with brilliant explanations, illuminating examples, lively polemics, and skillful exploitation of the discussion format to resolve any doubts, misconceptions, and questions of the reader. Galileo also touches on many interesting tangential matters and gives us some surprising insights. For instance, he proceeds to calculate the time it takes a body to fall to the Earth from a height as large as the distance to the Moon; that is, he assumes that gravity extends as far as the Moon.[19] This insight is usually attributed to Newton, but Galileo clearly anticipates him, although he fails to recognize that the strength of gravity decreases with distance, and his number for the time of fall (2 hours, 22 minutes, and 4 seconds) is far too short.

IN THE SECOND CHAPTER, Galileo introduces the fundamental concept of the relativity of motion. He illustrates this concept with examples drawn from experience, and he explores its significance for the motion of the Earth. By direct observation of everyday phenomena, Galileo had recognized that it is impossible to detect uniform translational motion, such as the motion of a ship moving with uniform speed along a straight line, except by looking outside the ship and checking its displacement relative to some landmark. In Galileo's own words, ". . . make the ship move with what velocity you please, so long as the motion is uniform and not fluctuating this way and that. You shall not be able to discern the least alteration in [any] of the forenamed effects, nor can you gather by any of them whether the ship moves or stands still."[20]

The *Dialogue* was published in 1632, but it is probable that Galileo thought about relativity much earlier, perhaps as early as 1592, when he held appointments at both the University of Padua and at the Arsenal in Venice, and often traveled back and forth between these cities. In those days, the preferred method of travel between Padua and Venice was by barge, or *burcio*, along the river Brenta, flowing from Padua into the Venetian lagoon. The banks of the river are lined with splendid villas, because of which the river has been called an extension of the Grand Canal. These villas were built by rich Venetians as summer homes, between the sixteenth and the eighteenth centuries, and many of them already existed in Galileo's time, among them Villa Sagredo, belonging to Galileo's Venetian friend Giovan Francesco Sagredo, whose name Galileo used in his *Dialogue* (it is known that Galileo occasionally stayed at Villa Sagredo, now an elegant restaurant).

So we can imagine Galileo on a warm summer day in 1593 comfortably sitting in the cabin of a barge, floating down the calm waters of the Brenta. He might have seen some boys running around the cabin, jumping over bales of cargo, tossing a ball to and fro, or rolling it along the deck. He might have seen flies buzzing about (probably very many flies) or fish swimming about in a bucket. He might have seen a stream of wine poured from a bottle into a goblet. And he would have noticed that all of these everyday phenomena happen on the moving barge in exactly the same way as when the barge is at rest, moored to the shore—the motion of the barge has no effect whatsoever on the behavior of bodies within it.

But in 1593, Galileo probably did not attach great importance to any of this, and, if he recognized that these observations implied a relativity of motion, he did not bother to set this down in writing. He did not publish this idea until 1632, and by then he had an agenda. He needed to defend the Copernican system against the criticisms by the adherents of Ptolemy. One such criticism was that if the Earth is orbiting around the Sun at a fearsome speed, then "beasts and men and buildings placed on the Earth would be precipitated from it."[21] To answer this criticism, Galileo proclaimed his new idea of the relativity of motion: he argued that the orbiting Earth is analogous to a moving ship, and we can't feel the speed of the Earth any more than we can feel the speed of the ship. The Earth has both a translational and a rotational motion, and everything on the Earth participates in this motion, so relative to the Earth everything behaves in exactly the same way as if the Earth were at rest: ". . . whatever motion comes to be attributed to the Earth must necessarily remain imperceptible to us and as if nonexistent, so long as we look only at terrestrial objects; for as inhabitants of the Earth, we consequently participate in the same motion."[22]

In the relativity of motion Galileo had found a decisive weapon to win the fight of the Copernican versus the Ptolemaic system. But Galileo did not fully understand the power of this weapon, and he failed to recognize that it held fatal consequences for his theory of ocean tides, his grand delusion, by means of which he thought he could overcome all opposition to the Copernican system.

IN THE FOURTH CHAPTER of the *Dialogue*, Galileo offers his theory of the ocean tides as the final, conclusive proof of the motion of the Earth. Galileo thought that this theory would provide not only a demonstration of the orbital translational motion of the Earth around the Sun but also a demonstration of the rotational motion of the Earth on its axis. He knew that high tides and low tides arrive on the seashores with clocklike regularity (in English, the word *tide* is in fact derived from the Anglo-Saxon word *tid*, meaning "time"; in German and in Dutch, the words for *tide* have a similar etymology). To explain this regular flux and reflux of the seas, he asked how the speed of an ocean basin—say, the Adriatic basin—changes in the course of a day because of the Earth's combined translational and rotational motion.

He thought of this combined motion as analogous to the motion of a dent on the rim of a rolling wheel—say, a dent on the wheel of a wheelbarrow—which also has simultaneous translational and rotational motions, that is, translational along the garden path and rotational around the axis of the wheel. The speed of such a dent is largest when it reaches the topmost position, where the translational and rotational motions of the wheel are both in the forward direction, so they add. The speed of the dent is smallest when it reaches the bottom, where the translational and rotational motions are in opposite directions, so they cancel. Likewise for the motion of an ocean basin on the surface of the Earth, the speed is largest at midnight (when the translational and rotational motions are in the same direction) and smallest at noon (when the translational and rotational motions are in opposite directions). From this, Galileo concluded that the ocean basins speed up and slow down each day, and he thought that such a speeding up and slowing down would cause the water to slosh back and forth and produce a flux and reflux of water on the shores. What gave him this idea was something he had observed in Venice during one of his visits: the Venetians transported drinking water in barges equipped with large holding tanks, and Galileo had noticed that whenever such a barge stopped suddenly, the water in the tanks would slosh forward.

But this theory of the tides is false: it is inconsistent, and it disagrees with the observational facts of the tides. The inconsistency in his theory is immediately obvious from the relativity of motion, according to which the translational motion of the Earth cannot have any effect whatsoever on the behavior of the oceans. The Earth can be thought of as a ship drifting through space, and as in the case of a ship drifting on a calmly flowing river, the uniform translational motion has no effect on anything that happens in the ship. This was Galileo's fatal mistake, the worst and most tragic scientific mistake in his long and distinguished career.[23]

Although Galileo can perhaps be forgiven for failing to recognize the inconsistencies in his theory of tides, he cannot be forgiven for dismissing or glossing over the obvious conflicts between his theory and observation. In his theory, water speeds up and slows down once per day, which would give one high tide and one low tide per day, whereas in fact there are two high tides and two low tides per day. But Galileo deluded himself that the second high and low tide each day must have something (what?) to do with peculiar conditions in the Mediterranean.

His theory also totally fails to account for the daily delay of the tides. In his theory, a high tide should occur at the same time each day; but in fact, at Venice, the tide comes 20 to 30 minutes later each day.[24] This delay must have been well known to Galileo from his visits. The tides at Venice are nearly a meter high (sometimes as much as 1.5 meters, when enhanced by adverse weather conditions), and Venice is a city attuned to the rhythm of the tides. Every Venetian of Galileo's days would have known that if high tide arrived at noon today, it will arrive 20 or 30 minutes after noon tomorrow. But Galileo does not mention this daily delay anywhere in his *Dialogue*—he probably decided not to reveal this problem, because he could not think of any explanation for it. If Galileo's Venetian friend

Sagredo had still been alive when Galileo wrote the *Dialogue*, he would have wondered what Galileo was playing at.

Finally, Galileo blithely dismisses as irrelevant the well-known correlation between the tides and the phase of the Moon. The tides "follow the Moon"—the tides are strongest at full moon and at new moon, and they are weak and erratic at first quarter and at last quarter. Kepler had commented on this correlation between the tides and the phase of the Moon, and he had concluded that the Moon somehow causes the tides. Since Galileo could not argue against Kepler's facts, he resorted to ridicule: "But among all great men who have philosophized about this remarkable effect, I am more astonished at Kepler than at any other. Despite his open and acute mind . . . he has nevertheless lent his ear and his assent to the moon's dominion over the waters, to occult properties, and to such puerilities."[25]

GALILEO COMPLETED THE *DIALOGUE* in 1631. All books were subject to censorship, and the book was examined by the pope's chief theologian and also by two Dominican censors of the Inquisition in Florence. The censors demanded some changes, such as the inclusion of the pope's sophist argument about the omnipotence of God and the limitations of the human mind, which Galileo duly inserted in the last pages. Simplicio presents the argument and attributes it to "a most eminent and learned person, and before which one must fall silent,"[26] and Salviati, for once, piously agrees. By means of the influence of the Tuscan ambassador, Galileo extracted the imprimatur, or permission to publish, from all the censors and hastily proceeded to print and publish the book in Florence.

The arrival of copies of the book in Rome caused an uproar. The pope flew into a rage, accusing Galileo of meddling with things that he ought not. He also bitterly accused Galileo of betrayal and of circumventing the censors, and he accused the censors of incompetence and collusion.

Exactly what brought on the papal rage remains somewhat of a mystery. Historians have conjectured that Galileo's Jesuit enemies incited the pope by accusing Galileo of defending the Copernican system or of holding subtle un-Catholic theological or philosophical ideas. But none of this seems sufficient to justify Urban's temper tantrum. In the book, Salviati speaks in defense of the Copernican system, but the censors, all of them well-trained Dominicans, had not objected to the extent of this defense. Neither had the censors discovered any other actionable theological defects in the book, and the printed book prominently displays the official stamps of approval from all three censors who had examined it. It seems likely that what provoked the pope's rage was not some theological or philosophical quibble, but some personal matter, known only to him and to Galileo.

A clue to the mystery may be provided by the discussions between Urban and Galileo some years before, which led to the writing of the book. We know that these discussions dealt with the "two systems." The pope undoubtedly emerged from these discussions thinking that he had won the argument, whereas Galileo emerged thinking *he* had won. The pope must have anticipated that Galileo's *Dialogue* would be a report of these discussions, and that it would reveal to Rome and to all the world, *urbi et orbi*, how cleverly the pope had defeated Galileo. It must have come as a very rude shock to His Holiness to find that the *Dialogue* claimed exactly the opposite: the pope (a.k.a. Simplicio) is defeated again and again by Galileo (a.k.a. Salviati). This conjecture is supported by what one of Galileo's friends wrote to him in a letter sent shortly after the book's arrival in Rome; he reports that Father Riccardi, the censor in Rome, told him that the error in the book was that: ". . . two or three arguments have been omitted at the end which were invented by our Lord's Holiness himself and with which, he says, he convinced Signor Galileo of the *falsity* of the Copernican theory. The book having fallen into His Holiness' hands, and these arguments having been found wanting, it was necessary to remedy the oversight."[27]

CONJECTURES ASIDE, the one thing that is certain is that something about Galileo's book offended Urban profoundly. The Tuscan ambassador reported that Urban ". . . was so incensed that he treated this affair as a personal one,"[28] and when he begged Urban to inform Galileo of the precise nature of the objections to his book, Urban answered violently, ". . . he knows well enough what the difficulties are, if he wants to know them; for We have discussed them with him, and he knew them all from Ourselves" and added, "I have used him better than he has used me, for he deceived me."[29] Hell hath no fury like a vainglorious pope spurned, and Urban's vendetta lasted as long as Galileo's life and beyond his grave. Upon Galileo's death, Urban refused to consent to the Grand Duke's wish to erect a funeral monument in the church of Santa Croce in Florence. Even many years after Galileo's death, when Urban was a hunched and wizened old man, his head sunk to the level of his shoulders, he was wont to repeat a long litany of furious complaints against Galileo.[30]

On orders from Urban, the Inquisition suspended the publication of the book and attempted to confiscate all the copies. But most copies were sold before they could be confiscated. The book circulated widely, it was received with much praise throughout Europe, and it was soon translated into Latin and into English. Galileo was summoned to Rome, to place himself in the hands of the Inquisition and to face trial for holding, defending, teaching, and discussing the Copernican system.

During the initial interrogation, Galileo put up a good defense, but under threat of worse accusations, Galileo finally agreed to what we would call today a plea bargain. He confessed that he had overstated the case for the Copernican system, ". . . a reader might have had reason to suppose that the arguments . . . were so expressed as to . . . compel conviction by their cogency" and that he had ". . . resorted to . . . devising, even in favor of false propositions, ingenious and plausible arguments . . . out of vainglorious ambition and pure ignorance and inadvertence."[31] In exchange for this confession, the commissary, or investigating judge, recommended leniency and hinted that Galileo might be allowed to rewrite and republish his *Dialogue* with suitable revisions and additions.

But Urban's rage was not appeased by Galileo's confession. He wanted Galileo crushed and humiliated, and he issued a decree instructing the inquisitors-general to subject Galileo to the threat of torture, to convict him of "vehement suspicion of heresy" for having held and defended the Copernican system, and to compel him to a public abjuration. The legal arguments that the inquisitors-general contrived for the preamble of the sentence were nothing but window dressing. Urban wanted Galileo convicted and sentenced, and his will was done.[32]

Galileo's sentence included a prohibition of his book, and its title was placed on the Index of forbidden books. This was a futile gesture, as Galileo had taken the precaution of sending copies to the Netherlands, where it was reprinted by the Elsevier publishing house and distributed widely, except in the papal dominions. The book remained on the Index until 1835, when the Church relented and permitted its republication.

In 1981, Pope John Paul II established a commission to re-examine the Galileo affair, and newspapers speculated that this might lead to an admission of judicial error on the part of the Church, with a retrial, a rehabilitation, or perhaps even a canonization of Galileo (Saint Galileo of Padua?). But John Paul soon revealed himself as an archconservative, and in 1992 the work of the commission was suddenly declared completed, and the blame for the conviction of Galileo was conveniently laid on the shoulders of low-ranking and unnamed theologians.[33] John Paul vaguely conceded that Galileo's judges had perhaps committed some subjective errors, and he issued a perfunctory apology, describing the whole matter as a "tragic, mutual misunderstanding." This is rather like calling the show trials of Stalin a "misunderstanding." Urban VIII and his cardinals did not suffer from any misunderstanding. They knew quite well what they were doing . . . and so did Galileo.

Although John Paul did not formally and explicitly rescind the condemnation of Galileo, it is generally recognized that the condemnation has been removed implicitly. With Pope Benedict XVI now in control of the Catholic nomenklatura, there is no chance that anything further will be done in the matter of Galileo anytime soon. John Paul was a conservative, but Benedict is a reactionary. As Cardinal "Panzer" Ratzinger, he was the prefect of the Inquisition (currently known by the more user-friendly name of the Congregation for the Doctrine of the Faith). One of his favorite maxims was said to be "Truth is not determined by majority votes"; this would maybe make a good motto for a scientist, but coming out of the mouth of the Grand Inquisitor it sounds rather ominous.

Ratzinger gave several speeches in which he quoted, with apparent approval, the controversial philosopher Paul Feyerabend, who had said: "The Church at the time of Galileo kept much more closely to reason than did Galileo himself . . . Her verdict against Galileo was rational and just, and the revision of this verdict can be justified only on the grounds of what is politically opportune."[34] After he became pope, these words came back to haunt him—he was forced to cancel a ceremonial visit to the Università "La Sapienza" in Rome in January 2008, because members of the faculty strongly protested against the visit, complaining that the quoted words were offensive.

The Galileo affair was a shameful episode in the relationship between science and the Church. It was shameful for science, because Galileo fraudulently attempted to prove the motion of the Earth by arguments that he knew, or should have known, to be false. Maybe his confession was feigned, maybe not; but the words he used in his confession describe his true crime

Galileo's villa in Arcetri. Here he spent his final years, under house arrest. The villa is now the property of the University of Florence, and is to be used as a small conference center. *(Photo Enrico Brunetti, INAF, Osservatorio di Arcetri, Firenze)*

quite accurately: he devised false propositions out of vainglorious ambition and pure ignorance and inadvertence. And it was doubly shameful for the Church, because the rash intervention of a vindictive pope prevented a reasonable settlement of the affair and the publication of a corrected version of Galileo's book. We have it on the authority of Dante that several popes wound up in the eighth circle of Hell. Urban VIII deserves to join them there for subverting the course of justice and of science.

AFTER THE SENTENCE was pronounced on Galileo at the Convent of Santa Maria Sopra Minerva, he was remounted on his mule and returned to his prison at the Palace of the Inquisition. Two days later he was released into the custody of the Tuscan ambassador, and then permitted to return to Florence. His sentence of perpetual imprisonment was commuted to house arrest in his villa in Arcetri, in the hills south of Florence.

In Arcetri, Galileo continued his scientific work. He resumed his study of mechanics, and described his experiments and principles of mechanics in the *Dialogue of Two New Sciences*. He completed this important book in 1636, and went blind the year after. This is how he was when John Milton passed through Arcetri in 1638: "There it was that I found and visited the famous Galileo, grown old, a prisoner of the Inquisition."[35] But not even blindness could stop Galileo's work, and he continued to dictate to his disciples further ideas about the development of mechanics until his final illness.

If you visit Arcetri, you can still find Galileos's house on Via Pian dei Guillari. A marble plaque, next to the front door, identifies the house as "Villa Galileo," and he died there in 1642, at age seventy-eight. The symptoms of his terminal disease suggest that he died of chronic lead poisoning brought about by consumption of lead-contaminated wine from his own cottage winery.[36] He was provisionally buried in a small room under the campanile of the church of Santa Croce in Florence. His favorite daughter, Suor Maria Celeste, a nun at the convent of San Matteo in Arcetri who had died some years before, was buried next to him. In 1737 his remains and those of his daughter were transferred to the nave of the church, which is the burial place of many illustrious Italians. Here, with the consent of Pope Clement XII, a splendid monument was erected to him, opposite the monument to Michelangelo. This was intended to suggest a spiritual continuity between Galileo and Michelangelo—it was widely believed that Galileo was born when Michelangelo died (they actually overlapped by a few days, so if they shared the same soul, they must have made do with half shares).

The fused bones of his left index finger were removed during the reburial and are now on view in a reliquary at the Museum of Science in Florence.[37] From this one might conclude that some Italians were seriously confused about the distinction between scientists and saints and failed to grasp that the characterization of Galileo as a martyr of science was merely a figure of speech.

3

"If I have seen farther . . ."

Isaac Newton in a letter to Robert Hooke, in which he acknowledges that his success in formulating the fundamental laws of physics depended on the contributions of his great predecessors, such as Galileo and Kepler.

The next step in the development of relativity was taken by Isaac Newton, the most awesome and the most awful physicist of all times. Newton gave relativity a completely new face. Whereas Galileo had presented relativity as an induction, that is, a generalization based on observation and direct experience, Newton presented relativity as a mathematical deduction from "Newton's laws," the fundamental laws of mechanics that he formulated and that came to dominate the development of physics for the next two-and-a-half centuries.

Newton's formulation of the laws of mechanics, his discovery of the law of gravitation, his application of these laws to a wide range of physical problems, and his brilliant contributions to mathematics rank among the greatest achievements of the Age of Reason. But Newton stood with one foot in the Age of Reason and with one foot in the Middle Ages. His atavistic medieval proclivities were revealed in his work on alchemy, to which he devoted much effort. Maynard Keynes, who gained fame as an economist but was also a distinguished Newton scholar, called him the last of the magicians.

Newton was a genius, the greatest genius ever seen in physics or in any other science. And he was mad, intermittently mad. Like Hamlet, he could have said, "I am but mad north-north-west. When the wind is southerly, I know a hawk from a handsaw." Newton suffered from hatter's disease, a peculiar form of madness that often afflicted hatmakers in the seventeenth century. The disease was caused by chronic poisoning with mercury compounds used for the stiffening of felt in the manufacture of hats, and its symptoms included tremors, irritability, loss of concentration, apathy, insomnia, depression, and loss of memory. Newton did not manufacture hats, but he used mercury in alchemical experiments, and he even drank some out of curiosity and routinely tasted the mercury compounds produced in his experiments (he described the taste of mercury as "strong, sourish, ungrateful").[1] Newton's mercury poisoning probably did not come from this reckless imbibing—mercury is poorly absorbed by the digestive tract. His poisoning probably came from inhalation of mercury vapors. In his alchemical experiments, Newton often heated mercury in open crucibles, and the mercury vapors produced under these conditions are readily absorbed through the lungs.

In a letter to John Locke, Newton himself described his severe insomnia and loss of memory, and he seems to have been vaguely aware of the cause of his symptoms: "The last winter by sleeping too often by my fire I got into an ill habit of sleeping and a distemper which this summer has been epidemical put me further out of order, so that when I wrote you I had not slept an hour a night in a fortnight and for five nights not a wink. I remember I wrote to you but what I said of your book I remember not . . ."[2] Besides, from samples of his writing, we know that he was also suffering from tremors. He himself attributed the premature graying of his hair to prolonged contact with mercury. His mercury poisoning was recently confirmed by forensic evidence: analysis of samples of his hair established the presence of elevated concentrations of mercury.[3]

NEWTON HAD TWO EPISODES of madness, one starting in 1677 and the other starting in 1693. Both episodes lasted more than a year and coincided with his alchemical experiments. The contemporary Dutch physicist Christiaan Huygens reported one of these episodes to Leibniz: "I do not know if you are acquainted with the accident which has happened to the good Mr. Newton, namely, that he has had an attack of phrenitis, which lasted eighteen months, and of which his friends have cured him by means of remedies, and keeping him shut up."[4]

Predictably, English historians pooh-pooh the notion that their greatest scientist was a lunatic. They claim that Newton's illness was much exaggerated by "foreign contemporary writers"[5] and that madness is "the charge that mediocrity hurls against great achievers."[6] Maybe so. But whether we describe Newton as mad or not, his behavior during the episodes was certainly bizarre, even if we make allowances for the usual eccentricities of Englishmen. In another letter to Locke and in a letter to Samuel Pepys, Newton wrote, "Being of opinion that you endeavored to embroil me with woemen & by other means, I was so much affected by it, as that when one told me you were sickly & would not live, I answered twere better if you were dead. I desire you to forgive me this uncharitableness"[7] and "I am extremely troubled at the embroilment I am in, and have neither ate nor slept well this twelvemonth, nor have my former consistency of mind but am now sensible that I must withdraw from your acquaintance, and see neither you nor the rest of my friends any more, if I may but leave them quietly."[8]

Newton's final alchemical experiments were performed in 1696, shortly before he moved from Cambridge to London. He displayed no madness after that date, but his mental abilities never recovered fully, and the work he did in his later years was not comparable to his early achievements. He became addicted to interminable theological speculations and he published only one important scientific work, his *Opticks*; but this work was an account of experiments he had performed much earlier, before his madness.

NEWTON WAS BORN in the hamlet of Woolsthorpe, near Grantham in Lincolnshire, in 1642, less than a year after Galileo's death. The Woolsthorpe farm held the privilege of a manor, which gave the owner the right to call himself "lord of the manor," and in his later years Newton was to attach great importance to his lordly title. But his father, who died before Isaac's birth, was a yeoman farmer of modest means and did not have the breeding that might be expected to come with such a manorial title. Both his father and his brother were illiterate, and his mother could barely write. When the young lad displayed an absolute lack of aptitude for farming, he was sent to a grammar school in Grantham, in the hope that he might ultimately make a living as a schoolteacher or a man of the

Woolsthorpe Manor near Grantham in Lincolnshire, the birthplace of Newton.
It is now a museum. The apple orchard with a descendant of Newton's apple tree
lies just beyond the right edge of the picture.

King's school, Newton's grammar school in Grantham, dates from 1528. It is
located next to the parish church of St. Wulfram's and is still used as a school today.

cloth. At first Newton was lazy, but then he quickly rose to first in his class, and one of his uncles, a Cambridge graduate, recommended that he be sent to the university.

Today, Cambridge is renowned for scientific research and other scholarship; but in Newton's day it had little reputation for anything, and it trailed far behind Oxford. It has been said that in those days the only thing a Cambridge education fitted anyone for was to be a competent priest or a bad doctor. Newton enrolled in Trinity College, Cambridge, as a "sizar," a position that excused him from payments for instruction, and room, and board, in exchange for working as a servant to the paying students of the college. As a servant, Newton was expected to perform menial tasks—fetch and carry beer and bread and firewood, polish boots, empty chamber pots.

The humiliations he experienced in this servile position left him permanently scarred, with a profound inferiority complex about his social position, for which he compensated with an aggressive assertiveness about his intellectual achievements. Neurotic, obsessive, hypochondriac, celibate, and a Puritan to boot, he was fierce in defending his claims of priority for his discoveries in mathematics and physics. Newton had what we might call the Rumpelstiltskin complex, after the pathetic little man in the Grimm fairy tale whose only joy in life was to hop around his campfire singing, "I am so glad that nobody knows that Rumpelstiltskin is my name." Like Rumpelstiltskin, Newton took insane pleasure in knowing something that nobody else knew. So it was with his demonstration of the elliptical planetary orbits, and especially with his knowledge of "fluxions" and "fluents," what we now call derivatives and integrals in differential and integral calculus. It is easy to imagine him skipping around his bedchamber at night cackling, "I am so glad that nobody knows what I know about ellipses and fluxions." He never revealed the details of his method of fluxions, and his papers on the fluxion calculus were published only after his death.

As a student at Cambridge, Newton focused on his own research and neglected the standard curriculum, with the result that he did poorly in his final exams ("he lost his groats," that is, he lost the small coins that by an ancient custom the student would place on the desk of the examiner and forfeit if he did poorly). In a close parallel, as a student at the Zurich Polytechnic, Einstein was equally neglectful of the standard curriculum, and did equally poorly in his exams. But Newton had impressed his professors with his mathematical abilities, so much so that after he graduated, the mathematician Isaac Barrow graciously yielded to Newton his chair as Lucasian Professor (a chair held today by Stephen Hawking). Whereas Einstein did not make much of a favorable impression on his professors, and he was left out in the cold.

At the time of his appointment as professor, Newton was only twenty-seven years old. He remained at Cambridge for the next twenty-seven years, and these were the most productive years of his scientific career. With his professorship he attained financial security; besides, upon the death of his mother, he inherited the Woolsthorpe farm, which made him a well-to-do country squire, so he could call himself "Isaak Newton of Woolstropp, Gentleman" (and, yes, lord of the manor).

NEWTON HELD THE ODD NOTION that whenever he discovered some new result in physics or mathematics, it became his personal property, which he was entitled to keep as a secret for as long as he chose, without any need to publish it to establish his priority. If another scientist later made the same discovery independently and published it first, Newton regarded this as trespass and as theft, and he would indignantly refuse to allow such a scientist any share of the credit. This odd notion conflicts sharply with the modern criterion: today, credit for a discovery is awarded strictly on the basis of priority of publication—the early bird gets the worm. In Newton's days, the criterion for credit for a discovery was not yet rigidly established. Claims for unpublished discoveries were sometimes accepted, especially if the scientist had the vociferous support of influential friends and patrons—sometimes the early bird got the worm, and sometimes the squeaky wheel got the grease.

Newton's secretiveness about his discoveries led him into many silly but savage disputes with other scientists about what they knew and when they knew it. Driven by his intense paranoia about his scientific accomplishments, he accused Robert Hooke, Gottfried Leibniz, and other scientists and mathematicians of stealing ideas from him. In his treatment of these scientists he was vicious and vindictive. Hooke was a talented scientist, best known for his investigations with microscopes, but he was a dwarfish man, with a stooped back. When Hooke asked for an acknowledgment that he had anticipated some of Newton's investigations of the colors in sunlight, Newton wrote a sarcastic refusal, in which he made an oblique reference to Hooke's diminutive size: "If I have seen farther, it is by standing on the shoulders of Giants."[9]

The German mathematician and philosopher Leibniz discovered the calculus independently, and, in contrast to Newton, he published his discoveries—by the modern criterion, Leibniz would have had full credit for the calculus, and Newton none. But when Leibniz asked a committee of the Royal Society to prepare an impartial report judging his share in the invention of the calculus, Newton not only packed the committee with his cronies, but he also wrote the report himself, and then wrote a favorable anonymous review of the report. In his private journal he gleefully recorded that he had bested Leibniz and "broke his heart."

English mathematicians paid the price for this shoddy treatment of Leibniz—for a long time thereafter their xenophobia and chauvinism compelled them to adhere to Newton's clumsy notations for derivatives and integrals, instead of adopting Leibniz's much clearer notation. Ultimately, Leibniz's notation won; it is now the standard notation for calculus, even in England (a vestige of Newton's notation remains in modern calculus in that time derivatives are sometimes still indicated by a dot placed over a mathematical symbol, for example $\dot{x}$; Newton used such a dot placed over the symbol and called it a "pricked" symbol).

NEWTON WAS THE VERY MODEL of the absentminded professor, the genius, and the mad scientist. When concentrating on a problem he would become oblivious to his surroundings, he would forget to take his meals, and he would forget to go to sleep. The cat he kept in his chambers at the university grew fat from gorging on the food Newton forgot to eat. A canard about him that circulated at Cambridge was that on one of the two occasions he proposed marriage, he fell into a reverie about a mathematical problem from which he was awakened by the screams and struggles of the young lady—he had absentmindedly taken hold of her small finger and used it to tamp the tobacco burning in his pipe. He never married, and at times displayed a phobia about women.

He was not a good teacher or lecturer; students found it difficult to understand him and few attended his lectures. Often, not even a single student would come to hear him, and he "would read to the walls," giving a lecture in an empty room (professors who failed to lecture, with students or without, were punished by a fine).

In 1696, at age fifty-four, Newton left Cambridge to assume a well-paid position as Warden of the Mint and, later, Master of the Mint. He was granted these positions by the patronage of the earl of Halifax, Lord of the Treasury, one of Newton's friends and admirers. But there is a delightfully scandalous tale connected with Newton's appointment. Newton had a very beautiful niece, whom the earl admired at least as much as he admired Newton, and who became the earl's mistress. Voltaire, always well-informed about juicy bits of gossip, commented, "I thought in my youth that Newton made his fortune by his merit . . . No such thing. Isaac Newton had a very charming niece, Madame Conduitt, who made a conquest of the minister Halifax. Fluxions and gravity would have been of no use without a pretty niece."[10]

Newton's appointment was intended merely as a comfortable and profitable sinecure. But, to the surprise of Halifax, Newton threw himself into the work at the mint with great zeal and enthusiasm. The mint was at that time engaged in an ambitious project for remaking the entire coinage of the realm. New coins were issued and old coins were withdrawn, because they had become untrustworthy, with wide circulation of worn, counterfeit, debased, and clipped coins. In this work, Newton's expertise in alchemy stood him in good stead, and he was able to manufacture coins of uniform quality with a standard composition of gold or silver. As Master of the Mint he was entitled to a percentage of all the coins he manufactured, from which he accumulated considerable wealth. And he was not above juggling the accounts at the mint, to gain a few extra pounds of profit.

Although he continued to collect his pay as a professor at Cambridge, he gave up his scientific and mathematical pursuits, and instead devoted much effort to theological writings, numerological prophecies about the resurrection of Christ and the end of the world, and biblical chronology, for which he calculated and recalculated the ages of the prophets, kings, and patriarchs listed in the books of the Bible. In this he anticipated the zany calculations of the age of the world performed by the Christian fundamentalists of today, who claim that the world is 11,000 years old and attempt to twist science into conformity with a literal interpretation of the Bible.

He also speculated on connections between science and religion, and he contemplated the charming concept that gravitation is communicated over the distance separating one particle from another by an incorporeal ether that consists of the spiritual body of Jesus Christ.[11] (Did he perform experiments to test whether the ingestion of Christ's body at Communion produces gravitational disturbances?) These private speculations contrasted sharply with his public attitude in his scientific writings, where he had avoided any such mumbo-jumbo and stuck to the observational facts, declaring, "I make no Hypotheses."

UPON HOOKE'S DEATH, Newton was elected president of the Royal Society, and he delighted in the pomp of presiding over its meetings. He treated the Royal Society as his fiefdom and used it to further the careers of his friends and supporters. Queen Anne knighted him, and he became *Sir* Isaac Newton, with a coat of arms displaying a pair of crossed bones, such as found in the skull-and-crossbones emblem of a pirate's flag. The knighthood was bestowed not for his glorious achievements in science, nor for his exemplary service at the mint, but merely to bolster his ebbing campaign for election to the House of Commons—but it did not help, and he was trounced by all the other candidates.

In the typical nouveau riche manner, he loved to make a display of his wealth, giving ostentatious dinners and decorating his entire London house in crimson: a crimson mohair bed, crimson bed curtains, crimson hangings, crimson settee—whenever a color is listed in the inventory of his possessions, it is crimson. He was generous to his poor relations, and enjoyed his role as the rich and powerful head of the Newton clan. But he was quite ungenerous to other petitioners begging for his help. As Warden of the Mint he persecuted and prosecuted coiners, clippers, and anybody found in possession of counterfeit coins. In this he displayed a ferocity bordering on sadistic cruelty. He often obtained convictions by the questionable practice of relying on mercenary informants, and he denied desperate pleas for clemency with contempt, haughtily declaring that "Criminals, like dogs, always return to their vomit."[12] It is passing strange that the greatest genius the world has ever known should have spent his later years in the manipulation of account books and the persecution of petty criminals. Perhaps he felt this was the way to get even with the world for the deprivations and humiliations he had suffered in his younger years.

Newton died in 1727, at age eighty-four. His funeral was all he could have wished for. He was the first scientist to be granted a state funeral, with the Lord Chancellor, two dukes, and three earls as his pallbearers. Voltaire commented that England honored a mathematician as other nations honor a king (in view of the execution of Charles I and the eviction of James II, he would have done well to add that how the English honor their kings depends on where the wind is blowing from). Newton was buried in the center of the nave of Westminster Abbey. There, at the choir screen, you can find an imposing baroque monument that his heirs erected to his memory.

WHATEVER HIS PERSONAL FAILINGS may have been, Isaac Newton was the greatest physicist of all times. The years from 1664 to 1666 can be reckoned as Newton's miracle years, analogous to Einstein's miracle year of 1905. In these years he made his great discoveries in mathematics, especially the differential and integral calculus, he completed his investigations on optics and the color spectrum of white light, he conceived of gravity extending as far as the Moon and beyond, and he concluded that the gravitational force that keeps the planets in their orbits must decrease with the inverse square of the distance.

In 1687, after long delays attributable to his reluctance to reveal all, he published his celebrated book *Mathematical Principles of Natural Philosophy*. Usually called the *Principia* from its Latin title, it is one of the glories of the Age of Reason.

At the beginning of his book, Newton proclaims the three fundamental laws of motion, now called Newton's laws. The following statements of the First and Third laws are Newton's own; but the statement of the Second Law is a modern reformulation of Newton's somewhat convoluted statement:

> Law I: *Every body continues in its state of rest or of uniform motion in a straight line, unless it is compelled to change that state by forces impressed upon it.*
> Law II: *The product of the mass and the acceleration of a body equals the force that acts upon it, or ma = F.*
> Law III: *To every action, there is always opposed an equal reaction.*

You may have been taught in some high school or college physics course that Newton's Second Law is $F = ma$. If so, you were taught by an engineer, not by a real physicist. Newton himself stated that the acceleration (or what he called the "change of motion") is proportional to the force, not that the force is proportional to the acceleration. Of course, both statements are true, but they differ in emphasis. For Newton, and for most physicists, the second law expresses how the force causes an acceleration, that is, the force is the known quantity and the acceleration is the unknown quantity. Accordingly, physicists prefer to write $ma = F$. Engineers prefer the opposite way of writing the Second Law, $F = ma$, because for them the force is often the unknown quantity, whereas the acceleration is the known quantity (for instance, an engineer may want to calculate what force will be exerted on the wheels of a car when it is racing around a curve at a known, high speed).

From his deceptively simple laws of motion, Newton deduced a wealth of momentous mathematical results. Among these results are ingenious proofs of Kepler's laws, which are shown to be consequences of Newton's laws of motion in conjunction with his Law of Universal Gravitation, according to which every particle attracts every other particle with a gravitational force directly proportional to the product of their masses and inversely proportional to the square of the distance. Newton had arrived at this law of gravitation as early as 1666, when, as in the preceding year, plague arrived in Cambridge, and the university again closed. Newton returned to Woolsthorpe, and, as he described it later, "I began to think of gravity extending to the orb of the Moon . . . & thereby compared the force requisite to keep the Moon in her Orb with the force of gravity at the surface of the earth, & found them answer pretty nearly. All this was in the plague years 1665–1666. For in those days I was in the prime of my age for invention & minded Mathematicks and Philosophy more than any time since."[13]

IN HIS OLD AGE, Newton told several friends and relatives that his thinking about gravity was inspired by an apple falling from a tree in the garden. John Conduitt, the husband of Newton's beautiful niece, retells this: "Whilst he was musing in a garden it came into his thought that the power of gravity (which brought an apple from the tree to the ground) was not limited to a certain distance from the earth but that this power must extend much farther than was usually thought. Why not as high as the moon said he to himself . . ."[14] The apple tree at Woolsthorpe Manor that inspired Newton's initial discovery of the law of gravitation survived until 1820, when it finally had to be cut down because of rot. The oldest of the trees that you can now find in the orchard in front of the manor is a replacement that incorporates cuttings from the original tree.

In retelling this story Conduitt misses the mark. The important new contribution that Newton brought to our understanding of gravity was not that gravity extends as far as the Moon—which had already been contemplated by Galileo, with whose work Newton was familiar—but that the strength of gravity decreases in inverse proportion to the square of the distance. Newton discovered this inverse-square law for the gravitational force by examining Kepler's Law of Periods, and he confirmed it by his calculation of the force of gravity affecting the Moon.

Newton was not alone in this discovery. Hooke held discussions with Edmond Halley, the preeminent astonomer of those days, and Christopher Wren, the famous English architect, and together they rediscovered the same law, after Newton's discovery but before Newton's publication. In a rare display of fairness, Newton consented to acknowledge their independent discovery in the first edition of the *Principia* (in later editions he reverted to true form, and deleted all mention of Hooke).

Newton's long delay in publishing his discovery of the inverse-square law of gravitation was partly due to his Rumpelstiltskin complex, and partly due to an error in his initial calculation of the magnitude of the force of gravity acting on the Moon. For his first calculation he had used the best available value for the radius of the Earth, but this value was in error by about 15 percent, and this threw off Newton's calculation and made him doubt the inverse-square law. Besides, there were two other difficulties that Newton had to deal with: he needed to prove that the combined gravitational forces of all the particles in the spherical body of the Earth or the Moon can be treated as a single gravitational force acting at the center of the Earth or the Moon, and he needed to prove that the inverse-square force produces elliptical orbits, in agreement with Kepler's laws. With the powerful mathematical methods available today, any reasonably competent undergraduate physics student can prove in twenty words or less that the gravitational action of a spherical body is concentrated at its center; but with the mathematical methods that Newton had available, this is a difficult thing to prove, and the demonstration he gives in the *Principia* is quite tedious.

BESIDES PRESENTING THE FORMULATION of the fundamental laws of motion and the deduction of Kepler's laws for planetary motion, the *Principia* also contains many other notable results, such as the deduction and calculation of the orbits of comets, equatorial bulge of the Earth, ocean tides, precession of the axis of rotation of the Earth, analysis of motion of fluids, etc. Stephen Hawking quite rightly called the *Principia* the most important single work ever published in the physical sciences. We could say that all physicists today are "people of the book," and the *Principia* is that book.

For two-and-a-half centuries Newton's mechanics ruled physics, and even when it was finally displaced by quantum mechanics in the first half of the twentieth century, it retained its hold over a homeland core, the domain of macroscopic physics and engineering, somewhat like the British Empire retaining its hold on the United Kingdom homeland when the imperial possessions were stripped off. To appreciate the great and lasting importance of Newton's work, it suffices to skim through any introductory physics textbook of today. Typically, in such a textbook, the first two or three chapters deal with results contributed by Galileo, and the next twelve or fifteen chapters deal with results contributed by Newton.

The brilliance and originality of the propositions and corollaries in the *Principia* is dazzling. The breadth and depth of Newton's discoveries in mechanics, optics, astronomy, and mathematics place him far above any other physicist. Not even Einstein can be compared to Newton. If we reckon net worth by the number of great discoveries, one Newton is worth several Einsteins.

In a preface to the *Principia*, Halley sang an extravagant ode to Newton:

> *Then ye who now on heavenly nectar fare,*
> *Come celebrate with me in song the name*
> *Of Newton to the Muses dear; for he*
> *Unlocked the hidden Treasuries of Truth:*
> *So richly through his mind has Phoebus cast*
> *The radiance of his own divinity.*
> *Nearer the gods no mortal may approach.*

And for once, this hype was justified. Newton showed us that motion of the heavens could be calculated from his laws, and that the laws that apply to the motion of the heavens were the same as those that hold on Earth. Kepler had merely described planetary motion, but Newton analyzed the motion and showed how the machinery of the heavens works. The fact that few could understand his calculations did not matter; on the contrary, it probably added to his cachet.

The *Principia* is a quite impenetrable book, except for mathematicians well trained in geometry. Few could read it in Newton's days, and today hardly anybody even tries. The Cambridge students said of Newton, "There goes a man that writt a book that neither he nor any body else understands."[15] The impenetrability of the *Principia* was not the result of a lack of talent for explanation. Newton was quite capable of clear explanations when he wanted, as illustrated by his illuminating discussion of the relationship between projectile motion and orbital motion, in which he imagined the motion of cannonballs fired horizontally from a cannon placed on a high mountain, with progressively larger and larger muzzle velocities. These cannonballs reach progressively larger and larger distances before they strike the ground, and, if the muzzle velocity is large enough, a cannonball can travel all the way around the Earth and return to its starting point, which means this cannonball achieves orbital motion. But Newton deliberately made most of his work as abstruse and unreadable as possible, "to avoid being baited by little Smatterers in Mathematicks."[16] Einstein's theory of general relativity has a reputation of being very difficult to understand, but in fact Einstein's work is much easier to read than Newton's *Principia*.

The *Principia* is a brilliant diamond, but it is a diamond with flaws. It is unreadable, it abounds in contradictions and inconsistencies, and it is festooned with a handful of gruesome mistakes. Some of these mistakes are outright errors in calculations and demonstrations, others are gaps in logic where Newton simply guessed what he could not prove.

The depth of Newton's intuition is often astonishing—many of his seemingly wild guesses hit the mark. Newton understood the physics underlying the equatorial bulge of the Earth and, in contrast to Galileo, he understood how the gravitational pull of the Moon causes the tides. He also understood how the gravitational pull of the Moon and the Sun on the equatorial bulge of the Earth causes a gradual drift of the orientation of the axis of rotation of the Earth (what astronomers call the precession of the equinoxes). But his mathematical methods were inadequate to treat these problems rigorously. He resorted to crude approximations, mixing good physics with poor mathematics. He did not admit to these mathematical deficiencies, and he passed off his crude guesstimates as exact calculations. During his lifetime, nobody challenged these errors, because nobody understood the book well enough to dispute him.

A CAREFUL EXAMINATION of Newton's writings has revealed that some of the errors in the *Principia* were deliberate and dishonest attempts to mislead. In the *Principia* Newton proposed that the exact quantitative agreement between theory and observation was the ultimate criterion of scientific truth. As he said in the preface, "He that works with less accuracy is an imperfect mechanic; and if any could work with perfect accuracy, he would be the most perfect mechanic of all."[17] And to convince his audience that he was the "most perfect mechanic," he proceeded to fabricate the required agreement between theory and observation, by fair means or foul. Newton faked some theoretical calculations and he engaged in flagrant cherry-picking of observational data, discarding those data that did not quite fit his calculations. Richard Westfall, one of Newton's most incisive biographers, called this "nothing short of deliberate fraud," and he labeled Newton a master of the "fudge factor."[18]

One example of Newton's fakery is seen in his theoretical calculation of the speed of sound. He correctly and perceptively identified sound as an oscillation arising from the elasticity of air, and he recognized that the speed of sound waves must depend on the ratio of the elasticity of air to its density. But when he calculated the value of the speed from this ratio, he found it was too low by about 20 percent, and to fix this discrepancy between theory and observation he conjured up two fudge factors, both of which were pure fiction.

First, he imagined that whereas sound travels at a finite speed in the space between the air particles, it travels at infinite speed through the body of any air particle it encounters, and he therefore inserted a correction factor that depends on the fraction of the air volume that is actually occupied by the bodies of particles. Then he imagined that not all of the air that surrounds us is "true air"; instead, some of it is "vapor," which does not participate in the propagation of sound and therefore, somehow, made sound travel even faster, and for this he inserted another correction factor. With these two fudge factors, his final theoretical value for the speed of sound came to 1142 feet per second, which—surprise, surprise—agreed exactly with the value of 1142 feet per second that had been recently measured in experiments by his friend William Derham.[19] "This passage," commented

Westfall, "is one of the most embarrassing in the whole *Principia*, since the adjustments rested on no empirical grounds whatever, and in their manifest hollowness served only to cast undeserved doubt on the basic analysis."[20]

Other examples of similar fakery are found in Newton's theoretical calculations of the precession of the equinoxes, the magnitude of the force of gravity acting on the Moon, the height of the tides, and the size of the equatorial bulge of the Earth.[21] In all of these cases, he had a good *qualitative* understanding of the underlying physics, but inadequate mathematical tools and/or inadequate observational data for an accurate *quantitative* analysis—and so he inserted fictional fudge factors into his calculations and/or cherry-picked the data.

These instances of fraud by Newton are somewhat reminiscent of the fraud that Galileo perpetrated with his theory of tides. But whereas we might give Galileo the benefit of doubt and charitably regard his mistaken theory of tides as an isolated case of self-delusion, no such excuse will serve for Newton. His acts of fraud occurred repeatedly, much too often to admit of self-delusion as a plausible explanation. In the perpetration of fraud, Newton was a recidivist, deserving of no charity. Besides, we have documentary evidence from letters between Newton and Roger Cotes, the editor of the second edition of the *Principia*, that they engaged in collusion to "mend" the numbers. Cotes would propose to Newton some fraudulent adjustment of observational data, "to make that Scholium appear to best advantage as to the numbers," and Newton would do Cotes one better by contriving some fudge factor that suited the occasion.

Newton's fraud did not receive wide attention because the *Principia* was much admired but little read, and its influence on the development of physics was indirect. Although most of the "Newtonian physics" found in today's physics textbooks is based on Newton's ideas and results, very little of it remains in the form presented by Newton in his *Principia*. Almost the entire content of his book was reorganized and rephrased by his followers, starting with the great Swiss mathematician Leonard Euler and continuing with the French savants and mathematicians of the eighteenth century: Pierre Maupertius, Jean Le Rond d'Alembert, Joseph Louis Lagrange, Pierre Simon de Laplace, Adrien Marie Legendre.

They all worshiped "le grand Newton," but they felt that his work needed to be remodeled with a touch of French elegance. They proceeded to tear down what Newton had built, keeping only the foundations, and on these foundations they built a new, remodeled version of mechanics, based on Leibniz's calculus and various new, auxiliary concepts, such as potential, moment of inertia, angular momentum, and, in a more sophisticated treatment, the principle of least action.

While translating Newton's work into the language of Leibniz's calculus, Euler reformulated Newton's Second Law in the form that is today found in all introductory physics textbooks: mass times acceleration equals force.[22] And then he went on to apply this reformulated version of Newton's laws to problems of the rotational motion of rigid bodies (such as the motion of a top or the rotational motion of the Earth), the vibrational motion of elastic bodies, and the motion of fluids, treating all of these as systems of pointlike particles. These were problems that Newton had failed to solve, because with Newton's geometrical methods the analysis of the three-dimensional motions of three-dimensional bodies is somewhere between difficult and impossible.

THE *PRINCIPIA* INCLUDES a statement of the principle of relativity of motion. Superficially, Newton's statement of this principle of relativity seems quite similar to Galileo's. This is not surprising—Newton was well acquainted with Galileo's work, probably from an English translation of the *Dialogue* published in London in 1661 or perhaps from a Latin translation. Galileo is one of the few scientists whom Newton acknowledges repeatedly in the *Principia*; oddly, he does not acknowledge him in connection with the relativity of motion.

In Corollary V of Book I of the *Principia*, Newton gave Galileo's somewhat vague idea of the relativity of motion a mathematically precise formulation. He carefully restricted it to uniform motion on a "right" (that is, straight) line, instead of Galileo's imprecise restriction to motions that "do not fluctuate this way and that":

Corollary V: *The motions of bodies included in a given space are the same among themselves, whether the space is at rest, or moves uniformly forwards in a right line . . .*

In imitation of Galileo, Newton illustrated this relativity of motion by the example of a ship: "A clear proof of this we have from the experiment of a ship, where all motion happens after the same manner, whether the ship is at rest, or is uniformly carried forwards in a right line."[23]

But despite the superficial resemblance, Newton's interpretation of relativity is drastically different—it is a complete makeover of Galileo's relativity. Newton provided a new theoretical foundation for the interpretation of relativity. He did not treat relativity as an independent law of nature, but as a mathematical consequence, or corollary, of the laws of motion. He was able to do this because in the *Principia* he had achieved a complete, coherent formulation of the laws of motion, from which all the properties of motion could be deduced by mathematical methods. The principle of relativity of motion expresses the mathematical fact that in Newton's theory velocity is relative, but acceleration is absolute; that is, the acceleration of a "movable" is exactly the same, whether we regard it from the point of view of a ship at rest or from the point of view of another ship in uniform motion. Since Newton's Second Law involves only this absolute acceleration (and not the relative velocity), the Second Law is unchanged by the motion of the ship, and "all motion happens after the same manner."

AFTER NEWTON, THERE WAS a long pause in the development of relativity. Newton's interpretation of relativity was accepted without challenge for three centuries—it seemed as though nothing else could be said or needed to be said about relativity. But in the later half of the nineteenth century, the investigation of the laws of electricity and magnetism and the investigation of the wave properties of light led to a re-examination of Newton's physics and his interpretation of relativity.

The laws of electricity and magnetism had been gradually discovered over a span of a hundred years, between 1760 and 1860 by a grand international team of physicists: the Frenchmen Charles Coulomb and André-Marie Ampère, the Dane Hans Christian Oersted, the Englishman Michael Faraday, and, finally, the Scot James Clerk Maxwell. The study of electric and magnetic forces led to the introduction of an important new concept into physics: Faraday recognized that these forces are communicated from one charge to another via electric and magnetic *fields*, a sort of modern, improved version of Newton's idea of communicating gravitational forces via the incorporeal body of Christ. In Faraday's view, each electric charge sits at the center of a web of electric fields, or electric disturbances (if the charge is in motion, it also has coexisting magnetic fields). This web of electric fields extends outward in all directions, like a spiderweb might extend outward from the location of a spider waiting for its prey. By means of this web of electric and magnetic fields, the electric charge pushes or pulls on other electric charges. This mechanism for the transmission of forces is called action-by-contact: the electric charge is in contact with electric fields that it creates in its vicinity, these electric fields are in contact with the electric fields a bit farther out, and so on, . . . and the electric field at some distance finally makes contact with another electric charge and pushes or pulls on it. The concept of fields plays a central role in modern physics. Today, at a fundamental level, all matter is described by fields. We might say that fields are the quintessence of matter.

NEWTON'S LAWS OF MECHANICS had ruled physics for two centuries, and the laws of electricity and magnetism were at first not viewed as in any way in conflict with Newton's laws. These laws were merely thought to provide the means of calculating the electric and magnetic forces that act on electric charges at rest or in motion. And it was believed that from these electric and magnetic forces the motion of the electric charges could be calculated in the usual way, by means of Newton's laws. But the merger of the laws of electricity and magnetism with Newtonian physics led to trouble.

In 1861, Maxwell finished assembling all the laws of electricity and magnetism in a system of four equations, Maxwell's equations. He discovered that these equations imply the existence of electromagnetic waves consisting of oscillating electric and magnetic fields that are created by oscillating electric charges. This is like what happens when you shake a spider violently back and forth and thereby break off some pieces of the spiderweb in which it sits; the loose pieces of spiderweb wafting away in the breeze represent a traveling electromagnetic wave pulse.

Maxwell calculated from his equations that the speed of traveling electromagnetic waves should be about 300,000 klicks per second, which is the same as the speed of light. From this he concluded that light must be such an electromagnetic wave, with very short wavelength. He also deduced that there must exist electromagnetic waves of longer wavelengths, and, indeed, such waves were discovered some years later by the German physicist Heinrich Hertz, in experiments with electric sparks. These waves came to be called radio waves, and their practical importance for communications was quickly recognized—the first radio transmitters were built just ten years after the discovery of radio waves.

The trouble with Maxwell's discovery was that his equations predicted a constant, immutable speed for radio waves—his equations gave a speed of 300,000 klicks per second, regardless of the speed of the reference frame or the speed of the transmitter. This was in contradiction to the intuitively self-evident addition rule for velocities valid in Newtonian physics, and this paradox motivated the experimental attempts to detect the effect of the motion of the Earth on the speed of light, among them the Michelson-Morley experiment. As described in the first chapter, these experiments gave negative results, that is, the motion of the Earth had no effect at all on the speed of light. This was good for Maxwell, but bad for Newton.

And there was another trouble: whenever an electric charge moves, its electric fields move with it; thus, whenever the push of a force accelerates an electric charge, it also has to accelerate the electric fields of that charge. Calculations from Maxwell's equations indicated that some of the force had to be allocated to accelerate the electric fields, therefore a force F acting on an electrically charged particle does not produce the acceleration a specified by Newton's law $ma = F$, but a somewhat smaller acceleration. Another way to express this is to say that the effective mass is larger than the particle mass m, and what is worse, the effective mass actually increases with the speed, because the "drag" of the electric fields increases.

This led theoreticians to several proposals of how the effective mass of an electrically charged particle would increase with speed, and experimenters tried to measure such increases of mass for high-speed electrons. Lacking the powerful accelerators that play such a large role in particle physics today, the maximum speeds that experimenters could give their electrons were modest by today's standards, and it proved difficult to extract quantitative results from the experimental data. But by 1900 it had become clear that there was indeed some increase of an electron's mass with speed, although the precise amount of this increase remained somewhat vague.

THE THIRD STEP in the development of relativity was taken by the venerable Dutch theoretical physicist Hendrik Antoon Lorentz, just a few years before Einstein published his own version of relativity. Like Newton, Lorentz treated relativity as a consequence of known laws of physics. But unlike Newton, the laws that Lorentz took as the basis for relativity were not Newton's laws of mechanics, but Maxwell's laws of electricity and magnetism. Lorentz was an authority on Maxwell's equations, and he was eminently qualified to explore the tangled relationships between relativity and the dynamics of electric and magnetic fields and charged particles.

Lorentz was born in 1853, in Arnhem, not far from where the Rhine crosses the Dutch-German border. Like Galileo and Newton, he displayed a precocious early talent for mathematics. At age nine, while his classmates struggled to memorize their multiplication tables, he taught himself how to do multiplication by using tables of logarithms. By age eighteen, he became a doctoral candidate at the University of Leiden. His mathematical expertise was illustrated by an incident that occurred during his candidacy exam, or entrance exam, in mathematics. The mathematics professor considered Lorentz's performance on the exam satisfactory, but not impressive, until he discovered that by some clerical error he had given Lorentz a wrong, much more difficult exam—the candidate had passed the final doctoral exam, rather than the candidacy exam. Instead of following the usual course of studies at Leiden, Lorentz elected to proceed by independent study, and, at age twenty, he earned his doctoral degree with a dissertation on Maxwell's equations, which were then considered to be the cutting edge of theoretical physics.

Lorentz became the leading expert on Maxwell's equations, especially in their application to electrically charged particles. Lorentz originated what came to be called the "electron theory" of matter, according to which the electrical, magnetic, and optical properties of matter are attributed to the presence of small electrically charged particles hidden within the atoms. This was considered a daring conjecture at the time; direct experimental evidence for the existence of such particles was not found until several years later. Lorentz interpreted the macroscopic electric charges and electric currents in Maxwell's equations in terms of distributions of small electrically charged particles at rest or in motion. He interpreted the magnetic forces acting between electric currents as resulting from the motion of electrically charged particles within the wires on which the currents flow, and he formulated the equation for the "Lorentz force," that is, the force that a magnetic field exerts on a moving charged particle.

At age twenty-four, after a brief stint of teaching high school, he was appointed to a chair of theoretical physics created for him at Leiden, where he lectured and pursued his research for the next thirty-five years. In 1912, he resigned his official position at Leiden to take up an appointment at the Teyler Institute in Haarlem (now the Teyler Museum), where he had been offered a well-equipped laboratory of his own. Although Lorentz is mostly known as a theoretical physicist, he was also fond of performing experiments. However, he continued to be connected with the University of Leiden in an extraofficial capacity, and he continued to present his celebrated "Monday lectures" there until the end of his life in 1928. These lectures dealt with new subjects in physics, and they were renowned for their clarity and for their sharp insights. Lorentz would review the latest developments "and then turn the subject round and round and over and over,"[24] shedding new light on it. The lectures were often attended by famous physicists visiting Leiden, Einstein among them in later years.

In contrast to Galileo and Newton, both of whom had a vicious and quarrelsome streak, Lorentz had a gentle and retiring disposition, and he never quarreled with anybody. Einstein said of him: "For me personally he meant more than all the others I have met in my life's journey. Just as he mastered physics and mathematical structures, so he mastered also himself—with ease and perfect serenity. His quite extraordinary lack of human weaknesses never had an intimidating influence on his fellow-men. Everyone felt his superiority; no one felt intimidated by it. For although he had a keen insight into human nature and human relationships, he had a charitable kindness toward it all . . . With all his devotion to scientific study he nevertheless was perfectly aware that the human intellect cannot penetrate very deeply into the essential core of things. It was not until my later years that I was able to appreciate fully this half skeptical, half humble disposition."[25]

For the first twenty years of his academic career, Lorentz preferred to work in isolation, in an ivory tower of his own making. Modest and indifferent to public acclaim, he rarely published in any language other than Dutch (although he was fluent in English, French, and German), and he thereby limited the circulation of his papers. A story narrated by his son illustrates Lorentz's isolationist tendencies. Leiden was then a small town, rarely visited by strangers. One day Lorentz was told that townspeople had seen a stranger wandering along the main street who seemed to be searching for someone, and that he had the looks of a foreign professor. Lorentz sighed, "I hope he will not turn out to be a physicist."[26]

LORENTZ ACHIEVED WIDE FAME for his interpretation of the Zeeman effect, a minuscule shift in the wavelength of light emitted by atoms when they are placed in a strong magnetic field. This effect was detected in 1897 by Pieter Zeeman, a colleague of Lorentz's at the University of Leiden. Lorentz had proposed this experiment as a test of his conjecture that light is emitted by electrically charged particles moving in the interior of the atoms. The magnetic field alters the motion of these particles, and thereby alters the wavelength of the light that they emit by an amount that depends on the ratio of the particle's electric charge to its mass. From the data on the Zeeman effect, Lorentz calculated the charge-to-mass ratio of the particles. Soon thereafter, the English physicist J. J. Thomson discovered electrically charged particles with this same charge-to-mass ratio in the "cathode rays" produced in electrical discharges in evacuated glass tubes. These particles came to be called *electrons*, and the Zeeman effect can be regarded as the earliest direct evidence that electrons are a constituent of atoms. With his interpretation of the Zeeman effect, Lorentz brought electrons into physics—Thomson *discovered* electrons, but Lorentz *invented* them.[27]

For the discovery of the Zeeman effect and its explanation, Zeeman and Lorentz were jointly awarded the Nobel Prize in 1902. This was only the second time a Nobel Prize had been awarded. The first had gone to Wilhelm Konrad Röntgen in 1901, for his discovery of X-rays (or Röntgen rays, as they are still called in Germany to this day). According to the terms of Alfred Nobel's bequest, the prize was to be awarded for discoveries that benefit mankind, and it was immediately evident to everybody that X-rays are indeed of great benefit in the practice of medicine. But what the benefit of the Zeeman effect might be was not clear to anybody at the time, and it remains unclear today—nobody has ever found any beneficial practical application of the Zeeman effect, except perhaps in astronomy, where it is routinely used to estimate the magnetic fields on stars and in interstellar and intergalactic clouds of gas. In 1902, not even this remotely "beneficial" application had been conceived, and the award of the Nobel Prize to Zeeman and Lorentz was somewhat of a stretch. But if Lorentz's explanation of the Zeeman effect was not beneficial to mankind, it was very beneficial to physics, and it certainly deserved a large prize of some sort.

With his rising fame, the world invaded Lorentz's ivory tower, and he was invited to give lectures in Germany, France, and the United States. He began to attend scientific meetings, and he found, somewhat to his surprise, that he enjoyed the contact with other physicists. In 1911 he was appointed chairman of the first Solvay Conference, a duty for which he was eminently qualified by his scientific reputation and his invariable tact and politeness. Besides, as a speaker of German, French, and English, and as citizen of a small and neutral country, he was politically acceptable to all. He continued as chairman of these conferences for the rest of his life, presiding over five conferences altogether. Einstein attended most of these conferences, and he had high praise for Lorentz: "H. A. Lorentz presided with incomparable tact and incredible virtuosity . . . He speaks all three languages equally well and is endowed with an exceptionally acute scientific mind."[28]

THE SOLVAY CONFERENCES were financed by Ernest Solvay, an immensely rich Belgian industrialist. He had started out as a chemical engineer and had invented a process for the manufacture of soda (sodium bicarbonate), a substance not only useful for treating indigestion but also for a wide range of industrial applications. He built a factory in Belgium, and began to manufacture soda on a large scale—and from this he accumulated wealth on a large scale. Solvay made generous donations to universities, and he founded the Solvay Institutes for medicine and sociology in Brussels. One of his pastimes was to dabble in physics, and he thought he had invented a theory of how gravity affects the "constitution of matter and energy."[29] It was a crackpot theory, but Solvay believed in it, and—in the usual manner of the rich and powerful—he would not take no for an answer.

When Solvay asked Walther Nernst, a renowned German chemist at the University of Berlin, for advice on how to spread his ideas about gravity, Nernst saw an opportunity to do a good turn to physics. He craftily suggested to Solvay that he should fund a conference on recent developments in physics. Some two dozen of the leading physicists were invited to each of the conferences. They were given posh lodgings at the Grand Hotel Metropole in Brussels, handed envelopes stuffed with ample amounts of cash as spending money, and wined and dined on sumptuous Belgian cuisine (Brussels is a gourmet's paradise, second only to Lyon).

Solvay would begin the conferences by lecturing the assembled captive audience of the world's best physicists on his nutty theory, and, after politely applauding Solvay, the physicists would then be free to hold their own discussions and lectures.

The Solvay Conferences continue to this day, at intervals of three years, but they have become less exclusive and less luxurious. The Grand Hotel Metropole still proudly displays group photographs of the famous participants of the first conferences in its lobby, and some of the Solvay Conferences (but not all) are still being held there.

In his final years, Lorentz presided over a committee established to analyze the hydrodynamic problems posed by plans to enclose the Zuiderzee, north of Amsterdam, and convert it into a freshwater lake, the IJssel lake. He devoted much effort to this, and he participated not only in the theoretical calculations but also supervised practical measurements of tidal currents and inspected the initial stages of the construction project.

When Lorentz died in 1928, he was deeply mourned by his countrymen, and thousands lined the route of his funeral procession through Haarlem. Some of the greatest physicists of the day spoke at his funeral, representing their countries: Paul Ehrenfest, his successor at Leiden, for the Netherlands; Lord Rutherford for England; Paul Langevin for France; and Einstein for Germany. Several statues were erected to his memory, among them a bust on the Lorentzplein in Haarlem, near the house where he lived. Streets and squares and schools were named after him throughout the Netherlands, and there is hardly a town where you will not find his name somewhere.

BETWEEN 1892 AND 1904, Lorentz gradually formulated a theory of relativity for electric and magnetic phenomena, on the basis of Maxwell's equations. He explained the negative result of the Michelson-Morley experiment by the length contraction, but took this explanation much further than FitzGerald, by showing how this contraction was a direct consequence of Maxwell's equations. For this purpose, Lorentz relied on a model for the microstructure of solid bodies: he assumed that at the microscopic level, a solid body consists of positively and negatively charged particles held in equilibrium by electric forces. Since his investigation of the Zeeman effect had established that atoms contain electrons, the presence of negatively charged particles was assured, but Lorentz had no actual evidence for positively charged particles. His assumption was no more than inspired guesswork, and it was not confirmed until ten years later, when experiments revealed the existence of positively charged particles, in the form of atomic nuclei. From Maxwell's equations, Lorentz calculated what would happen to the array of positive and negative charges in a solid body when this body is in motion relative to our reference frame—and he found that this would produce a contraction of exactly the magnitude needed to explain the Michelson-Morley result.

Furthermore, Lorentz investigated what would happen to electric and magnetic fields if the electric charges and the electric currents that generate these fields were in motion relative to our reference frame. From this he found the Lorentz transformation equations for length and time, and he found the transformation equations for the electric and magnetic fields. With these he was able to show that all electric and magnetic experiments in all inertial reference frames occur in the same way—and thus observers in different inertial reference frames are incapable of detecting the motion of their reference frames by means of such experiments.

Lorentz completed his theory of relativity for electricity and magnetism in 1904, a year before Einstein developed his own theory, which was more broadly based than that of Lorentz's, encompassing not only electricity and magnetism, but all of physics. But Lorentz's theory anticipated many of the results of Einstein's theory, with the exception of the time dilation and the formula for aberration. This prior discovery of some of Einstein's results by Lorentz did not lead to any dispute between them—Lorentz was much too gentle a scholar to become embroiled in a dispute over priority. With a characteristic sense of balance and fairness, Lorentz conceded, "It would be unjust not to add that, besides the fascinating boldness of its starting point, Einstein's theory has another marked advantage over mine . . . by which the theory of electromagnetic phenomena in moving systems gains a simplicity that I had not been able to attain,"[30] and he explained that the chief cause of his failure was in not recognizing the true physical significance of the synchronization of clocks.

LORENTZ'S WORK ON RELATIVITY was improved and expanded by the famous French mathematician and physicist Henri Poincaré, a professor at the Sorbonne and president of the French Academy of

Sciences. Henri was a cousin of Raymond Poincaré, who became president of France just before the beginning of World War I and who was much hated by the Germans as a warmonger—the French called him Poincaré-la-guerre and the Germans called him Poincaré-Schweinskaree (piece of pig, or pig cutlet). After the war, the Germans hated him even more for his insistence on the payment of harsh and ruinous reparations. But cousin Henri avoided hate-by-association—he died in 1912, two years before the beginning of the war.

Henri Poincaré was a graduate of the legendary École Polytechnique, which teaches a unique combination of mathematics and engineering. Poincaré's greatest work was an encyclopedic treatise on the mathematics of celestial mechanics, but in the tradition of the Polytechnique he also pursued a great many practical problems, from mining to longitude determinations to the reform of time zones and the reform of angular measures (Poincaré unsuccessfully proposed that a circle be divided into 400 degrees, instead of 360).

Poincaré wrote several popular books on philosophical issues in science. The first of these books, *Science and Hypothesis*, was in the reading program of the "Olympia Academy," and one of the members of the academy recalled that it "profoundly impressed us and kept us breathless for weeks on end."[31] This book includes a discussion of the meaning of time and the meaning of synchronization that exerted a crucial influence on Einstein's thinking about the role of time in relativity.

In 1904, in a lecture at St. Louis University in Missouri, Poincaré presented his thoughts on relativity, and he formulated a Principle of Relativity for all the laws of physics, anticipating Einstein by a whole year. However, it seems Einstein did not know of this lecture, and he formulated his own version of the Principle of Relativity independently. Many years later, Einstein told his hagiographer, Carl Seelig:

> There is no doubt that the special theory of relativity, if we regard its development in retrospect, was ripe for discovery in 1905. Lorentz had already observed that for the analysis of Maxwell's equations the transformations that were later known by his name are essential, and Poincaré had penetrated even deeper into these connections. Concerning myself, I knew only of Lorentz' important work of 1895 . . . but not of Lorentz' later work, nor the consecutive investigations of Poincaré. In this sense my work of 1905 was independent.[32]

Poincaré was well acquainted with Lorentz's work, and in 1905 he published a revised version of Lorentz's treatment of relativity. He corrected a mistake that Lorentz had made in the transformation of the electric current from one reference frame to another, and he took some first tentative steps to extend the treatment of relativity to gravitational forces. He recognized that gravitational effects should propagate at the speed of light, and he initiated the earliest speculations about gravitational waves.

This work by Poincaré was published a few weeks before Einstein's 1905 paper on relativity. This timing has led to some claims that Lorentz and Poincaré invented relativity first and Einstein plagiarized their work. Thus, Jules Leveugle—like Poincaré, a graduate of the École Polytechnique—recently accused Einstein of plagiarizing the paper that Poincaré presented to the Academy of Sciences in Paris on June 5, 1905. However, in view of the delays in printing and shipping Poincaré's paper, it is next to impossible for Einstein to have received it in time to prepare a plagiarized version by the end of June, when he sent his own paper to *Annalen*.[33]

Plagiarism from Lorentz's 1904 paper might have been possible, because Einstein might have had access to that paper. But the internal evidence from the papers speaks against plagiarism: Lorentz and Einstein used completely different methodologies, and, besides, their papers contain discordant technical mistakes. Lorentz makes a mistake in his calculation of the transformation of the electric current (which Einstein does not imitate), and Einstein makes a mistake in his calculation of the transverse relativistic mass (which he would not have made if he had imitated Lorentz). Concordance of these mistakes would have been strong evidence of plagiarism; discordance is not strong evidence of anything, but it entitles Einstein to the benefit of doubt.

More recently, Leveugle proposed a daft conspiracy theory, according to which a cabal of German physicists led by Max Planck in Berlin and the mathematician David Hilbert in Göttingen contrived to deny credit to Poincaré by perpetrating a pernicious fraud. Supposedly, these German malefactors had access to Poincaré's paper and they plagiarized it, composing Einstein's 1905 paper and then publishing it under his name, with his consent, to hide their misdeed. Supposedly they selected Einstein as their front man because Planck had noted his proclivity to stealing ideas from others. In the absence of any actual evidence for this conspiracy, we might equally well suppose that Einstein's paper was written by Clever Hans the horse.

FOR MANY YEARS, the theory of special relativity was called the theory of Lorentz and Einstein—this is the way it was usually referred to in the physics literature until the end of World War I. But then Einstein's fame skyrocketed because of the "blitz" of publicity generated by his theory of general relativity, and Lorentz's name disappeared from view, except in connection with the Lorentz transformations.

Although the equations and the results found by Lorentz and by Einstein were the same, the interpretations they gave to these were quite different. Lorentz continued to believe in the ether to the end of his life. He understood that his and Einstein's results concerning the length contraction and the transformation equations for coordinates and for electric and magnetic fields implied that the ether was not detectable by any direct mechanical or electromagnetic experiment, but he adhered to the nineteenth-century view that light waves had to be waves in something— they had to be oscillation of an "ether." And Lorentz obstinately insisted that the reference frame of the ether establishes a criterion for absolute rest—he needed the ether as a mental crutch. Einstein dispensed with the ether altogether.

Throughout his life, Lorentz kept aloof from Einstein's theory of relativity. He never opposed Einstein's theory, he even praised Einstein's "boldness," but he never committed himself to Einstein's theory. Lorentz's lack of commitment pained Einstein. He admired Lorentz and venerated him as somewhat of a father figure, and he craved his approval. He said, "The four men who laid the foundations of physics on which I have been able to construct my theory are Galileo, Newton, Maxwell, and Lorentz."[34] He visited Lorentz several times in Leiden, and he always enjoyed these visits very much and even considered accepting a position at Leiden, to be near Lorentz. But their tacit disagreement remained.

4

"A storm broke loose in my mind"

*Einstein, in a letter to his son, describing the sudden inspiration
that led him to the invention of special relativity.*

E instein had thought about relativity and the puzzle of the speed of
light for many years. During his student years at the Zurich Polytechnic he talked about this puzzle with Mileva, and in a letter to
her, he once wrote enthusiastically "I'll be so proud and happy when we
jointly will bring to conclusion our research on relative motion."[1] This
statement has sometimes been misinterpreted as showing that Mileva
participated in the formulation of relativity. But it merely indicates an
emotional partnership, not a scientific partnership. Mileva's own letters to her relatives always speak of "Albertle's theories," and she never
made any claims to have contributed to these theories, not even after
the bitter divorce of later years.[2] By 1905, Mileva had lost all interest in
physics; she had become Einstein's wife and cook, and the mother of his
children (*Kinder, Küche*, without *Kirche*), but not his collaborator.

Einstein often discussed the puzzle of the speed of light with friends
and with the members of his "Olympia Academy," and together they read
some of the books and papers that explored relativity, in particular the

writings of Lorentz and of Poincaré. However, before his miracle year, he failed to make progress with this problem, and he felt frustrated that his long brooding had led to no conclusive results. As he tells us in his autobiography, "By and by I despaired of the possibility of discovering the true laws by means of constructive efforts based on known facts. The longer and the more desperately I tried, the more I came to the conviction that only the discovery of a universal formal principle could lead to assured results . . . After ten years of reflection such a principle [of constant speed of light] resulted from a paradox upon which I had already hit at an age of sixteen years: If I pursue a beam of light with the velocity c (velocity of light in vacuum), I should perceive such a light beam as an . . . electromagnetic field at rest. However, such a thing does not seem to exist, neither on the basis of experience nor according to Maxwell's equations."[3]

Finally, on a beautiful spring day in Berne, in the middle of May 1905, Einstein suddenly came upon the idea that the solution to the puzzle of the speed of light lay hidden in the procedures used to measure time. He was, once again, discussing this puzzle with his friend Michele Besso, an engineer also working at the patent office, when he was struck by an inspiration. He later recalled, "A storm broke loose in my mind,"[4] and "Suddenly I understood where the key to the problem lay."[5] The next morning, when he again met Besso, he was so excited that instead of greeting Besso with the usual "Grüezi," he blurted out, "Thank you. I've completely solved the problem." And he explained, "An analysis of the concept of time was my solution. Time cannot be absolutely defined, and there is an inseparable relation between time and signal velocity."[6] He pointed at one of Berne's clock towers and then to the distant clock tower of the neighboring town of Muri to exemplify for his friend his crucial idea about the synchronization of clocks at different locations.

IT WAS NOT AN ACCIDENT that Einstein came upon the idea that synchronization plays a crucial role in the speed-of-light puzzle. An obsession with clock synchronization had swept across Europe. This was in part motivated by the very practical concerns of scheduling the transmissions of telegraphic messages and scheduling the departures and arrivals of trains. Telegraph lines and railways linked all of Europe into a gigantic clockwork in which the operation of telegraphs and of trains at different locations had to be synchronized to permit the smooth flow of messages, passengers, and cargo.

But the obsession with synchronization went beyond practical concerns —it was sometimes commercial opportunism and sometimes a fad. The technology for synchronization of clocks had come under commercial development, and manufacturers of synchronization equipment were eager to sell, whether the public needed such equipment or not. In Paris, one company sold synchronization equipment operating with pulses of compressed air transmitted by underground pipes; others sold synchronization equipment operating with electrical pulses transmitted on wires. When well-to-do Parisians installed such synchronization equipment in their homes, linking their clocks to a central station, they were responding to the fascination exerted by this gadgetry, not to any real need for precise synchronization. For ordinary use in a home, a pendulum clock was more than adequate to keep time to within better than a minute a week, and synchronization could be checked by listening to the sound of the bell of a nearby church (church clocks and other public clocks were regulated by electrical links with the Paris Observatory).

The Swiss of those days were more obsessed with synchronization than anybody else. All the major Swiss cities had recently installed electrical links between the clocks on their railroad stations, post offices, clock towers, church spires, etc., to ensure that all these clocks stayed in exact synchronization. In Berne, more than two dozen clocks located in public places were linked in this way—including the clock at the picturesque *Zeitglockenturm*, a clock tower with an astronomical clock, replete with dancing bears, in lieu of cuckoos, which pop out each hour to entertain the tourists, in precise synchronization with other clocks throughout the city. This clock tower stands at the beginning of Kramgasse, where Einstein had his apartment. By leaning out of the window he would have been able to see the large dial on the clock tower.

These installations for the synchronization of public clocks were in tune with the long tradition of manufacturing of clocks and watches in Switzerland. The Swiss were and are famous for making expensive and accurate timepieces. Only a few of the great names in watchmaking are not Swiss, and there is perhaps a touch of envy behind the remark of the cynic who said that the only contribution of the Swiss to civilization is the cuckoo clock. And if you make the best watches in the world, you cannot tolerate that the clocks displayed over the streets of your cities are out of synch.

At the patent office, Einstein was the expert for electrotechnology, and he reviewed patent applications for electromagnetic devices used for the operation of citywide networks of synchronized clocks: devices for transmitting time signals, devices for resetting remote clocks, and even devices that operated with radio signals. Einstein must have had synchronization problems on his mind during many days at the patent office, and it is no wonder that he thought of synchronization in connection with the speed-of-light puzzle. And he was undoubtedly also influenced by his readings about relativity, especially the writings of Poincaré.

AFTER HIS INSPIRATION, Einstein completed his paper on relativity in just a few feverish weeks, and he sent it to *Annalen* by the end of June. The paper has signs of having been written in a hurry. There is a feeling of breathless excitement and incoherence about the whole thing. For instance, in an early part of the paper, Einstein promises to calculate the length contraction of a rod, but he never delivers on this promise, and, instead, he calculates the length contraction of a sphere. His cumbersome, clumsy mathematical treatment leading to the Lorentz transformations takes more than four pages of messy calculations to achieve what any competent physics student today can do in less than one page. Einstein was in a rush to get to the results, and he didn't bother to make the effort to find a clear, elegant argument. As he said sometime later, "I adhered scrupulously to the precept of that brilliant theoretical physicist L. Boltzmann, according to whom matters of elegance ought to be left to the tailor and the cobbler."[7]

In the first part of this paper, he presents a lengthy discussion of his ideas about clock synchronization and the implications for the puzzle of the speed of light. He argues that to measure the speed of light, we must let a light signal race from one location to another in our reference frame, over a known distance, and determine the "time of flight" by taking the difference between the readings of the clocks at the starting point and at the finish point. Synchronization plays an essential role in this, because the two clocks used to measure the time of flight must be synchronized— otherwise the difference between their readings would be meaningless.

And hereby hangs a tale: it is not immediately obvious how we can accomplish an *exact* synchronization of two clocks at different locations. For two clocks at the same location, synchronization simply means to bring the hands into the same positions ("Gentlemen, synchronize your watches!"). But for two clocks separated by many meters or many kilometers, we need to invent some synchronization procedure, and it has to be more accurate than the synchronization procedure by electric signals used in the city of Berne. This procedure did not take into account the travel time of the electric signals, and it therefore did not give an *exact* synchronization.

Einstein proposed to achieve an exact synchronization of clocks at different locations by sending light signals back and forth between the clocks. For instance, consider a clock in Berne and a clock in Lucerne, 60 klicks to the east. Then Einstein's proposal is to send a light signal at, say, 12:00 exactly, on a straight line from Berne to Lucerne (difficult to do because of all those quaint Swiss mountains, alive with the sound of music . . . but never mind), and bounce it back to Berne immediately. The Berne clock will tell us that the light signal returns at 12:00 plus 0.0004 seconds; that is, the light signal took 0.0004 seconds for this round trip. Obviously, it must then have taken 0.0002 seconds to get to Lucerne, and if the Lucerne clock is synchronized with the Berne clock, the Lucerne clock must have shown 12:00 plus 0.0002 seconds when the light signal arrived. If it did not, its synchronization is wrong, and the clock must be advanced or retarded by the amount it deviated from 0.0002 seconds.

Physicists today call this the Einstein synchronization procedure, and it is often described as one of his original and brilliant ideas. Einstein gives us no hint of whether he invented this synchronization procedure on his own, or whether he copied it from elsewhere. His paper lacks footnotes and references. Nowhere in the entire paper is there any hint that anybody else had contemplated synchronization methods before Einstein, and an uninformed reader might get the impression that Einstein created his synchronization procedure out of nothing. What is more, nowhere is there any hint that anybody else had contemplated relativity before Einstein—nowhere are the important contributions of Lorentz and of Poincaré mentioned. Today, the absence of such acknowledgments of earlier works by other scientists is unacceptable; if Einstein had submitted his paper to a modern journal, it would have been sent back to him with a peremptory request to supply appropriate references. *Annalen* had a somewhat lax policy regarding references, and it permitted Einstein to

get away with his cavalier way of skipping the contributions of his predecessors in relativity.

In a perceptive criticism of Einstein's writing, the historian Peter Galison has remarked that the absence of footnotes and references makes
Einstein's early papers look more like patent applications than scientific
writings.[8] And it is quite possible that Einstein drew his stylistic inspirations
from the patent applications that crossed his desk. Reportedly, Einstein
once said, "The secret to creativity is knowing how to hide your sources."[9]
This sounds like the credo of a plagiarist, but it is probably merely Einstein's realistic appraisal of the style of patent applicants. Patents are
investments by the inventor, and commercial considerations take precedence over intellectual honesty—the patent applicant needs to convince
the examiner that his invention is totally original, and the applicant will
never reveal what, if any, parts of his inventions were adopted or adapted
from other sources. Einstein had been examining patent applications for
two years, and in his omission of references he was imitating the style of
the applicants. This perhaps explains Einstein's style, but does not excuse
it. Einstein had also been reviewing scientific writings for *Annalen* for several years, and he should have known that the ethical standards of a scientist are supposed to exceed those of a commercial entrepreneur. This
omission of references also afflicts his other publications of the year of
miracles, 1905. Only several years later does Einstein begin to add some
references to his papers—he abandons the mindset of the patent clerk
and acquires that of the scientist.

IN FACT, EINSTEIN'S TREATMENT of synchronization was
by no means original. The question of the meaning of synchronization of
clocks at different locations had been discussed by Poincaré, with whose
writings Einstein and his friends at the Olympia Academy were familiar.
The "Einstein" procedure for clock synchronization had also been discussed by Poincaré. Furthermore, it had been widely used by surveyors and
astronomers for many years, since the early days of telegraphy. Soon after
the first long-distance telegraph lines were laid in the 1840s, it had been
recognized that by sending a telegraph signal back and forth along the
telegraph line, it was possible to synchronize clocks over intercontinental
distances by the method later imitated by Einstein.

Such clock synchronization plays a crucial role in the determination of geographic longitude, and all the early transcontinental and transatlantic telegraph lines were exploited for the determination of exact longitudes. The basic idea is quite simple: In one day, or 24 hours, the Sun moves westward across the sky by 360 degrees, returning to its starting point. With synchronized clocks in, say, London and New York, we can determine how much later the Sun crosses the meridian (the north-south line) at New York than it does at London—5 hours or, more exactly, 4.9 hours. By simple proportions, we can then calculate that the longitude difference between New York and London must be a fraction 4.9/24 of a full circle, that is, 360 degrees × 4.9/24, which equals 74 degrees.

The principles of this method for longitude determination were known in antiquity to Greek astronomers and philosophers, but they did not have synchronized clocks to put it into practice. Longitude determinations by synchronized clocks did not become practical until the eighteenth century, when the ingenious English clockmaker John Harrison succeeded in constructing portable clocks that could maintain their synchronization for several months while being transported by ship over transatlantic distances. The tale of Harrison's long struggles to build an accurate portable clock is charmingly told in Dava Sobel's book *Longitude*. But Sobel does not reveal the reason why Harrison's first three clocks—beautiful brass machines, about the size of suitcases, now on display at the Greenwich Observatory—proved inadequate. These clocks were only moderately accurate, because they were too large. If Harrison had understood the physics of clocks better, he would have recognized that a small clock, ticking at a fast rate, is less affected by the bouncing motion inflicted by transport in a ship. By happenstance, his fourth and last clock, Harrison IV, was small, and it remained "on time" to within 1 or 2 minutes during transatlantic voyages lasting as long as 5 months. This finally earned Harrison the prize of £20,000 offered by the English Parliament for finding a practical method for the determination of longitude at sea.

Much more precise longitude determinations became possible when intercontinental and transoceanic telegraph lines were laid in the second half of the nineteenth century. When the first transatlantic cable was completed from Ireland to Newfoundland in 1866, telegraphic synchronization of clocks reduced the uncertainty of older longitude determinations by a factor of about 20. The uncertainty in the geographical position of the American continent was thereby reduced from about 1 mile to 1/20 of a mile.[10] Telegraphic synchronization of clocks was also used extensively for longitude determinations in the survey of Siberia in 1875 and in the Great Survey of India in 1877. And telegraphic synchronization of clocks remained the most accurate synchronization procedure until the development of radio, after 1904.

Einstein undoubtedly knew about the technical details of the telegraphic synchronization method, and he must have been thoroughly familiar with various automatic synchronization devices through his work in the patent office, where he specialized in the examination of electric and magnetic equipment. It is also likely that he had some practical experience with the telegraphic equipment manufactured in his father's factory. All of this is clear evidence that Einstein's synchronization procedure was in no way an original idea; he merely adapted it from other sources—without giving credit where credit was due.

EINSTEIN'S SYNCHRONIZATION PROCEDURE relies on the implicit assumption that the speed of light between the locations of the two clocks is the same in each direction: same speed from Berne to Lucerne and from Lucerne to Berne. And Einstein emphasized this meant that with the clocks synchronized by his procedure, it was not possible to perform a logically meaningful test of the constancy of the speed of light. As he told Besso, there is an "inseparable relation between (synchronized) time and signal velocity." Any attempt to verify the constancy of the speed of light with these synchronized clocks would, of course, verify it. But the constancy is built into the synchronization, and therefore such an attempt would be circular reasoning—like a cat chasing its own tail, it would prove nothing. Einstein had deliberately designed his synchronization procedure to hide the effect of the speed of the Earth on the speed of light, because he thought that this was the clever and right thing to do. In essence, Einstein's synchronization procedure was a parlor trick to make the speed of light appear constant, regardless of what the speed "really" is.

But Einstein's trick had a fatal defect, and this is what Donald Crowhurst, drifting in mid-Atlantic on June 24, 1969, recognized when he exclaimed, "You can't do THAT!" and "swindler!" Crowhurst recognized that we can expose Einstein's parlor trick by testing the synchronization of the clocks by other means, for instance, by transporting a clock from one location to another.

To understand clearly the defect in Einstein's trick, consider the following fable: While the *Titanic* is racing through the calm waters and the calm air of the frigid North Atlantic, dinner is being served in the Grand Salon. A scientifically inclined passenger mentions to the captain that, due to the high speed of the ship relative to the air, the speed of a sound signal along the deck is slower when proceeding from stern to bow than from bow to stern. This displeases the captain—he would prefer that orders he yells from the stern to the bow arrive without extra delay. So he calls for the First Engineer and commands: "Forthwith on the deck of this ship the speed of sound from bow to stern shall be the same as the speed from stern to bow. See to it."

The First Engineer thinks the captain has gone bonkers, but he bites his tongue—onboard, the captain's word is law. Fortunately he remembers that some years ago some Swiss (or maybe German?) patent clerk had invented a trick for solving a similar problem with the speed of light. After some quick calculations, the engineer decides that if he sets the clock at the bow back by a fraction of a second, a sound signal sent from the stern will appear to arrive at the bow a bit sooner, and the speed will appear to be a bit higher, so as to satisfy the captain's command. He resets the clock at the bow by the required amount, and he resets all the clocks elsewhere on deck by smaller amounts, in proportion to their distance from the stern. The scientifically inclined passenger finds this puzzling. She decides to use her own pocket chronometer to check the clocks. She compares her chronometer with the clock at the stern, and walks forward along the deck, to the bow. She discovers that the clocks at the bow are out of synch. She perceives that the First Engineer is tricking the captain, and she wonders what the officers of this ship are smoking. Meanwhile the captain issues another order, "Forthwith there shall be no icebergs in the path of this ship. Make full speed ahead . . ."

Crowhurst found himself in the role of a scientifically inclined passenger on spaceship Earth racing through the calm ether. The chronometer he carried on his boat played the role of the passenger's pocket chronometer. Crowhurst was an experienced celestial navigator, and he knew that his chronometer had been synchronized to standard time, or Greenwich time, when he began his trip in England. He knew that his chronometer would remain in synchronization with the clock at the Greenwich Observatory, and that, in principle, his chronometer could be used to synchronize clocks placed at different locations on spaceship Earth and to check the "time of flight" of a light signal sent from Greenwich to any other location. In practice, his own chronometer was not accurate enough to detect the time of flight of such a light signal, but as an electronics engineer, Crowhurst probably knew of the portable atomic clocks recently developed by Hewlett-Packard, which were sufficiently accurate to detect the travel time of light signals between two locations on the Earth (and had a price tag larger than the entire cost of his boat). Thus, it must have been quite clear to Crowhurst that, in principle, and perhaps even in practice, a chronometer transported from one location to another does provide synchronization of the clocks at these locations and thereby permits an unambiguous measurement of the one-way speed of light.

And this was Einstein's big mistake: He forgot that besides synchronization with light signals there are other synchronization procedures—such as synchronization with transported clocks—by means of which it is possible to detect his trick and expose it as fraudulent. Synchronization by light signals does not permit us to check whether the one-way speed of light is *really* constant. But synchronization by other procedures permits us to check whether the one-way speed of light is *really* constant.

HOW COULD EINSTEIN have failed to see something so obvious? The explanation is that Einstein was not thinking like a physicist, but like a patent clerk. The most accurate available method for clock synchronization in 1905 was synchronization by electric signals (equivalent to light signals). This is what Einstein had learned in the patent office, and so he assumed that this was the right method to use for synchronization. He forgot that in the discussion of the foundations of relativity, he needed to pay attention to principles, not practice. He forgot that he was writing a paper about physics, not a patent application.

When Einstein wrote that the constancy of the one-way speed of light is "in reality neither a *supposition nor a hypothesis* about the physical nature of light, but a *stipulation* which I can make of my own free will in order to arrive at a definition of simultaneity,"[11] he was stipulating something that was not subject to stipulation. In everyday language, the distinction between *hypothesis* and *stipulation* is sometimes blurred, but the way Einstein used these words can be made clear by a simple example: when your rich aunt says that next year the exchange rate is going to be four dollars per euro, she is making a hypothesis (this may or may not turn out to be true; we won't know for sure until we check by a "measurement," next year); whereas when she says that next year she is going to buy euros from her European relatives at four dollars per euro, she is making a stipulation (she is making this stipulation of her own free will, and, furthermore, she can change this stipulation anytime she wants, so any "measurement" we do next year will merely confirm whatever stipulation she has finally adopted).

Einstein was entitled to make a hypothesis about the speed of light, but not a stipulation. The speed of light is either constant or not, and only measurement can decide what it is. The experimentally testable facts of physics and the computable facts of mathematics cannot be decided by stipulation. Einstein's attempt to stipulate the constancy of the speed of light was as silly as the notorious attempt of the Iowa legislature in 1888 to stipulate a value of 22/7 for π—the value of π is either equal to 22/7 or not, and only computation can decide what it is ($\pi = 3.14159\ldots$ versus $22/7 = 3.14285\ldots$). Sometimes a genius is as daft as a legislator. No wonder that Einstein's stipulation drove poor Donald Crowhurst into madness.

THE FIRST VAGUE DOUBTS about Einstein's resolution of the puzzle of the speed of light were raised in 1907 by Arnold Sommerfeld, a highly respected professor of theoretical physics at the University of Munich, much admired for the excellence of his teaching (his complete lectures on theoretical physics, published in a six-volume set, are still found on the bookshelves of many physicists today). Sommerfeld admired Einstein's work on relativity, but he felt uneasy about the abstract, stipulational treatment of the puzzle of the speed of light, and about the purely formalistic deductions of time dilation, length contraction, and other physical consequences of relativity, without any explanation of what mechanism might possibly produce such a slowing of clocks or contraction of solid bodies. Einstein solves the puzzle of the speed of light by a questionable trick, and the deductions of time dilation and length contraction are mathematically rigorous, but they lack concrete mechanical explanations. Exactly what is the mechanism that slows down the rate of a moving clock, and exactly what is the mechanism that causes a contraction of the length of a moving body? Einstein remains silent on this; he merely says that it has to happen, somehow, to keep the speed of light constant in all reference frames.

Sommerfeld complained to Lorentz about Einstein's derivations: ". . . As ingenious as they are, it seems to me that there is something almost unhealthy in their nonconstructive and nonvisualizable dogmatics. An Englishman would hardly have put forth such a theory; perhaps . . . what is expressed here is the conceptually abstract nature of the Semite. I hope you will be able to breathe some life into this ingenious conceptual framework."[12] These remarks have often been interpreted as anti-Semitic, but such an interpretation does an injustice to Sommerfeld, who was certainly no anti-Semite. He was deeply impressed by Einstein, and wrote to him expressing his admiration. He later became a friend of Einstein's, and later still, when Nazi threats forced Einstein to leave Germany, he was the only German professor who dared to fulminate against the Nazi's treatment of Einstein openly and bluntly. For his defense of Einstein, Sommerfeld (along with Planck and Heisenberg) was pilloried as a "White Jew" and as a "lieutenant" of Einstein's in the official newspaper of the SS.[13]

Sommerfeld's remarks are merely a frank, objective appraisal of Einstein's theory, and that Jews are intellectually inventive and inclined to abstruse, convoluted arguments is a widely held view, even held by Jews themselves—it may be prejudicial, but it is not derogatory. Einstein himself described one of his Jewish colleagues in Zurich as a *Rabbinerkopf*, a rabbinical thinker. Much later, Max Born—a Jew, a Nobel Prize winner, and a friend of Einstein's—in one of his lectures was to describe "Jewish physics" as an attempt "to get hold of the laws of nature by thinking alone."[14] Einstein readily conceded the point, with the comment, "I am confident that 'Jewish physics' is not to be killed," to which Born replied, "I have always appreciated your good Jewish physics."[15]

Sommerfeld's characterization of Einstein's ideas as conceptually abstract and dogmatic was on target, and perhaps so was his speculation about a Semitic influence in these ideas. From the reading list of the Olympia Academy, we know that Einstein had recently read Spinoza's *Ethics*, and the axiomatic-deductive treatment in the first section of Einstein's paper on relativity is reminiscent of the first pages of Spinoza's book, with its definitions, axioms, and deductions. Einstein revered Spinoza; after a visit to Spinoza's house during a visit to Leiden in 1920, he wrote a long poem in his honor beginning with the line: "How much do I love that noble man, more than I could say with words . . ."[16]

Sommerfeld's remark about the contrast between Einstein and Englishmen was also on target. The so-called Cambridge school of physicists was well-known for its emphasis on mechanistic explanations. For instance, Maxwell was of that school, and he initially contrived a complicated mechanistic model of the ether with little gyroscopes linked by springs, to explain the properties of electromagnetic waves that are represented mathematically by his equations. Such mechanistic explanations are diametrically opposite to Einstein's approach to relativity. Einstein had the faith of a mystic—he had an absolute trust in his solution of the puzzle of the speed of light, and he thought that any explanations in the English style were superfluous.

IT IS NOT KNOWN what Lorentz replied to Sommerfeld, but it is clear that Sommerfeld remained dissatisfied with Einstein's trick for achieving a constant speed of light by stipulation. In his own teaching, Sommerfeld preferred to treat the constancy of the speed of light as a consequence of Maxwell's equations, and he declared in his lectures, "The path taken by Einstein in 1905 in his discovery of the special theory of relativity was steep and difficult. . . . the path which we shall take is wide and effortless. *It proceeds from the universal validity of the Maxwell equations . . .* and it ends almost inadvertently at the *Lorentz transformations.*"[17] But Sommerfeld never put his finger on exactly what was wrong with Einstein's trick.[18]

Einstein's mistake was first identified explicitly in the physics literature by Sir Arthur Eddington, a brilliant and influential astronomer, director of the Cambridge Observatory. In 1918, Eddington confirmed the prediction of general relativity for the bending of light by the Sun, causing a sensation in the popular press and elevating Einstein to the rank of an international celebrity. Five years later, in 1923, he published a textbook called *The Mathematical Theory of Relativity* (still in print), which includes a careful analysis of the conceptual foundations of special and general relativity.

Eddington showed that in special relativity synchronization can be achieved by the slow transport of chronometers, and he showed that, in principle, this permits measurement of the one-way speed of light. Thus, the constancy of the one-way speed of light can be tested by experiment, although Eddington conceded that ". . . owing to the obvious practical difficulties, it has not [yet] been possible to verify it directly."[19] However, Eddington refrained from any overt criticism of Einstein, and he did not label Einstein's mistake as such. This restraint has also been observed by later writers, who proposed synchronization by clock transport or by mechanical signals and implicitly recognized Einstein's mistake, but always treated him as a sacred cow.[20]

THE DISTINCTION BETWEEN regarding the speed of light as constant by stipulation or as constant by experiment might seem like a splitting of hairs. And, indeed, if you are calculating some of the consequences of the theory of relativity, it makes no difference whether the speed of light in your calculations is constant for one reason or for the other. But if you are planning an experimental test of some of the strange, counterintuitive predictions of the theory, then you need to know what part of relativity is a stipulation and what part is hypothesis, so you know what you can test—any part of relativity that is a (legitimate) stipulation cannot be tested in any meaningful way.

The 1960s ushered in a new era of experimentation in relativity. Physicists exploited gamma rays, lasers, and masers (that is, microwave lasers) to perform new tests of relativity. In part, all of this experimental activity was opportunistic. It was not driven by any real doubts about relativity, but by the availability of new techniques—physicists had new gadgets at their disposal, and they wanted things to do with them. And in part it was driven by a fresh insight into how the Doppler shift of light could be used to measure the one-way speed of light, in contradiction to Einstein's claim that tests of the one-way speed of light are meaningless.

You have probably had occasion to experience the Doppler shift of sound when standing on the sidewalk while an ambulance speeds by, with its siren wailing its dire warnings of disaster. The pitch of the siren is higher than normal while the ambulance approaches you, but then becomes lower than normal when the ambulance recedes from you. This increase or decrease of pitch, to higher than normal or to lower than normal, is called the Doppler shift. It was discovered by the Austrian physicist Johannes Doppler in 1842, who noticed it standing near a railroad track while a whistling train rushed by. He promptly confirmed his observation quantitatively by a charming musical experiment: he placed trumpeters on an open train car, so that musically trained listeners on the ground could tell him how many notes sharp or flat the trumpets sounded while the train was approaching or receding (at a speed of 70 klicks/hour, the change of pitch is about one semitone). Doppler also gave the explanation of the Doppler shift: When the trumpets are approaching the listener, each succeeding pulse of sound, or each succeeding wavecrest of the sound wave, has to travel a somewhat shorter distance than the preceding pulse, so the pulses arrive at the listener's ear closer together than when the trumpets are at rest. The opposite is true if the trumpets are receding.

In the case of light emitted by a moving source, the Doppler shift manifests itself as an analogous change in the "pitch" of the light wave, that is, a change in the frequency, which, for a light wave, means a change of color (red light has the lowest frequency, blue light the highest). The magnitude of this Doppler shift depends on the ratio of the speed of the source to the speed of light. From a quantitative measurement of the Doppler shift, we can calculate the speed of the source; this is what a police officer does when he measures the speed of a car with a "radar gun" (but the gun uses a back-and-forth radar signal, whereas we are interested in a one-way signal originating at the source, or the car). Alternatively, if we know the speed of the source, we can calculate the speed of light. And the speed of light we obtain from a measurement of the Doppler shift is the one-way speed, that is, speed in the direction from the source to the receiver. This was the new insight that launched the Doppler-shift experiments of the 1960s. All these experiments confirmed that the one-way speed of light is indeed constant, the same in all directions, to within better than one part in a billion.[21]

THE 1960s ALSO LED to the development of high-precision atomic clocks, controlled by the oscillations of molecules or atoms, such as rubidium or cesium. In 1967, the cesium atomic clock was adopted as the fundamental standard for the definition of the second of time, replacing the earlier standard based on the rotation of the Earth. Atomic clocks were installed in all the main metrology laboratories and also in astronomical observatories, especially radio-astronomy observatories.

The wide availability of high-precision atomic clocks made it possible to perform direct measurements of the one-way speed of light and to test that the one-way speed is the same in all directions. To avoid disturbances to the atomic clocks during transport, the experimenters adopted a clever trick, which can be illustrated by a simple example. Suppose we want to test the equality of the speeds of light signals traveling from Berne to Lucerne and back. For this purpose, we send a light signal from Berne to Lucerne, but instead of immediately sending the signal back, we hold it up in Lucerne for *exactly* half a day.[22] We then send the signal back to Berne and calculate its speed from the round-trip distance and the total travel time registered by our clock in Berne, with, of course, an allowance for the 12-hour delay in Lucerne.

At first sight this seems a silly complication compared with simply sending the signal back immediately. But there is method in this madness: Lucerne is almost exactly east of Berne, and in 12 hours the rotation of the Earth turns the Berne-to-Lucerne track of the signal through 180 degrees, and consequently the returning signal is actually traveling in *exactly the same direction through space* as the outgoing signal. For instance, if the Berne-Lucerne segment was originally parallel to the direction of motion of the Earth, then, 12 hours later, the Lucerne-Berne segment is parallel to this direction of motion. Thus, with the 12-hour delay, this test gives us the one-way speed of the light signal along the direction parallel to the motion of the Earth. And by performing this test once more, on the next day, with a signal that starts in Lucerne, it gives us the one-way speed in the opposite direction, antiparallel to the motion of the Earth.

For this test, the clocks at Berne and Lucerne do not even have to be synchronized—all that matters is that the Lucerne clock be capable of counting off exactly half a day. This makes it clear that Einstein's contention that there is an "inseparable relation between (synchronized) time and signal velocity" is false. Synchronization is not needed for one-way speed measurements. Einstein failed to consider all possible variants of the use of light signals and clocks for measurements of the speed of light. Einstein was very inventive, but he had a one-track mind, and after he was struck by his mystical inspiration about clock synchronization by light signals he ceased to think about alternatives. If he had thought about it a bit longer, maybe he would have come up with a dozen alternative methods for measuring the one-way speed of light.

The most recent and most rigorous tests of the one-way speed of light were performed by analyzing radio signals sent back-and-forth between stations on the ground and satellites of the Global Positioning System (GPS), with various time delays. These GPS satellites carry high-precision atomic clocks, well suited for "counting off" delays between arrival of a radio signal and its later retransmission. The tests confirmed that the one-way speeds of light in all directions every which way across the sky are equal to within about one part in a billion. Einstein was lucky—more lucky than the Iowa legislators who had to withdraw their legislation about the value of π. What he had asserted by stipulation actually was confirmed by experiment. In the end, he turned out to have been right for the wrong reason.

EINSTEIN'S INTERPRETATION of the constant speed of light as a stipulation was not widely followed by his contemporaries. Only Max Born, a professor at Göttingen who became a close friend of Einstein's and later gained the Nobel Prize for contributing the crucial probabilistic interpretation of quantum mechanics, wholeheartedly imitated Einstein's approach in a book about relativity he wrote in 1920. Most other physicists treated the constant speed of light as a hypothesis subject to experimental verification. This was the approach adopted by the Swiss physicist Wolfgang Pauli, who wrote an influential article on relativity for the *Mathematical Encyclopedia* in 1921. The article was immediately reprinted as a separate book, and it became a classic—for the next fifty years it remained one of the best introductions to relativity (both special relativity and general relativity).

Remarkably, Pauli was just a twenty-one-year-old physics student when he wrote this comprehensive article. He was studying physics in Munich, under the direction of Sommerfeld, who was the editor of the *Encyclopedia* and invited the young Pauli to write the article on relativity. Pauli went on to outshine his teacher—he became a professor at the Zurich Polytechnic and a winner of the Nobel Prize.

In 1921, Pauli did not overtly criticize Einstein's approach to relativity, although he later was to become famous for his trenchant and witty comments about the failings of his colleagues. In the 1930s, Pauli began to treat Einstein with the same lack of respect he showed for everybody else. When Einstein constructed a new version of his unified theory of gravitation and electromagnetism, Pauli mocked him in a letter: "All that remains is to congratulate you (or should I rather say: offer condolences?) for having joined the ranks of the pure mathematicians."[23] And when Einstein published a critique of quantum mechanics, Pauli wrote to a colleague, "Einstein has again made a public announcement about quantum mechanics . . . As is well known, this is a catastrophe each time it happens."[24] But in 1921, the only comments that the student Pauli made about the celebrated professor Einstein were highly favorable. He did not emphasize the differences between his and Einstein's way of establishing the foundations for the constant speed of light.

Most physicists followed Pauli's lead. They did not criticize Einstein openly, but nevertheless adopted an empirical foundation for relativity, taking the constant speed of light as a law of physics, subject to experimental tests. They regarded the constant speed of light not as a stipulation, but as a fundamental law of nature, rooted in experiments, just like the principle of relativity is rooted in experiments. And they regarded Einstein's synchronization procedure with light signals as merely one possible synchronization procedure, which had some practical advantages over synchronization by slow clock transport, but was in principle equivalent to slow clock transport. Thus, they ignored Einstein's inspiration: instead of treating the speed of light as something to be stipulated, they treated it as something to be measured, and instead of treating the second postulate as a stipulation, they treated it as an experimentally demonstrable fact about nature.[25]

This is the attitude found in almost all modern textbooks on relativity today. Some writers even go a step farther. They do not bother to state the constant speed of light as a separate principle, but treat it as a corollary of the Principle of Relativity. And at first sight this seems plausible: the Principle of Relativity states that all laws of physics are the same in all inertial reference frames, and it would seem we can then say that Maxwell's equations and the speed of light implied by them must be the same in all inertial reference frames. But there is a pitfall in this, and even the very learned Sommerfeld fell into this pit in his lectures.[26] The deduction of a constant speed of light from the Maxwell equations hinges on knowing that the Maxwell equations are absolutely exact—and we can't be absolutely sure of this. Maybe the Maxwell equations deviate from the true equations obeyed by nature by some small amount, as of yet undetected, that might alter the speed of light under some circumstances. To rule out this possibility, we need to retain the second postulate, or some variant of it.[27]

EINSTEIN'S SECOND BIG MISTAKE in relativity was his fail-ure to consider the implications of the Michelson-Morley experiment. In view of Einstein's own fuzzy recollections about the experiment, we can't be sure that he knew of it in 1905—but ignorance is no excuse. If he didn't know of it, he should have known; and if he knew, he should have understood that it sets limits to the role of synchronization in the puzzle of the speed of light. The Michelson-Morley experiment does not depend on synchronization—it compares the *round-trip* speeds of light signals parallel and perpendicular to the motion of the Earth, and the null result of the experiment provides unambiguous evidence that these *round-trip* speeds are equal. This equality cannot be achieved by any resyn-chronization trick. (In terms of the fable about the *Titanic*: if the First Engineer had been ordered to make the speeds of sound equal not only fore and aft, but also athwartships, he would not have been able to do so, no matter how much he messed with the clocks.)[28] The outcome of the experiment is independent of any stipulation that Einstein chose to make or not to make—the experiment provides some important direct empiri-cal evidence about the behavior of light.

Einstein had a persistent mental block about the Michelson-Morley experiment. Abraham Pais, who was acquainted with Einstein and wrote a laudatory scientific biography of him, said that perhaps his inspiration about synchronization in 1905 "was so overwhelming that it seared his mind and partially blocked out reflections and information that had been with him earlier."[29] For a mystic in the throes of a revelation, such a "sear-ing" of the mind is indeed a typical reaction. And ten years later, when Einstein wrote the book that was to seal the fate of Donald Crowhurst, he perhaps still suffered from the aftereffects of this searing. By then he knew of the Michelson-Morley experiment, but he still failed to appreciate its full significance.

In the book, Einstein discusses the Michelson-Morley experiment only cursorily. He felt he had no need of this experiment—he *knew* relativity was right, and he had no need for any experimental support. His treat-ment of the Michelson-Morley experiment is unfriendly and adversarial, like that of a lawyer facing a hostile witness. He regards the experiment as a challenge to be overcome by his theory of relativity, and he shows how his theory survives this challenge by means of the length contraction.

He could, and should, have treated the Michelson-Morley experiment as a friendly witness, giving aid and support to his theory. The evidence from the experiment can actually be exploited in two ways: in the reference frame of the Solar System, the Earth and the experimental apparatus are in motion, and the null result gives evidence for length contraction; but in the reference frame of the Earth, the apparatus is at rest, and the null result gives evidence for equal, constant round-trip speeds of light in the two perpendicular directions.

Einstein's misappreciation of the experiment was a serious mistake. How serious did not become clear until the 1940s, when a new analysis of the implications of the Michelson-Morley experiment by the American physicist H. P. Robertson showed that this experiment, in conjunction with a similar experiment using an interferometer with arms of unequal lengths, can be used to provide an empirical basis for the Lorentz transformation equations and thereby establish the constancy of the one-way speed of light. This means that the constant round-trip speed implies the constant one-way speed.[30] Thus, the experiment provides quantitative confirmation for one of the basic principles of the theory of relativity and for one of its important predictions: the constant speed of light and the length contraction. Until the development of the new experiments in the 1960s, it remained the most precise quantitative evidence for relativity.[31]

EINSTEIN'S THIRD MISTAKE in the 1905 paper is in his calculation of the relativistic mass. He recognized, quite correctly, that the alterations imposed on the acceleration of a particle by length contraction and time dilation alter Newton's Second Law, and effectively make the mass m in the equation $ma = F$ into a variable quantity. The mass increases with the speed of the particle, so a fast-moving particle effectively has more inertia. Einstein also recognized that this inertia depends on the direction of the acceleration: it is larger for a longitudinal acceleration, parallel to the velocity (as in the case of a particle accelerating along a straight line), than it is for a transverse acceleration, perpendicular to the velocity (as in the case of the centripetal acceleration of a particle traveling at constant speed around a circle).

But when Einstein calculated this speed-dependent mass for the transverse case, he made a bizarre misstep. To find the mass, he needed to compare the acceleration and the force. But he compared the acceleration in one reference frame with the force in *another* reference frame: he used the experimenter's reference frame (at rest in the laboratory) for the former and the particle's own reference frame (moving with the particle) for the latter. This gave him a mass that was larger than it actually is. It was a mistake like that of a tailor who interchanges the waist and inseam measurements of a pair of trousers, and presents his customer with trousers wide at the waist and short at the ankles.

Einstein's mistake was not accidental. He knew he was using two different reference frames for acceleration and for force, and he did so deliberately, for some mysterious and perhaps mystic reason of his own. He was dealing with a force that had different magnitudes in different reference frames, and he seems to have thought that the choice between these different magnitudes had to be made by definition, by stipulation. In this rush to stipulation Einstein was making the same kind of mistake he made in the one-way speed of light: instead of relying on measurements he was again relying on his "free will."

When an experimenter measures the inertia of a particle by pushing on the particle with a calibrated force and observing the response, she does not care what the force is in any reference frame other than her own—the force in her reference frame is a measurable quantity, and the force acting on a particle is whatever measurements in her reference frame say it is. There is no freedom of will or of stipulation in this. Years later, Einstein somewhat sheepishly conceded that his calculation of the transverse mass from his definition of force "is not advantageous, as was first shown by M. Planck."[32] This was rather an understatement: Dear customer, I concede that the fit of the trousers I tailored for you is not advantageous.

How did Planck become involved in this miscalculation? Among his other duties as professor at the University of Berlin and president of the German Physical Society, Planck was also an editor of *Annalen*, and Einstein's paper crossed his desk before it appeared in print. Planck immediately perceived that it was an important new contribution to relativity, and he also immediately perceived that the calculation of the mass was wrong.

One reason why this mistake must have been obvious to Planck was that he knew of calculations of the relativistic mass done a year earlier by Lorentz and also of older calculations done by Poincaré and by J. J. Thomson, the discoverer of the electron. These earlier calculations lacked the generality of Einstein's calculations; they applied only to "bodies" made of electric and magnetic fields. The disagreement between Lorentz's formulas and Einstein's set off warning bells in Planck's mind, and when he checked Einstein's calculations he immediately spotted Einstein's mistake.

Planck was the perfect gentleman. Instead of crowing that he had found a mistake in Einstein's calculations, he simply published an entirely new and elegant recalculation of the mass, with which he obtained the correct formula. By rights, this formula should thereafter have been associated with Planck's name. But history often treats scientists unfairly, and today only few physicists are aware that the formula for the relativistic transverse mass in Einstein's paper was wrong, and that the correct formula is due to Planck.

Einstein can be faulted not only for the mistake in his calculation, but also for his ignorance of Lorentz's work. But in this there are some extenuating circumstances: Lorentz published his paper (in English) in the *Proceedings of the Royal Academy of Amsterdam*, a journal that did not enjoy very wide distribution, so Einstein would have found it difficult to locate a copy. It seems he did not get around to reading Lorentz's paper until much later.

5

"Motions of inanimate, small, suspended bodies"

Einstein in a letter to his friend Conrad Habicht,
describing the random, restless motion of small bodies
suspended in a liquid, which he interpreted as evidence for
the existence of atoms.

Einstein's celebrated work on relativity tends to overshadow some of his other contributions to physics, especially his seminal contributions to the controversy surrounding the atomic picture of matter in the early years of the twentieth century. Einstein published several highly original works on how the microscopic properties of atoms explain the macroscopic behavior of matter in various phenomena. The results of these investigations played a crucial role in securing the acceptance of the atomic picture of matter, and it was through these results that Einstein's name was first brought to the attention of fellow physicists. In the literature of physics, two of Einstein's works on the atomic question—his papers on the size of molecules and on Brownian motion—have been cited more often than his papers on relativity. These two papers helped to convince physicists and chemists of the existence of atoms.

Today we have available powerful microscopes that give us direct visual evidence for the existence of atoms. These microscopes—atomic-force microscopes and electron-tunneling microscopes—do not operate with light, but with a small probe that explores the surface of a sheet of metal by touch and thereby senses individual atoms as bumps, much as your finger senses the bumps on the surface of a coin. Electronic circuits register what the probe senses and create a graphic display of the shape of the atomic bumps, giving us visual evidence for the existence of atoms.

No such visual evidence for atoms was available in Einstein's days, a hundred years ago. Then, the evidence for atoms was purely circumstantial, derived from chemistry and from the physics of gases. Early in the nineteenth century, the English chemist John Dalton had formulated an atomic model for chemical reactions that neatly explained the proportions in which diverse chemical elements must be mixed to form chemical compounds. This atomic model was further developed by the Swedish and English chemists Jöns Berzelius and William Prout, and it quickly gained wide acceptance among chemists.

But physicists were slower to follow suit. Most of them viewed the atomic model of the chemists as purely hypothetical, and they were hesitant to accept atoms as real physical entities. As one of them declared: "The atomic theory has been of priceless value to chemists, but it has more than once happened in the history of science that a hypothesis, after having been useful in the discovery and coordination of knowledge, has been abandoned and replaced by one more in harmony with later discoveries. Some distinguished chemists have thought that this fate may be awaiting the atomic theory, and that in future chemists may be able to obtain all the guidance they need from the science of the transformations of energy."[1]

IN 1811 THE ITALIAN physicist Amadeo Avogadro concluded from chemical experiments and the known properties of gases that standard volumes of 22.4 liters (a large bucketful) of any kind of gas at a standard temperature and pressure contain the same number of molecules. Chemistry does not tell us the value of this number, which came to be called Avogadro's number, but evidently the atoms must be very small and the number in a bucketful must be very large, so that gases, liquids, and solids appear homogeneous, rather than lumpy. To make the atomic model of matter more concrete, physicists needed to find a way to count the atoms. This proved difficult, and the first determination of Avogadro's number from the study of gases was not achieved until half a century later.

In 1845 the Scotsman John Waterston contrived an atomic model that accounted for the physical properties of gases. He envisioned a gas as a swarm of atoms (or molecules) performing a restless dance, that is, a fast random motion, with atoms zipping through space in all directions and bouncing off each other and off the walls of the container; and he attributed the pressure of the gas to the repeated impacts of the gas molecules on the walls. But when Waterston sought to publish this idea in the *Proceedings of the Royal Society*, his paper was harshly rejected, a referee declaring that "this paper is nothing but nonsense, unfit even for reading before the Society."[2] According to the standard procedure of the society, the rejected paper was filed in its vaults, and Waterston—who was then living in India and had not kept any copy of his paper—made no effort to recover the original and seek publication elsewhere. The paper was rediscovered some fifty years later by John William Strutt, Lord Rayleigh, professor of experimental physics at Cambridge and successor to Maxwell. Rayleigh noticed a reference to the paper in some of Waterston's later writings, and he rescued it from the vaults of the society and arranged for its publication, including a generous and heartfelt apology to the late Waterston.

By that time other physicists had independently discovered the same idea. The credit for the first publication on the atomic theory of gases belongs to Maxwell, who is best remembered for his cardinal contributions to electrodynamics, but who also was the founder of "kinetic theory," that is, the statistical study of the restless dance of the atoms with the aim of explaining the gross, average features of large congregations of atoms in gases, liquids, and solids. With the atomic picture of matter, Maxwell and his followers were able to explain and to predict various properties of gases and liquids. For instance, from kinetic theory, Maxwell extracted the prediction that the viscosity of gases must *increase* with temperature. This is completely counterintuitive—from our experience with liquids we know that viscous liquids, such as honey or oil, become *less* viscous when heated, and we intuitively expect the same for gases. Maxwell was so surprised by this theoretical prediction, that he immediately performed an experimental test of the viscosity of a gas and found that indeed the viscosity increases with temperature (for a gas held at constant density).

Starting in the 1860s, the Austrian physicist Ludwig Boltzmann, professor of theoretical physics at Vienna, skillfully exploited statistical methods for treating the large congregations of atoms found in samples of gases, liquids, and solids, and he became the leading advocate of atomic theory in physics. Perhaps his greatest discovery was the proportionality between probability and entropy, with a constant of proportionality called "Boltzmann's constant." The thermodynamic entropy is a measure of the disorder in a physical system, and thus Boltzmann's proportionality says that when a system is in a state of high probability, it is in a state of high disorder (for instance, for a more or less uniform distribution of air molecules over the volume of a room, the probability and the disorder are high; but for a nonuniform distribution, with an unusually large concentration of molecules in one corner of the room, the probability and the disorder are lower). However, Boltzmann maintained that his constant of proportionality was not directly measurable, because in the final steps of the calculation of any macroscopic, average property of a large congregation of atoms— such as a calculation of pressure, density, or volume of a gas—the constant is always multiplied by Avogadro's number, so only the product of these two appears in the final result, and it becomes impossible to determine them separately.[3]

Despite the success of Boltzmann's statistical methods for calculating the average behavior of large numbers of atoms, nineteenth-century chemists and physicists regarded atomic theory as hypothetical and controversial. Robert Bunsen (he of the Bunsen burner) avoided talking of atoms, and Hermann von Helmholtz and Gustav Kirchhoff took the same stance.[4] Around the turn of the century, the leading opponent of atomic theory was Wilhelm Ostwald, a highly regarded professor of physical chemistry at Leipzig, who insisted on describing physical phenomena and chemical processes entirely in terms of energy. Boltzmann also faced strong opposition from his well-known Viennese colleague Ernst Mach, professor of natural philosophy, who held the opinion that all hypotheses or theoretical models in physics must be avoided, and that physics must be limited to the extraction of general laws from observational and experimental data. Since atoms were invisible and undetectable, Mach thought it was futile to speculate about them. He would make fun of whoever spoke of atoms by asking, in Austrian patois, "Ham's eins g'sehn?" ("Have you seen one?").[5] And he accused atomic theorists of attempting "to make into a fundamental tenet of physics a notion as thoroughly naïve and crude as that which holds [atomic] matter to be absolutely unchanging."[6]

Boltzmann was much troubled by Mach's hostility, as is evident from his description of one of their clashes: "I once engaged in a lively debate on the value of atomic theories with a group of academicians, including councilor professor Mach, right on the floor of the academy . . . Suddenly Mach spoke up and said tersely, 'I don't believe that atoms exist.' This sentence went round and round in my head."[7] In 1902, Mach had to give up his professorship to Boltzmann because of ill health, but he persisted in his stubborn opposition to atomism for the rest of his life, and after his death a note was found amongst his papers: "I can accept relativity just as little as I can accept the existence of atoms."[8]

By an odd coincidence, both Waterston and Boltzmann were driven into severe depression by the harsh criticism that their theories received, and both committed suicide. Waterston walked out of his house in Edinburgh on a summer day in 1883 and disappeared without a trace; the police investigation suggested that he drowned himself in the Firth. Boltzmann hanged himself from the crossbars of a window of his hotel room while vacationing in Duino, near Trieste, in 1906.[9] The Italians called this hanging *alla Condé*, because some years earlier the last prince of the ancient French family Condé had committed suicide in this manner. Boltzmann was buried in Vienna, in the Central Cemetery. His formula for probability and entropy is engraved on his tombstone.

SOME ROUGH GUESSES for Avogadro's number had been made in the first half of the nineteenth century, but the first solid, quantitative determination was achieved by the Austrian chemist Johann Loschmidt around 1860. He started with a simple idea: when a cubic centimeter of, say, air is compressed and cooled so it condenses and becomes liquid, all the air molecules will be closely packed together, and the volume of the condensed, liquid air must then equal the product of the number of molecules and the volume per molecule. Hence the volume of the condensed, liquid air gives us a relationship between the number of molecules and the volume per molecule, that is, a relationship between the number of molecules and the size of a molecule—if we know one of these quantities, we can calculate the other. Loschmidt then searched for a second relationship between these quantities, so that he could calculate both of them (algebra tells us that if we want to calculate two unknowns, we need two equations). To get such an additional relationship, he exploited the theoretical formula for the viscosity of a gas, which depends, among other things, on the size of the molecules. From the measured value of the viscosity of air, he extracted the size of a molecule, and he then calculated the number of molecules in 1 cubic centimeter of air.

The number Loschmidt found is very, very large, and to make it a bit easier to grasp, it is best to re-express it as the number of molecules in 1 cubic millimeter of air—according to Loschmidt, 2 billion billion molecules. Today we know that Loschmidt was off by a factor of about 10—according to modern measurements, the number of molecules in 1 cubic millimeter of air is actually 27 billion billion. By a pretty coincidence, the number of molecules in a cubic millimeter of air is not very different from the number of grains of wheat in the charming Indian folktale about the king, the philosopher, and the chessboard. As a reward for services rendered, the philosopher asked the king to give him enough grains of wheat to fill a chessboard—one grain in the first square, two in the second, four in the third, eight in the fourth, etc. This adds up to 18 billion billion grains of wheat, just 30 percent short of Loschmidt's number—the two numbers are in the same ballpark (or, in the jargon of physics, they are "of the same order of magnitude"). The king ordered the philosopher executed to avoid paying this exorbitant reward. This was a needless breach of promise and judicial murder. If the king had been more quick-witted than is customary in the ranks of inbred hereditary monarchs, he could simply have ordered the philosopher to count the grains himself. Even at a counting rate of several grains per second, day and night, the philosopher would have taken more than 100 billion years to complete the counting—and this gives us a good way to grasp the enormous magnitude of Loschmidt's number.

A few years later, Maxwell refined and improved Loschmidt's calculation, and obtained a value closer to the modern value. Unfortunately, nineteenth-century physicists could think of no other reliable method for the determination of Avogadro's number. Several methods involving surface tension and capillarity of liquids had been tried; but these methods gave only rough estimates, because the microphysics involved in the behavior of liquids was poorly understood.[10] This meant that adherents of the atomic picture could offer no corroboration—they were in the position of a detective who has found one witness to a crime (or one witness to the presence of weapons of mass destruction), but needs to find several more to establish the fact beyond reasonable doubt.

It was not until 1901 that Max Planck found another good way to determine Avogadro's number, as a by-product of his formula for the energy distribution in blackbody radiation, that is, thermal radiation emitted by hot, glowing bodies (see Chapter 6). His result, 28 billion billion molecules per cubic millimeter, was within a few percent of the modern value. But Planck's result did little to convince the skeptics opposing atomic theory. Planck's theory of blackbody radiation was viewed with even more suspicion than kinetic theory, and it was deemed to be a highly questionable witness for validating the atomic picture.

EINSTEIN FIRST BECAME INTERESTED in atomic theory during his high school years, and later, at the Zurich Polytechnic, he read Boltzmann's book on heat and statistical mechanics. His first published paper, which appeared in *Annalen* shortly after his graduation, deals with the implications of atomic theory for molecular forces. He later declared this paper to have been the "worthless work of a novice,"[11] but at the time he was quite naturally proud to get it published. He thought a publication would be his ticket to an academic position. He sent a copy to Boltzmann, and also to Ostwald. He hoped for a position as an assistant to Ostwald, and in a letter enclosed with his paper he asked, "whether you perhaps have employment for a mathematical physicist . . ."[12] When he received no reply, he again tried to contact Ostwald with a transparent excuse: "I am not sure whether I previously included my address . . ."[13] And this was soon followed by a pathetic letter to Ostwald from Einstein's father—sent without Einstein's knowledge—in which he wrote, "My son is deeply unhappy in his present state of unemployment, and each day the idea becomes more strongly implanted in him that his career has taken a wrong turn and he will find no companionship . . . I take the liberty to beg you politely that you read the paper published by him in *Annalen* and perhaps send him a few lines of encouragement, so he recovers his joy in life and in work. If it were possible for you to obtain for him a position of assistant now or next fall, my gratitude would know no limits . . ."[14]

Ostwald, who undoubtedly was swamped with such begging letters, never replied; but, with Germanic thoroughness, he stored the letters in his files, where they were rediscovered long after Einstein's death. It is amusing to speculate what the consequences would have been if Ostwald, the leading opponent of atomic theory, had taken on as an assistant Einstein, who was a firm believer in atomic theory and was soon to publish several crucial papers in support of this theory. Given that the young Einstein was an incorrigible and tactless loudmouth and know-it-all, an assistantship with Ostwald would have led to mayhem. Einstein was temperamentally unfit to act as assistant to anybody. When he was briefly employed as assistant to a private teacher in Schaffhausen in 1901, before gaining the appointment at the patent office, he promptly rebelled against his employer, demanding more independence, and then simply went AWOL, never to return.

In his late years, Einstein recognized that the appointment at the patent office was ideal for him. But in 1900 his failure to obtain an assistant position after graduation made him feel rejected, and this rejection rankled. In a paranoid mood, he blamed Weber, his professor at the Polytechnic, accusing him of undermining his efforts to secure an academic position and plotting against him; "I am convinced that Weber is to blame . . ."[15] And he wrote to Grossman: ". . . I am seeking a position of assistant at a university, and I would have found such a one long ago, were it not for Weber's foul play against me."[16] Maybe this accusation was born of unrequited academic admiration—in earlier years he had said about Weber, "I look forward from one of his lectures to the next."[17]

It was absurd to suppose that Weber would have taken the trouble to plot against Einstein, whose mediocre grades at the Polytechnic were quite sufficient to abort any hope of an assistant position. Of the four graduates in physics in the class of 1900, he had the worst grades (the others all got assistantships). When Weber died in 1912, Einstein viciously commented, "The death of Weber is good for the Polytechnic."[18] And even as late as 1918, some lingering resentment surfaced in his reply to a generous offer with which the Polytechnic hoped to entice the then-famous Einstein back to Switzerland: "How happy I would have been 18 years ago, if I could have become a humble assistant at the Polytechnic! But I could not achieve that! The world is a madhouse, fame is everything."[19]

IN 1902–4 EINSTEIN published three papers in which he attempted to fill in some gaps in Boltzmann's arguments. These papers received little attention, because, unbeknownst to Einstein, the American physicist Willard Gibbs had already published a book with a much more comprehensive treatment. Einstein later declared, "Had I known Gibbs' book at that time, I would not have published those papers, and instead limited myself to the treatment of a few points."[20] There is nothing much memorable about these papers, except that they contain a mistake, stubbornly repeated twice.

Einstein tried to deduce the Second Law of Thermodynamics from the laws of mechanics. This law of thermodynamics asserts that heat always flows from a hot body to a cold body, but never in the opposite direction; alternatively, it asserts that the disorder in any isolated physical system always increases. For instance, if the heating system of your house deposits some hot air in one corner of your room, the hot air pretty quickly spreads all over the room, that is, the heat spreads all over the room. This can be viewed as a flow of heat from the hot corner into the colder regions of the room, or it can be viewed as an increase of disorder (initially the hot and cold masses of air are segregated in a somewhat orderly arrangement; later they are mixed and disordered). To deduce this law from mechanics, Einstein introduced the assumption that in any physical system the probability of the configuration always increases. For a room left to itself, the concentration of heat in one corner of the room has low probability, whereas the distribution of the heat all over the room has a higher probability, and Einstein's assumption of an increase of probability "explains" the flow of heat demanded by the Second Law. But his argument fails to relate the change of probability to the laws of mechanics—there is nothing in mechanics to prevent a reversed flow of heat, that is, a flow of heat from the cold regions of the room into one hot corner. In fact, it is a basic tenet of mechanics that for every motion there is a possible reverse motion, and for every process there is a possible reverse process—the laws of mechanics work equally well in the forward and the backward directions.

Einstein's attempted derivation of the Second Law was severely criticized. One reviewer wrote, "If one assumes, like Einstein, that more probable follow less probable configurations, one thereby introduces a special assumption that is not in any way endowed with self-evidence and absolutely requires a demonstration."[21] Einstein conceded that this was a fair criticism, and he admitted, "My derivation had not satisfied me even at the time, which is why I gave a second derivation shortly thereafter . . ."[22] But this second derivation was no better than the first, and Einstein abandoned the topic. It is somewhat of a mystery why he opened this can of worms, because the derivation of the Second Law of Thermodynamics from the laws of mechanics had already been formulated by Boltzmann in his celebrated H theorem. Admittedly, Boltzmann also had to rely on a "special assumption," his controversial *Stosszahlansatz*, which, in essence, says that the initial positions and velocities of the molecules are random, or uncorrelated. It was rather presumptuous—but quite typical—of the young Einstein to imagine that he could improve on Boltzmann's work. Although Einstein's papers on the foundations of statistical mechanics have faded into the obscurity they richly deserve, the mathematical techniques that Einstein learned during their preparation provided the springboard for some of his later, highly successful forays into atomism and the determination of Avogadro's number.

IN 1905, IN QUICK SUCCESSION, Einstein invented three or four new methods for the determination of Avogadro's number. The first of Einstein's methods was presented in the dissertation he submitted to the University of Zurich as a candidate for the doctoral degree. He had already submitted a dissertation on a different topic three years earlier, but he had quickly withdrawn it when it became clear that it was about to be rejected (the withdrawal was a typically Swiss financial maneuver—it permitted him to obtain the refund of the SF 230 filing fee; rejection would have blocked the refund).

Einstein's immediate reaction to this fiasco was sour grapes; in a letter to a friend he wrote that he was giving up on the idea of gaining a doctorate, as it would be of little use to him, and besides "the whole comedy has become tiresome for me."[23] But some time later he reconsidered, and proposed to write a dissertation on electrodynamics, the topic that later was to lead to his groundbreaking paper on relativity. The one and only physics professor at Zurich, an experimentalist, failed to appreciate the profound significance of this topic and found the theoretical treatment too bewildering. So Einstein selected another, more experimentally oriented topic from the several pieces of work he had then in progress. Einstein finally settled on the presentation of a new method for the determination of Avogadro's number. This dissertation was accepted, it gained him the doctorate, and it was published a year later in *Annalen*.

Einstein's calculation of Avogadro's number and the molecular size relies on experimental data on the viscosity of solutions of sugar in water and the diffusion of sugar in water.[24] As shown by Loschmidt, the determinations of the molecular size and of Avogadro's number go hand in hand, and Einstein accordingly gave his dissertation the title "A New Determination of the Molecular Size." He dedicated the dissertation to his friend Marcel Grossmann, who had often helped him with mathematical matters, and who had been instrumental in securing for him the position at the patent office.

To relate the macroscopic properties of sugar solutions to the microscopic size of the sugar molecule, Einstein performed a complicated hydrodynamical calculation of how the presence of one or several sugar molecules affects the viscosity of water. This calculation goes on and on for twelve pages, and it forms the bulk of his dissertation. In the words that Mileva once used to describe the work of one of her professors, we might say that Einstein "calculated and calculated, set up equations, differentiated, integrated, substituted . . ."

In his calculation, Einstein displays a good grasp of the principles of hydrodynamics, and he also displays some good physical insights.[25] But his execution of the mathematical program immediately turns into a comedy of errors. The editors of the modern edition of Einstein's *Collected Papers* attached more than thirty footnotes to the reprint of the dissertation, listing all the mathematical errors. These range from simple mistakes of sign, to typos in mathematical symbols, to wholesale omissions of terms in equations and errors in coefficients. It looks as though Einstein raced through his calculation and never bothered to review it and check it.

SOME OF THE MATHEMATICAL ERRORS in the original dissertation lead to results that are patently absurd. For instance, by a concatenation of mathematical mistakes, he obtains the odd result that the presence of one single sugar molecule in water *decreases* the viscosity of the water (although the presence of many such molecules increases it). This should have set alarm bells ringing in Einstein's mind. But he does not stop to reflect on this oddity and blindly accepts it as true. He even highlights it for the reader, calling it "remarkable." Evidently, Einstein would have benefited from a rule that John Wheeler, a professor at Princeton and one of the great physics teachers of the twentieth century, recommended to his students: *Never make a calculation before you know the answer.* This rule has come to be called Wheeler's First Moral Principle, and it means that you should first make a preliminary rough estimate, so you know what to expect, and don't fall into the trap of blindly believing an erroneous calculation (unfortunately, Wheeler never revealed to his students his Second or Third Moral Principles).

Maybe the frenetic pace of Einstein's work during the year of miracles had taken a toll on his mathematical performance. But maybe the heavy infestation of mathematical mistakes in the dissertation is a good gauge of Einstein's mathematical incompetence. He was not fond of mathematics; as he once said to a French colleague: "As for me, I do not believe in mathematics."[26] The calculation in his dissertation was the most complicated calculation he had attempted up to this time, and it remained the most complicated calculation in all his works until he began his investigations in general relativity (for which he again secured the mathematical help of his friend Marcel Grossmann). In later years, when he could afford it, he always kept a hired mathematical assistant at hand to perform calculations for him. During his academic career, he went through a total of ten assistants, to which might be added a handful of "collaborators," who calculated for him, but were not his dependents.[27] Thus, many of the complicated calculations published in his later work were not done by him.

A few years after the publication of this dissertation, Einstein was informed that experimental measurements of the viscosity of water containing small spheres in suspension contradicted his theoretical formulas, and he decided his mathematical calculations needed checking. He first made a vague effort to do the checking himself, and he quickly concluded, "I have checked my previous calculations and arguments and found no error in them."[28] It is a telling sign of Einstein's own views about his mathematical talents that he then begged one of the students in the class he was teaching to recheck everything, promising that the student "would gain much credit in this matter."[29] Nobody ever gains credit by confirming a correct calculation, so we can take this as a hint that Einstein expected that mistakes would surface. The student promptly found several serious mistakes. Einstein finally published a corrected version of the dissertation in 1922.[30]

BESIDES THE ABUNDANCE of mathematical mistakes, there is also an abundance of mistakes in the physical assumptions underlying the mathematical calculations (out of kindness, or out of inadvertence, the editors of the *Collected Works* did not list any of the latter mistakes). The mistakes in physical assumptions range from oversimplifications to the outright zany. They are reminiscent of the anecdote about the physics professor who was asked to calculate the milk production of a cow and replied, "It's easy. Assume the cow is a sphere . . ." Einstein must have been aware that there was no way for him to calculate the actual forces in the flowing water arising from the presence of the microscopic sugar molecules, and so he calculated—in wonderful detail and with much mathematical virtuosity—the forces arising from the presence of smooth, macroscopic spheres, and he pretended that this was the same thing. Perhaps he expected that his professors would be dazzled by his mathematical fireworks and overlook his questionable physical assumptions. And in this he was right: one thesis examiner commended him for dealing with a calculation that is "among the most difficult in hydrodynamics," and the other declared that Einstein demonstrated "fundamental mastery of the relevant mathematical methods."[31]

The thesis examiners not only failed to catch Einstein's mathematical mistakes, but they also failed to notice that he had an inadequate grasp of the basic physics involved in his treatment, and that his physical assumptions were deeply flawed. His assumptions that that sugar molecules can be treated as rigid spheres and that water can be treated as a fluid were inconsistent—a sugar molecule is only four times as large as a water molecule, and therefore the sugar molecule experiences individual collisions with water molecules, rather than the average pressure and viscosity that would be experienced by a larger body immersed in water. Any child who has jumped into a "ballpit" filled with plastic balls at an amusement park could have explained the difference to Einstein: the ball-molecules filling the ballpit are too large compared with the child's body to permit smooth flow, and therefore the motion of these ball-molecules around the child's floundering body is nothing like the fluid motion of water in a swimming pool. Furthermore, the sugar molecule is not spherical, and it has a surface that is not at all smooth (it consists of forty-five atoms arranged in a quite lumpy structure).

These flaws might be considered naïve oversimplifications.[32] But Einstein's final assumption descended to the zany: for his hydrodynamical calculations, he assumed that the sugar molecule is at rest in the water and does not rotate. This is nonsensical, because according to kinetic theory sugar molecules must have a random thermal velocity of about 140 meters per second and a comparable rotational velocity. It is a total mystery why his thesis advisors overlooked this glaring mistake. They were quite ordinary, dull professors at what was then a dull, second-rate university, but even the dullest of dull physics professors should not have been this blind. Einstein's dissertation should have been rejected.

APART FROM THE UNREASONABLE physical assumptions, there is something else that is odd about Einstein's dissertation: for the purpose of determining Avogadro's number and the size of the sugar molecules this complicated hydrodynamical calculation is totally superfluous. Like Loschmidt, Einstein needed two mathematical relationships to extract Avogadro's number and the molecular size. For one of these he used data from diffusion of sugar in water. For the other, instead of using the result of his complicated calculation, he could have used the simple fact that when sugar molecules are tightly packed in solid, undissolved sugar, the volume of 1 gram is 0.61 cubic centimeters. He even mentions this fact in his dissertation—but then makes no use of it. Loschmidt had used a fact of just this sort about tightly packed air molecules in his method, and it is unclear why Einstein did not exploit the corresponding fact about tightly packed sugar molecules. He hints that a sugar molecule in water might be surrounded by a layer of attached water molecules, which might change the dimension of the molecule and introduce uncertainties in a simple calculation of this kind. But this is not much of an argument, because his complicated calculation has similar, and worse, uncertainties.

The real reason for Einstein's insistence on retaining the complicated calculation is probably that without it the dissertation would have been too flimsy—if he had discarded this fancy calculation and extracted the size of the molecule from the simple fact about the volume of solid sugar, his dissertation would have been cut down from seventeen pages to a mere six pages. It would then have been too short, inadequate even for a second-rate university. In fact, his professor had already complained about its insufficient length. Einstein later gleefully told one of his biographers that when the professor sent the dissertation back to him, he added one sentence—and it was accepted without further comment.[33]

The most that can be extracted from Einstein's highly idealized and faulty mathematical model of the motion of sugar molecules is a crude estimate of Loschmidt's number. With the known value of the viscosity of sugar solutions, Einstein found a value of 9 billion billion molecules per cubic millimeter. This is in the right ballpark—the actual value obtained by modern methods is 27 billion billion molecules per cubic millimeter. Here, as in much of his earlier and later work, Einstein was lucky. By some fortuitous cancellations among all his errors, he got a better value than he deserved.

In spite of its flaws, Einstein's hydrodynamical calculation has found a niche in the literature of physics. It is a poor description of the behavior of sugar molecules, but it is a good description of the behavior of small, but not microscopically small, spheres suspended in a fluid. Einstein's formula for the viscosity was verified some years later by the French physicist Jean Perrin and his collaborators in experiments with small droplets of gamboge resin suspended in water (but this did not help to determine Loschmidt's number—for that we need experiments with *molecules*). Thus, Einstein's fancy mathematical calculation, after elimination of its mathematical mistakes, actually gives the correct viscosity for a liquid containing small spheres in suspension. Einstein's formula has a wide variety of practical, technological applications; it has been used to calculate the behavior of sand in cement mixes, casein droplets in milk, aerosols in clouds, etc.[34]

EINSTEIN'S NEXT METHOD for finding the Loschmidt number was much better—and completely free of mistakes. Sometime in 1904, Einstein recognized that Boltzmann's statistical methods could be applied not only to invisible atoms and molecules, but also to much larger bodies, such as grains of matter large enough to be seen with a microscope. It occurred to him that if such grains of matter are suspended in a liquid, the impact of molecules on the grains might lead to pressure fluctuations on the grain. For a large body immersed in liquid—such as, say, your finger immersed in water—the impacts of water molecules from opposite directions average out quite precisely, and there is no net force pushing your finger to one side or another. But for a very small body, there are not enough impacts for such a precise averaging, and sometimes there is an excess of impacts on one side or another, which causes an unbalanced force that accelerates the body in one direction or another. This gives the body a random, restless motion, proceeding in zigzags in unpredictable directions, like a drunk wandering about the town square.

Motion of this sort had been noticed long ago in suspensions of small bodies, such as grains of pollen immersed in water. It was called Brownian motion, after the English botanist Robert Brown who had described this motion in detail in the 1820s. Before Brown, the motion had usually been attributed to some kind of "life force" of the pollen, but he disproved this notion by showing that small grains of nonbiological materials—such as grains of gold, silver, or mercury—behaved in the same way. Several physicists proposed that the explanation was to be found in the dancing motion of atoms and molecules communicated to the grains by collisions, but in the absence of quantitative calculations this explanation remained controversial.

Einstein recognized that the connection between the invisible molecular motions and the visible Brownian motion could provide direct observational evidence for molecules, and he proceeded to calculate the Brownian motion by statistical methods. According to kinetic theory, water molecules at room temperature have an average speed of about 600 meters per second, but grains of pollen, with a mass much larger than water molecules, have lower speeds—a grain of pollen of diameter 1/1000 millimeter should have an average speed of a few millimeters per second. This speed is not directly observable, because, within a small fraction of a second, the grain of pollen will be stopped by the impacts of the surrounding water molecules, and these impacts will then change its motion at random in some other direction. Thus, the velocity does not persist unchanged long enough to be measured.

However, although the particle has no directly detectable velocity and no steady motion in any given direction, it has a zigzag dancing motion that gradually causes it to drift farther and farther away from its starting point, somewhat like the motion of a dancing couple on a tightly packed dancing floor, when the couple does not try to progress in any selected direction, but merely takes advantage of the random openings that are available.

Einstein calculated that this dancing motion would cause a grain of pollen to wander about 1/100 millimeter in a minute, 2/100 millimeters in 4 minutes, 3/100 millimeters in 9 minutes, etc. (the distance increases in proportion to the square root of the time, rather than in proportion to the time, because the grain sometimes takes backward steps, canceling some of its progress). Although these distances are small, they can be readily measured when the grain is observed with a microscope, and Einstein proposed that experimenters attempt such measurements. Einstein showed that the average distance traveled by a small grain performing Brownian motion is related to Avogadro's number, so measurements of this distance permit an accurate determination of Avogadro's number. He ended his paper on a wistful note: "May some scientist soon succeed to decide the issue here set forth, of such importance for kinetic theory!"[35]

A YEAR AFTER EINSTEIN'S PAPER, the Austrian physicist Marian von Smoluchowski published another theoretical treatment of Brownian motion, in which he arrived at results similar to Einstein's. Smoluchowski had obtained his results some years before, but had not made any efforts to publish them. The case of Smoluchowski-Einstein is similar to the case of Waterston-Maxwell: in both cases the scientist who first made the discovery was "scooped" by a latecomer who published promptly. Although the credit for the first publication of the theory of Brownian motion belongs to Einstein, Smoluchowski's work has not been forgotten, because his treatment of the theory of Brownian motion was much simpler than Einstein's. With some further improvements by the French physicist Paul Langevin in 1908, it has become the standard treatment. In modern expositions of the theory of Brownian motion, physicists today genuflect in the direction of Einstein and then invariably proceed à la Smoluchowski-Langevin.

The paper on Brownian motion was the first of Einstein's papers to attract wide attention. Today we think of this paper as the least spectacular of the five great papers of the year of miracles. But in 1906–7 it was the paper that physicists found easiest to understand and that presented the most concrete and experimentally accessible results. Several physicists and chemists contacted Einstein with questions about aspects of the paper, among them Wilhelm Konrad Röntgen, the first Nobel Prize winner, who wanted to know whether the generation of Brownian motion by molecular impacts might violate the Second Law of Thermodynamics.

Experimenters took up Einstein's challenge to test his theoretical predictions. The first measurements produced somewhat ambiguous results, but in 1908 the French physicist Jean Perrin and his students confirmed Einstein's predictions by a series of precise experiments. This not only gave clear evidence for the atomic theory of matter, but also led to an accurate determination of Avogadro's number, in good agreement with the values previously obtained by Planck from his theory of blackbody radiation and by Maxwell from the viscosity of gases. For his experimental work on Brownian motion, Perrin later received the Nobel Prize.

The direct visual evidence for the dance of the atoms revealed in Brownian motion convinced many skeptics of atomic theory. Ostwald conceded in 1909, and he was a good loser—he later twice recommended Einstein for the Nobel Prize. But it did not convince Mach. Einstein sent him a copy of one of his papers on Brownian motion and included a letter in which he explained, "Under the microscope one directly sees, in a sense, a portion of the thermal energy in the form of mechanical energy of moving particles,"[36] but Mach stubbornly remained in opposition to atomic theory to the end of his life—he wanted *visual* evidence. When he was shown the scintillations produced by individual alpha particles emitted by radioactive materials, he exclaimed, "Now I believe in atoms."[37] But his belief did not last, and he quickly reverted to his skepticism. He died an unbeliever, and, as a relapsed heretic, he presumably resides with other heretics in the sixth circle of hell—unless some of his other valuable contributions to physics and to positivist philosophy saved him from that fate.

IN 1908, EINSTEIN continued his investigations of fluctuation phenomena in a paper on opalescence. By this is meant the strong scattering of a beam of light sometimes observed in a very dense gas or in a mixture of two liquids. When a beam of light is aimed into a tank filled with an opalescent gas or liquid, the beam is broken up, and light is scattered, or deviated, in all directions, so the entire volume of the tank fills with a luminous glow, an "opalescent" glow. Such scattering of a light beam is analogous to the scattering of ocean waves on submerged boulders near the shore; the wavefronts are deflected by the boulders, and they fill the entire water surface near the boulders with a seething, irregular mix of waves. Smoluchowski had suggested that opalescence arises from irregularities that have formed within the gas or liquid. In a gas compressed to high density, such irregularities form when some of the gas assembles into patches with a density almost as large as that of liquefied gas, and in a mixture of two liquids such irregularities form when one of the liquids assembles into patches within the other.

Einstein adopted Smoluchowski's idea and developed it quantitatively. He calculated the size of the density fluctuations in a gas, and then calculated the scattering of a light beam incident on these density fluctuations. As a by-product of his calculations, he found the explanation for the blue color of the sky. We are so used to the blue color of the sky that we hardly ever wonder, Why is the sky blue? Why not green, or yellow? Or why not black, as the sky in the interplanetary space above the Earth?

The first explanation of the blue color of the sky had been attempted by Lord Rayleigh. He had calculated the scattering of sunlight by air molecules and had concluded that blue light is scattered by these molecules much more than any other color. This scattered light then bounces around in the atmosphere, and it fills the entire sky with the familiar beautiful blue glow that we see on sunny days. However, Rayleigh's explanation had a serious flaw: a simple calculation shows that if the air is uniformly distributed throughout the atmosphere, the scattering of a beam of light by the air molecules should cancel exactly. Whenever a beam of light encounters a molecule and suffers some scattering, it will encounter another molecule later and suffer some more scattering that exactly cancels the first (this is a case of destructive interference of the scattered waves; the second scattered wave cancels the first). Rayleigh had overlooked this subtle cancellation effect, and it ruined his simple explanation for the blue color of the sky.

Einstein rescued Rayleigh's explanation from this disaster. He showed that although the scattering cancels for a perfectly uniform distribution of air, the actual distribution of air in the atmosphere has small irregularities in density, because sometimes there are a few more air molecules in a cubic centimeter and sometimes a few less. By an arithmetical coincidence, the small failure of cancellation arising from this small excess or defect of molecules restores the net amount of scattering to exactly the value originally calculated by Rayleigh. Whenever you look up at a beautiful blue sky, let your thoughts drift to Einstein for a moment, and remember that he gave us the explanation for this blue color.

Quantitative measurements of the scattering of light confirmed Einstein's calculation, and led to yet one more determination of Avogadro's number. In 1909, Perrin summarized the evidence for the existence of atoms and molecules. He argued that the concordance of the values of Avogadro's number obtained by a wide variety of completely different and independent methods establishes the reality of atoms and molecules beyond all reasonable doubt: "I think it is impossible that a mind free of preconception can reflect upon the extreme diversity of the phenomena which thus converge to the same result without experiencing a strong impression, and I think it will henceforth be difficult to defend by rational arguments a hostile attitude to the molecular hypothesis."[38]

6

"What is the light quantum?"

Einstein continued to ask this question fifty years after
he launched the idea of light quanta in 1905,
which gained him his Nobel Prize.

oward the end of the nineteenth century, physicists lived in a state
of blissful delusion. They thought that most of the big problems
of physics had been solved, and that the remaining problems were
merely minor details. When Max Planck, upon graduation from his high
school in Munich at the young age of sixteen, asked two professors for
advice on whether to pursue a career in music or in physics, the musician
said to him, "If you have to ask, go study something else!" and the physicist urged him to find a different career because everything worthwhile
in physics had already been discovered, and only a few insignificant gaps
remained to be filled.[1] In his autobiography, Einstein described the pre-
vailing beliefs of the physicists of those days: "In the beginning (if there
ever was one), God created Newton's laws of motion and the necessary
masses and forces. That is all; everything else follows by deduction upon
the development of suitable mathematical methods."[2]

But by 1899, physicists must have wondered whether something was
wrong with God, Newton's laws, or their mathematical methods. The long-

festering problem of the ether and the speed of light remained unsolved—it was not to be solved until 1905, by Einstein. And physicists faced another acute problem in connection with the investigations of the radiant heat, or thermal radiation, emitted by hot, glowing bodies, such as hot pieces of coal in an oven, iron in a blacksmith's forge, or pottery in a kiln.

Physicists had first become interested in this thermal radiation forty years earlier, when they recognized that the laws of thermodynamics place serious constraints on how this thermal radiation is emitted. For a body with a hollow cavity accessible only by a small hole—for instance, a small hole in the side of a hot, closed oven—the laws of thermodynamics require that the radiation emerging from the hole be entirely independent of the material of the cavity. It makes no difference at all whether the cavity is lined with iron, brick, pottery, or Meissen porcelain: the emitted thermal radiation depends on the temperature of the cavity, and on nothing else. Thus, the formula for this radiation is a universal law.

Thermal radiation of this kind is called *cavity radiation* or *blackbody radiation*. (The latter name arises from the peculiar circumstance that when such a cavity is cold, the hole will look absolutely black, more black than any black paint you can imagine. You can easily check this by taking a closed cardboard box and making a small hole in its side; the hole will look pitch black, regardless of the color of the interior of the box.) Physicists were eager to discover the universal law of blackbody radiation, because they expected it would reveal some fundamental insights into the behavior of radiation and matter.

PROGRESS WAS SLOW. Experimental physicists needed to develop accurate techniques for the quantitative measurement of thermal radiation, and theoretical physicists needed to achieve a clear understanding of the principles of statistical mechanics and the laws governing the production of radiation by the (presumed) electric charges in the atoms. Finally, toward the end of the nineteenth century, experimenters invented new techniques for the measurement of thermal infrared radiation. They proceeded to measure not only the total intensity of radiation emitted by glowing bodies but also its spectrum, that is, the intensity of radiation at each frequency or each wavelength; and they investigated how this radiation changes with temperature. For instance, when you heat a piece of iron to higher and higher temperatures, it first becomes red hot, then white hot, then blue hot—the frequency of the predominant radiation shifts from low (red) to high (blue) as the temperature increases.

Most of this experimental work was done in Germany in the 1890s, by the well-funded physicists in Berlin. Kaiser Wilhelm II, "All-Highest," may have been a moron in matters political and military who bears a large share of the responsibility for failing to halt the mad rush toward World War I, but he deserves some credit as a keen yachtsman and as an enthusiastic supporter of German science. He could tell the difference between luff and leech, and he allocated generous grants to the Imperial Physical-Technical Institute and, later, donated a large tract in his hunting preserve of Dahlem on the outskirts of Berlin for the foundation of the Kaiser-Wilhelm Institute (today called the Max-Planck Institute).

The data obtained in these experiments looked deceptively simple. At a fixed temperature, the data fell on a smooth curve when displayed in a graph of intensity versus frequency—the points representing the intensity started near zero at low frequencies then climbed steeply upward to reach a peak at an intermediate, predominant frequency, and finally descended on a gentle slope toward zero intensity for high frequencies. The intensity curves for higher temperatures were higher, but, qualitatively, they still had the same shape.

This suggested to the experimenters that there should exist a simple mathematical formula describing these intensity curves. And, as experimenters are wont to do, they attempted some trial-and-error curve fitting in the hope of finding a good mathematical formula. Today, you can do such curve fitting with nifty programs on your PC—you tell the computer what points you want to fit, and it instantly spits out a mathematical formula that fits the points to perfection. But in those days, curve fitting involved laborious calculations done by hand, mostly by the hands of the assistants of the professors. The best fit was achieved by a formula contrived by the German physicist Wilhelm Wien. He had earlier deduced some general features of the radiation law from thermodynamics, and all these important contributions to the radiation problem later earned him the Nobel Prize, in 1911.

But curve fitting is only the first step in the amalgamation of experiment and theory. After the experimenter fits his curves, the theoretician has to explain them. To their consternation, the Berlin physicists found that their experimental curves were in conflict with theoretical calculations based on Newtonian physics, which predicted that the radiation should increase with frequency—no gentle rise, peaking, and downward slope, but a steep rise that becomes steeper and steeper. This means that most of the radiation at any temperature should be high-frequency radiation, that is, ultraviolet radiation. This conflict of experiment and theory came to be called the "ultraviolet catastrophe."[3]

THE CONFLICT BETWEEN theory and experiment was resolved within less than a year by Max Planck, by then a professor at the University of Berlin. Against the advice he had been given upon graduation from high school, Planck had pursued a successful career in physics, and he had become a leading authority in thermodynamics. He had published many papers and a highly regarded textbook on this subject—and he had paid close attention to the problem of blackbody radiation for several years.

Planck relied on his expertise in thermodynamics—his "incomparably delicate thermodynamic sensitivity,"[4] as von Laue called it—to make an intelligent guess for the radiation law. After a few weeks of tinkering, he found a formula that seemed a perfect fit to the experimental data. He presented this formula, now known as Planck's law, at a meeting of the German Physical Society in October 1900. New data on thermal radiation obtained by experimenters showed that, at low frequencies, Wien's law deviated from the data. The usually dry-as-dust reports of the meeting state that this led to a "lively discussion," and Planck's proposal of his new law generated much interest. Overnight, the experimenters checked their data against Planck's formula, and, the very next morning, they informed him that he had a perfect fit. But at this point, Planck had no theoretical justification for his law; as he admitted later, it was merely "an interpolation formula, guessed by good luck."[5]

Planck immediately began an intense effort to find the missing theoretical justification. Within eight weeks he found the answer: "After a few weeks of the most strenuous work of my life, the darkness lifted and an unexpected vista began to appear."[6] On December 14, 1900, at a memorable meeting of the German Physical Society, he announced a revolutionary hypothesis of quantization of energy. This marks the birthday of quantum theory.

In his theory, Planck treated the emitters of radiation in the walls of the cavity as small oscillators, that is, small electrically charged particles attached to springs. Although this is a rather crude and sketchy model for an atom, Planck knew from his study of thermodynamics that this model was a valid theoretical substitute for a more realistic, and more difficult, model of an atom. The random thermal disturbances in the hot walls of the cavity make these oscillators vibrate with various amounts of energy— some vibrate more, some less—and when they vibrate, they emit radiation that ultimately emerges from the hole in the cavity.

Planck then calculated what average energy the oscillators must have to yield the correct formula for the radiation law. He introduced a small, fundamental "energy packet," or energy quantum, into his calculations, and assigned to each oscillator either one such energy packet, or two, or three, etc., but never a fraction of an energy packet. He found that if the energies are allotted to the oscillators according to this quantization rule, the calculated thermal radiation is in accord with his radiation law; but if the energy were allotted in any smaller fractional amount, the calculated thermal radiation would not match the radiation law. This means that the oscillators are incapable of emitting energy smoothly and continuously—

Planck's constant h inscribed on a plaque on the wall of the Prussian Academy of Sciences. *(Courtesy of Kaihsu Tai, copyright © Kaihsu Tai)*

they can emit energy only in quanta of a fixed magnitude. An oscillator can emit one quantum, or two quanta, or three quanta, etc; but it cannot emit a fraction of a quantum. This is analogous to the way a chicken lays eggs—it will lay one, or two, or three eggs, but it cannot lay half an egg.

According to Planck's theory, the magnitude of the energy quanta of an oscillator equals the product of the oscillation frequency and a fundamental constant, later called Planck's constant, although Planck himself never called it by that name—he called it the "action" constant. If you stroll along Unter den Linden, the old main street of Berlin, you will find this constant, represented by the symbol h, inscribed on the wall of the university, at the building where Planck first announced his discovery. It is also carved on Planck's tombstone, in the Göttingen cemetery. Apart from the speed of light, Planck's constant is the most important constant in physics; it determines the magnitude of all the quantum effects that govern the realm of the atom and the nucleus.

AS AN IMMEDIATE BY-PRODUCT of his quantum theory, Planck was also able to determine the magnitude of Boltzman's constant, which Ludwig Boltzmann had introduced into his theoretical formula relating entropy and probability but had believed to be beyond the reach of measurement. From the value of Boltzmann's constant, Planck also determined a new value of Avogadro's number, far more accurate than previous estimates from the properties of gases, and he determined the value of the fundamental electric charge of electrons and ions. Planck had always attached much importance to the identification and the determination of the fundamental constant of physics, and he was delighted with these concrete results he was able to extract from this quantum theory. But most of his contemporaries ignored these results—his calculations of Avogadro's number and the fundamental electric charge from data on thermal radiation seemed too far-fetched to be believable. Planck's value for the fundamental electric charge was not considered credible until 1908, when Rutherford obtained an almost identical value by measurements on alpha particles emitted by radioactive sources.[7]

Planck's energy quanta are very small, and this is why they had never been noticed before. For instance, for a large oscillator, such as a child bouncing up and down on a pogo stick at the rate of, say, one bounce per second, Planck's energy quantum is so small that a bounce with only one energy quantum would be imperceptible. The typical bouncing motion of a child on a pogo stick involves an enormous number of energy quanta, more energy quanta than there are grains of sand in all the Earth, even if all the Earth were made of sand. In everyday life we do not notice the energy quanta—we typically deal with so many energy quanta, that the energy distribution seems smooth and continuous. According to Planck's picture, the smooth distribution of energy is an illusion that arises from the small size of the energy packets, just as the smooth mass distribution in a sample of water is an illusion that arises from the small size of the "mass packets," or atoms, that make up the sample. Planck's quanta are, in effect, atoms of energy.

The proposal of energy quantization was a radical innovation. It was in direct conflict with the well-established principles of Newtonian physics, according to which energy is smoothly allocated to the particles within matter, to be emitted or absorbed in a continuous manner. In fact, Newton's laws forbid the quantization of energy, because whenever a particle has one of these quantized energies, a small push, perhaps from a breath of air, can alter the energy by a small amount and leave the particle with the "wrong" energy. Planck was well aware that by his quantization of energy he was undermining the laws of mechanics that had ruled physics for two and a half centuries. His son Ernst, a young boy at the time, later recalled that during long walks in Berlin's Grunewald Park, his father told him he had made a discovery comparable to those of Newton.[8]

WITH HIS HYPOTHESIS of quantization of energy, Planck launched a quantum revolution that was to last for thirty years—"thirty years that shook physics," as George Gamow, one of the quantum physicists who participated in this revolution, called them. When Einstein discovered the first clues that launched his search for the theory of general relativity, he commented, "Nature shows us only the tail of the lion. But I do not doubt that the lion belongs to it, even though he cannot reveal himself directly to our sight because of his enormous extent."[9] This comment applies equally well to the search for the quantum theory. Planck had found the tail of the quantum lion, and it took a long and intensive effort by the best and brightest theoretical and experimental physicists to reveal the full extent of this lion, a ferocious, voracious beast, that even today springs surprises on us and has not yet been fully tamed.

The quantum revolution devoured the foundations of Newtonian mechanics and gave us an entirely new quantum mechanics, where all the energies in atoms and nuclei and molecules are quantized, where electrons make unpredictable jumps from one state of motion to another, where position and velocities are afflicted with mysterious indeterminacies, and where the certain and firm predictions of Newtonian mechanics are replaced by mere probabilistic assertions. Today, Newtonian mechanics is called classical mechanics, and it is viewed as an approximate form of quantum mechanics, applicable only to very large, very massive bodies, for which quantum effects are imperceptible.

But Planck was quite uncomfortable in the role of a revolutionary. He was the quintessential German professor, with an obsessive devotion to law and order, not only in physics but also in his personal and public life. For him, the quantization of energy was "an act of desperation," into which he was forced because "a theoretical explanation had to be found, whatever the cost."[10] At first, most of Planck's contemporaries did not take his theory of energy quanta very seriously. They assumed that sooner or later a more conventional explanation of the blackbody spectrum would be found. Only Einstein fully grasped the enormity of what Planck had done; he declared that "it was as if the floor had been pulled out from under the feet, with no firm ground to be seen anywhere upon which one could have built."[11]

Some historians have claimed that in his later work Planck retreated from his quantum hypothesis. This is a canard. In his later work, Planck did not remove the quantization; he merely tried to fix some inconsistencies in the details. Planck understood that Newton's physics was broken, irretrievably broken. Humpty Dumpty Newton had taken a big fall, and all the king's horses and all the king's men could not put Newton's physics together again. As he said in his address to the Solvay Conference in Brussels in 1911, "The framework of classical dynamics . . . is obviously too narrow to account also for all those physical phenomena which are not accessible to our coarse senses."[12] At best, Planck hoped to achieve some damage control—he hoped that the break in Newton's physics could be limited.

One inconsistency in Planck's original work involved the treatment of the interaction between the oscillators and the thermal radiation in the blackbody cavity. All such thermal radiation is electromagnetic radiation, mostly ordinary light and infrared light, but with some admixture of ultraviolet light and microwaves. In his original calculation, Planck had assumed that although the oscillator energies are quantized, the interaction between the oscillator and the electromagnetic radiation could be described by Newton's laws of motion and Maxwell's equations. As Einstein pointed out in a paper some years later, this is inconsistent, because quantization demands deviations from Newton's laws, and the emission of radiation by a quantized oscillator demands deviations from Maxwell's equations. Planck was probably aware of this problem even before Einstein rubbed his nose in it, but perhaps Einstein's criticism provided some extra incentive to attempt a resolution and to find a deeper explanation of quantization. Planck explored the possibility that the oscillators emit energy in quantized amounts, whereas they absorb energy continuously, in nonquantized amounts, as expected according to Maxwell's electrodynamics. But this provided only a partial answer to Einstein's objection.

IN 1905, IN THE FIRST of the four papers he wrote in that year, before his paper on relativity, Einstein took Planck's quantization of the energy of oscillators one step further, by proposing that the energy of light is also quantized. Einstein was led to this proposal by an analysis of the blackbody radiation law. Einstein found that the radiation in the cavity had some of the thermodynamic properties of a gas—the radiation behaves as though it were made out of pointlike particles, each of them with an energy equal to one of Planck's quanta. He called these particles light quanta, or "Lichtquanten." Today we call these particles photons, but this name was not given to them until the late 1920s, and Einstein continued to call them light quanta throughout his life.

In a letter to his friend Habicht, Einstein described his proposal of light quanta as "very revolutionary."[13] But in fact, this proposal was rather less revolutionary than the original quantization of energy by Planck—it was merely a natural outgrowth of Planck's idea. Furthermore, some of Einstein's courage abandoned him when he selected a title for his paper. He called it "On a Heuristic Viewpoint Concerning the Production and Transformation of Light." The word *heuristic* is one of those erudite Greek waffle words that physicists and philosophers are wont to use when they don't want to stick their necks out too far. It announces that a viewpoint is adopted tentatively, as an investigative tool, in the hope that it might lead to further discoveries. And it warns the reader that the author isn't willing to put his money where his mouth is. Einstein could equally well have entitled his paper "An Investigative Viewpoint . . ." or a "A Working Hypothesis . . .", but that would have sounded a lot less impressive.

At first sight, Einstein's hypothesis of light quanta seems a necessary consequence of Planck's quantization of the energy of the oscillators that emit the light. Since an oscillator can emit only one, or two, or three, etc., energy quanta, the emitted light necessarily has one, or two, or three, etc., energy quanta; thus, it would seem self-evident that the light must be quantized. Although it is true that light is quantized immediately after its emission, it is not self-evident that it will stay that way. Planck and his contemporaries thought light was a wave, and when a wave strikes a small obstacle, such as a small dustgrain, it is scattered; that is, it breaks up into small wavelets that spread out into various directions (you can readily see such scattering in a water wave, when it strikes a rock, piling, or a buoy). By such scattering, the energy in the wave is subdivided, and there might be a corresponding subdivision of the energy quanta, which would erase the quantization of the wave. In terms of the quantum versus egg analogy: the eggs might break up during scattering, so they become a distributed, amorphous mass of well-stirred eggs.

Einstein assumed this would not happen. He adopted the "heuristic viewpoint" that the energy of the light wave remains quantized forever. The eggs laid by Einstein's quantum chickens do not split into smaller pieces or merge into larger pieces—they remain as individual eggs forever, until they are absorbed by another chicken somewhere else. As he wrote in his paper, "According to the assumption here to be kept in view, the energy in a light ray propagating outward from a source is not distributed continuously over . . . space, but instead consists of a finite number of energy quanta localized at points of space, which move without dividing themselves and can be absorbed and produced only as a whole."[14]

Thus, Einstein thought of light quanta as tiny corpuscles, and he thought of a light beam as a stream of such corpuscles. The small magnitude of the energy in the individual light quanta implies that ordinary light, from the Sun or from a lightbulb, contains very many quanta. For example, in bright sunshine, each cubic centimeter of sunlight contains about 10 million quanta. The small magnitude of the quanta and their high density make ordinary light seem smooth and continuous.

A corpuscular picture of light had been proposed centuries before, by Newton. He had expressed his proposal as a rhetorical question: "Are not the rays of light very small bodies emitted from shining substances?"[15] However, this picture was abandoned in the early nineteenth century, when investigations by the English physicist Thomas Young and the French physicist Augustin Fresnel revealed that light displays interference effects and scattering effects. Such effects are characteristic of waves, and they flatly contradict any simple corpuscular picture of light. For instance, in the case of destructive interference, two light waves meet crest to trough, and they thereby cancel each other. But such a cancellation makes no sense whatsoever for two streams of corpuscles—corpuscles can't cancel one another. When you add eggs to eggs, you get more eggs, not less eggs.

Planck felt that the compelling experimental evidence for the wave nature of light ruled out any corpuscular picture, and in his treatment of blackbody radiation he had carefully avoided a direct quantization of light—in his theory, quantization only affected light indirectly, by the "imprint" that the quantization of the oscillators places upon the average characteristics of light that they emit and absorb in the blackbody cavity. Planck could not avoid the quantization of the oscillators, and he felt constrained to accept the serious consequences for Newtonian mechanics implied by the quantization of the oscillator energies. But he tried his best to avoid the quantization of light, which would lead to equally serious consequences for Maxwell's theory of electromagnetism.

Einstein delighted in the role of the revolutionary. He was a devotee of Maxwell's, but he had fewer inhibitions than Planck about dumping Maxwell, if this would lead to a profitable payoff. And the payoff for Einstein's "heuristic viewpoint" was bountiful. Einstein applied his ideas with great success to the explanation of several phenomena, among which the photoelectric effect was the most remarkable—and this ultimately gained him the Nobel Prize in 1922.

THE PHOTOELECTRIC EFFECT had been discovered many years before, in 1887, when Heinrich Hertz noticed during his early experiments with radio waves in his laboratory in Karlsruhe that electric sparks between adjacent metallic terminals could be triggered more readily when the terminals were illuminated with ultraviolet light. Evidently, the electrons absorbed energy from the light, and this caused their emission from the terminals, producing the electric current revealed in the sparks. The effect was investigated in much detail by Philipp Lenard, a skillful experimental physicist, who began his career as an assistant to Hertz and became a professor at the University of Kiel and later at Heidelberg.

After World War I, Lenard was to transmogrify into a virulent anti-Semite and a vociferous enemy of Einstein. But in the early years, Lenard admired Einstein and his theories, and, vice versa, Einstein admired Lenard and his work. In 1901 he wrote to Mileva, "I just read a most beautiful paper by Lenard about the generation of cathode rays (electrons) by ultraviolet light. Under the influence of this beautiful piece I am filled with such happiness and such joy, that you must absolutely have some of that too."[16] Einstein's letter was written in response to Mileva's letter informing him she was pregnant; ever the scientist, Einstein first discusses Lenard, and only toward the end of his letter does he discuss the awkward matter of pregnancy. Mileva, also, was a fan of Lenard's; she had attended some of his lectures at Heidelberg when she spent a semester there, and in an earlier letter to Einstein she had gushed, "Oh, it was so pretty yesterday in the lecture of Prof. Lenard, who is now talking about the kinetic theory of gases . . . it was established that the molecules of O [oxygen] move with a speed of 400 m per second, then the goodly professor calculated and calculated, set up equations, differentiated, integrated, substituted, and finally it emerged that indeed the molecules move with this speed, but that they advance over a path of only 1/100 of a hair's breadth."[17]

In his investigations of the photoelectric effect, Lenard found that the energy absorbed by the electrons from the ultraviolet light increased with the frequency of light, but it did not depend on the intensity of light. If the frequency was high, the electrons absorbed enough energy from the light to be ejected from the terminals; but if the frequency was low they did not absorb much energy, and they stayed put. This behavior of the electrons was at variance with what was expected on the basis of Maxwell's equations. A strong light wave, of high intensity, was expected to give the electrons large energies. And a high-frequency wave, exerting a quick succession of alternating pushes in opposite directions, was expected to give the electrons less energy than a low-frequency wave.

IT SEEMS THAT as far back as 1901, these conflicts between Lenard's experimental results and classical theory led Einstein to contemplate the need to abandon some consequences of Maxwell's equations and the conventional description of light as a wave. And in 1905, in one of his flashes of inspiration, he recognized that Planck's quantization held the answer, if the quantization is applied not only to the oscillators but also to the light itself. He assigned to each light quantum an energy equal to that given by Planck's quantization rule for oscillators: the energy of a light quantum equals the product of the light frequency and Planck's constant. And he assumed that in the photoelectric effect, electrons absorb these light quanta from the light wave, one quantum at a time. This means that each electron that captures a light quantum acquires an energy proportional to the frequency of the light—in agreement with Lenard's experimental results.

This relation between the electron energy and the frequency of the light came to be called Einstein's photoelectric equation. An increase of intensity of the light merely means that more light quanta rain down on the electrons, and more electrons will capture a light quantum; but this does not increase the amount of energy captured by each individual electron. It is like throwing sunflower seeds to a large flock of hungry birds, each of which flies away as soon as it captures a seed—if you throw more seeds, you will not increase the mouthfuls taken by any individual bird, but you will increase the number of birds that do get a mouthful.

Unfortunately, Lenard's data on electron energies were not very precise. They indicated that the electron energy increases with frequency, but they did not show exactly how much it increases. At the end of his paper, Einstein had to be satisfied with a rather weak claim: "As far as I can see, our concept is not contradicted by the properties of the photoelectric effect observed by Mr. Lenard."[18]

Soon after the publication of Einstein's paper, Lenard, who had just received the Nobel Prize for his experiments with cathode rays, initiated a cordial correspondence with Einstein, sending him a reprint of some of his newest work. Einstein replied that he had studied this new work "with the same feeling of admiration as your earlier works,"[19] and in a letter to a colleague he described Lenard as "not only a skillful master of his future, but a true genius."[20] But despite these expressions of mutual respect, Lenard and Einstein only agreed to disagree. Lenard did not accept Einstein's interpretation of the photoelectric effect, and he continued to view the emission of electrons as some kind of resonance phenomenon that supposedly arises from Maxwell's equations.

Precise data confirming Einstein's interpretation of the photoelectric effect did not become available until 1916, when the American physicist Robert Millikan reported his very meticulous measurements of the electron energies in the photoelectric effect. For his experiments, Millikan placed the metallic terminals in an evacuated glass bulb, and he contrived a method for shaving off the top layer of the terminals with a knife blade installed within the evacuated bulb, so as to obtain a perfectly clean, uniform metallic surface, uncontaminated by air.

With his accurate measurements, Millikan confirmed Einstein's equation for the electron energy, and he confirmed that electrons, indeed, absorb light quanta of the expected magnitude. For this work, and for his earlier work on measurements of the electric charge of the electron, Millikan was awarded the Nobel Prize in 1923, a year after Einstein. However, although Millikan confirmed Einstein's equation for the photoelectric effect, he refused to believe Einstein's theory of light quanta, and he said, "Despite . . . the apparently complete success of the Einstein equation, the physical theory, of which it was designed to be the symbolic expression, is found so untenable that Einstein himself, I believe, no longer holds to it."[21]

In 1923, the American physicist Arthur Compton provided further evidence for Einstein's light quanta (which by then were called photons). He found that when X-rays are incident on a sample of atoms—say, a piece of metal—the X-rays make the electrons in the atoms recoil, in the way expected for elastic collisions between the incident X-ray photons and the electrons. The collision between an incident X-ray photon and an electron is rather like a collision between an incident high-speed billiard ball and a stationary billiard ball; in such a collision the initially stationary billiard ball recoils, acquiring some of the energy of the initially moving billiard ball. Compton found that the measured recoil energy agreed exactly with the energy calculated for an elastic photon-electron collision, in which the photon is treated as a high-speed, extremely relativistic particle. For this discovery, Compton gained the Nobel Prize a few years later.

PLANCK, ALSO, WAS UNHAPPY with Einstein's idea of persistent quanta of light. He insisted that only the emission and the absorption of light should be quantized, but that light would not retain its quantization after emission. This was also the view held by Lorentz, who supported "Planck's hypothesis of the energy elements," but disapproved of Einstein's "light-quanta which retain their individuality in propagation."[22] And indeed, if light is a wave, the persistence of the quanta of light would seem to make no sense whatsoever. If you have a light wave whose energy is one single quantum, it would seem you could fabricate a light wave with, say, half a quantum by simply chopping this light wave in two, for instance, by closing the shutter of a camera when the light wave is halfway through. Planck was willing to accept (reluctantly) the quantization of the energies of oscillators, because he thought that perhaps there are some unknown modifications of Newton's laws of mechanics on the atomic and subatomic scale, just as there are relativistic modifications of Newton's laws at high speed. But he regarded a persisting quantization of the energy of light waves as nonsensical.

Planck felt so sure the notion of light quanta was wrong that when in 1913 he nominated Einstein for membership in the Prussian Academy of Sciences, he thought it necessary to defend Einstein in a speech to the academicians: "One should not count it too heavily against him that in his speculations he occasionally might have missed the mark, as e.g., in his hypothesis of light quanta; because without taking a risk, it is not possible to introduce any real innovation even in the most exact of the natural sciences."[23] There is much irony in this: nine years later Einstein was to receive the Nobel Prize for just the idea that Planck held up for criticism.

The conflict between the corpuscular and the wave aspects of light was not resolved until many years later, when quantum mechanics gave us a new conception of light. What quantum mechanics teaches us is that light has both wave and particle properties. Sometimes the wave aspect is dominant, sometimes the particle aspect—it depends on what experiment we perform on the light. A scattering experiment emphasizes the wave aspect, whereas a photoelectric experiment emphasizes the particle aspect. Because of its dual nature, light might be called a "wavicle," that is, a combination of wave and particle.

But Einstein never accepted this modern quantum-mechanical picture of light. For Einstein, the conflict between the corpuscular and the wave aspects was never resolved. In his later years he complained, "A full fifty years of deliberate brooding have not brought me any closer to the question: What is the light quantum? Today every clod thinks he knows it, but he deceives himself."[24]

THE CONFLICT BETWEEN Einstein's corpuscular view and the known wave properties of light was a defect in Einstein's theory of light quanta, but it was not a mistake. Paraphrasing Planck, we can say it was the price that Einstein paid for "introducing real innovation . . . in the most exact of the natural sciences." However, Einstein's paper on light quanta contained a genuine, glaring mistake in its analysis of the blackbody radiation law. Einstein had arrived at the interpretation of blackbody radiation as a gas of light quanta by means of an analysis that relied on Wien's law for the blackbody radiation rather than Planck's law. Wien's law agrees with Planck's law at high frequencies, but it deviates from Planck's law (and from observation) at low frequencies. Thus, Einstein's interpretation of blackbody radiation as a gas of light quanta worked only for the high-frequency quanta, but not for the low-frequency quanta. In his paper, Einstein emphasized the good fit at high frequency and pretended not to notice the misfit at low frequency, like a tailor who tells the customer how beautifully the jacket fits at the shoulders, and pretends not to notice that the sleeves are much too long, ending somewhere near the knees.[25]

And there is another mistake in Einstein's paper. In a footnote, Einstein claimed to show that the relation between pressure, volume, and temperature of the radiation is the same as that for an ordinary gas of atoms or molecules. This footnote was intended to lend additional support to his contention that blackbody radiation can be regarded as a gas of corpuscles. But in his derivation of the pressure-volume-temperature relation Einstein overlooked a subtle effect: When the volume of the radiation is compressed by an external force, the frequency of the quanta of light increases (in the same way that the frequency, or pitch, of a trombone increases when the player shortens the slide). This adds an additional term to the pressure, which does not exist in the case of an ordinary gas and which modifies the pressure-volume-temperature relation by a factor of 3. It is a puzzle why Einstein did not immediately spot this mistake—the factor of 3 in the expression for the pressure of radiation confined in a container was well known in those days, from Maxwell's equations. It is even more of a puzzle why this mistake was not spotted by Einstein's contemporaries, or by anybody else until recently.[26]

Neither of these mistakes was ever corrected by Einstein, but both mistakes became moot in 1924, long after Einstein's idea of light quanta had become accepted (and he had received a Nobel Prize). At that time, Einstein's original treatment of the gas of light quanta was superseded by an entirely new statistical treatment by the Indian physicist Satyendranath Bose, who discovered that, in quantum mechanics, gases behave in a surprisingly different way. Bose was an instructor at the University of Dacca; he had an overabundance of original and wild ideas, and was described as rather "flaky" by an Indian colleague. In June of 1924, Bose sent Einstein a paper with a new derivation of Planck's law. The paper was in English, and Bose asked for Einstein's help in its translation and publication in a German journal. He wrote, "If you think the paper is worth publication, I shall be grateful if you arrange its publication in *Zeitschrift für Physik*," and he added, "Though a complete stranger to you, I do not feel any hesitation in making such a request. Because we are all your pupils though profiting only by your teachings through your writings . . ."[27] Bose had previously sent the same paper to *Philosophical Magazine*, the leading English physics journal, which had rejected it. Einstein immediately understood the profound significance of this paper, and he replied by a postcard: "I have translated your paper and given it to *Zeitschrift für Physik* for publication. It means an important step forward and pleased me very much."[28]

Bose had recognized that the statistical treatment of a gas of quantummechanical light quanta required a drastic modification of the statistical methods of Boltzmann. The classical particles in Boltzmann's statistical treatment of a gas are identical (all are, say, molecules of oxygen), but they are nevertheless distinguishable because we can, in principle, keep track of them by observing them at all times (for instance, we might film their motion, and thereby know exactly which particle went where and which particle did what). In contrast, the quantum-mechanical light quanta in blackbody radiation are absolutely indistinguishable. When two such quanta collide, we have no way of knowing which is which after a collision, because the quantum-mechanical uncertainties in their motion prevent us from tracking the motion to discover which went where during the collision—two light quanta enter the scene of the collision, and two emerge, but we can't tell which is which. Bose showed how to take this indistinguishability into account in a new and clever statistical treatment, and he showed how this leads directly to Planck's law. Thus, with the new Bose statistics, Einstein's interpretation of blackbody radiation as a gas of light quanta worked for the high-frequency quanta and also for the low-frequency quanta.

IN 1906, IN ANOTHER FLASH of insight, Einstein found a brilliant confirmation of Planck's quantization of the energy of oscillators. The atoms in a solid body, such as a metal or other crystalline solid, are held in place by elastic bonds that link them to adjacent atoms. These elastic bonds can be thought of as springs, so each atom is effectively an oscillator, that is, a mass attached to a spring, which can oscillate around its equilibrium position within some limits. By his insight, Einstein recognized that the energy of oscillation of each atom must then be quantized, according to Planck's quantization condition for oscillators.

The individual energy quanta for this oscillatory motion are quite small and they were not directly measurable by the techniques available in those days. However, when the metal is placed in an environment at a given temperature, the random thermal disturbances will cause each atom to oscillate with some average amount of energy, and for a large number of atoms (such as the number of atoms in, say, a 1-kilogram sample of metal), the total combined energy can be quite substantial. This meant that the quantized oscillator energies would show up indirectly in measurements of the thermal energy stored in metals, what is called "latent heat."

From Planck's quantization condition, Einstein deduced a formula for the decrease of latent heat with decreasing temperature, and when he compared his formula with experimental data, he found good agreement. His result was all the more impressive because previous theoretical calculations, based on Newtonian physics, had given totally wrong results—at low temperatures, the latent heats calculated from Newtonian physics were much larger than what was indicated by the experimental data.

Planck and his physicist-chemist colleague Walther Nernst attached great significance to Einstein's theoretical results about the small value of the latent heat near the absolute zero of temperature because this confirmed the experimental observations that had led Nernst to formulate the Third Law of Thermodynamics. Stated in the simplest way, this law says that the absolute zero of temperature can never be reached. And, indeed, not even our modern techniques of cooling have permitted us to reach the absolute zero of temperature. The lowest temperature ever attained in a laboratory experiment is a tenth of a billionth of a degree above absolute zero—almost zero, but not exactly zero.

In spite of these early successes of Planck's quantum theory, Einstein had serious misgivings because of the absence of a solid physical foundation for quantization. To a friend he wrote, "*Theory* is too presumptuous a word—it is only a groping without correct foundation. The more successes the quantum theory has, the sillier it looks. How nonphysicists would scoff if they were able to follow the odd course of development."[29] The later development of quantum mechanics did not allay Einstein's misgivings. He always remained dissatisfied with the explanations offered by modern quantum mechanics, and he was to become involved in lengthy disputes with Niels Bohr (more on this in Chapter 11) and other quantum theorists.

PLANCK SUCCEEDED IN EXPLAINING blackbody radiation by quantization. This replaced the puzzle of blackbody radiation by a deeper puzzle: what causes the quantization? As Nernst said, "At present, quantum theory is essentially only a rule for calculation, of apparently very strange, one might even say grotesque, nature; but it has proven so fruitful by the work of Planck, as far as radiation is concerned, and by the work of Einstein, as far as molecular mechanics is concerned . . . that it is the duty of science to take it seriously and to subject it to careful investigations."[30]

The explanation of Planck's quantization was to occupy physicists for the first thirty years of the twentieth century. It led to the development of quantum mechanics at the hands of the Dane Niels Bohr, the Germans Werner Heisenberg and Max Born, the Austrian Erwin Schrödinger, and the Englishman Paul Dirac, all of whom were to earn Nobel Prizes for their contributions. The long interval between Planck's discovery of quantization in oscillators and the formulation of a complete and coherent theory of quantization for all physical systems is a measure of the profoundly revolutionary nature of Planck's idea. None of Einstein's discoveries required such a long interval for complete assimilation into physics. And none of Einstein's discoveries sent such shock waves throughout all aspects of physics.

Planck was still the dominant figure in physics until 1918, when Einstein was suddenly propelled to fame by stories about general relativity in the popular press. In 1918, Planck was awarded the Nobel Prize for his discovery of quantization. Why the Nobel committee waited eighteen years before awarding Planck the prize can only be explained by the rule of the least common denominator: a committee is always as dumb as the dumbest of its members. Maybe the Nobel committee was hoping that Planck's unconventional ideas would fade away. In 1911, they had awarded the Nobel Prize to Wilhelm Wien, whose law for the blackbody radiation predated Planck's law, but did not quite agree with experiment. The Nobel committee probably liked it better than Planck's law, because is rested (mostly) on conventional Newtonian physics. But by 1918, theoretical investigations by Niels Bohr and other quantum theorists had established that quantization afflicts not only oscillators but also the internal mechanics of atoms, and it must have become clear even to the dumbest member of the committee that quantization was here to stay.

Planck was forty-two when he discovered quantization. This is an advanced age for making a great discovery. Most great discoveries in science are made by scientists in their twenties. By comparison with these frisky youngsters, the older scientists are dinosaurs—big and powerful, but slow to move. Planck's discovery of quantization was an exception to the rule. But he was too old for active participation in the subsequent developments of quantum theory. When quantum mechanics was finally fully formulated by the work of Heisenberg and Schrödinger in 1925, Planck was in his late sixties—he was then a very old professor approaching retirement (and Schrödinger became Planck's successor as professor of theoretical physics in Berlin). Although Planck was not able to participate in the development of quantum mechanics, he will always be remembered as the father of quantum mechanics.

We have it on the authority of Hans Bethe that the greatest scientific discoveries of the twentieth century were quantum theory, DNA, and relativity theory, in that order.[31] In December 1999, *Time* magazine named Einstein the man of the century. They would have done better to award this honor to Planck and to the other physicists who developed quantum mechanics. The twentieth century was primarily the century of the quantum, and only secondarily the century of relativity. Electronics became our dominant technology; almost all the gadgets we use run on electronics—from computers to cell phones, radios, CD players, and TVs. The transistors, integrated circuits, processor and memory chips in our electronic devices all rely on quantum physics. Electronics represents the single largest sector of our economy—whenever the stocks of computer chip manufacturers sink, the entire stock market is dragged down. In contrast, relativity plays hardly any direct role in modern technology. The Global Positioning System (GPS) is an exception: corrections from relativity need to be incorporated into the clocks on the GPS satellites to attain the required accuracy, but, of course, the entire GPS system relies on electronics.

As the founding father of quantum theory, Max Planck might therefore be considered a greater physicist than Einstein. Planck took the first revolutionary and courageous step toward quantum theory with his discovery of the quantization of energy. This was a discovery of greater moment than any of Einstein's discoveries. But Einstein supporters can argue that he made several other great discoveries besides relativity, and the points he gains from these extra discoveries place him ahead of Planck.

"The argument is jolly and beguiling"

*Einstein, in a letter to his friend Conrad Habicht, describing his discovery of
the relation between mass and energy.*

The equation $E = mc^2$, relating the energy of any physical system to
its mass, has been variously hailed as the world's most famous equation; Einstein's greatest discovery; the equation that changed the
world and gave us nuclear fission and the atomic bomb; the equation that
holds the secret to the stars and drives the universe, etc., etc. The equation
has been inscribed on monuments to Einstein, on the covers of books and
magazines, on postage stamps, on a €10 coin, on T-shirts, and on coffee
mugs; it is highlighted in science museum displays; there is a whole book
that purports to be a "biography" of this equation; and there was a television documentary on Einstein's 1905 discoveries that used this equation as
its title. For the man on the street, $E = mc^2$ is practically synonymous with
Einstein—if you ask any Tom, Dick, or Harry about Einstein, the response
is likely to be, "Ah, yes, eh equals em ce squared."

All of this is hype and nonsense. It illustrates that 50 million Frenchmen can be wrong, and they often are. Einstein was not the discoverer
of $E = mc^2$. The equation was known several years before Einstein, and it

was used by Lorentz in some early attempts at a theoretical description of the mass and energy of electrons.[1] The equation played only a marginal role in the discovery of nuclear fission and the development of the atomic bomb. The physicists who worked on this project—Enrico Fermi, Robert Oppenheimer, Edward Teller—were nuclear physicists, and the tools of their trade were primarily the laws of quantum physics. To achieve control over nuclear fission, they needed an understanding of the nuclear forces and the quantum mechanics of the nucleus. $E = mc^2$ was of little concern to them—although they occasionally used $E = mc^2$, they could have done just as well without it.

And last but not least: the paper that Einstein published in 1905 with the discovery (or rediscovery) of $E = mc^2$ contained a hidden mistake, a fatal flaw in the argument. The proof by which Einstein first attempted to establish $E = mc^2$ was incomplete, and so were several other proofs that he published in subsequent years. The first complete proof of $E = mc^2$ was not found by Einstein, in 1905 or at any other time. It was found by Max von Laue, in 1911. Von Laue was a talented German physicist, best remembered for his experimental work with X rays, which established that these rays were electromagnetic waves, of the same kind as light, but of much shorter wavelengths. For his work with X-rays, he was awarded the 1914 Nobel Prize, whereas his work on the proof of $E = mc^2$ is hardly ever remembered.[2] Nobody was ever awarded the Nobel Prize for $E = mc^2$, perhaps because the members of the Swedish academy in charge of these awards never managed to sort out their confusion as to who should be given credit for this equation.

IN A LETTER to his friend Conrad Habicht, Einstein describes his first four grand papers of 1905, his year of miracles:

> What are you doing, you frozen whale, you smoked, dried, and canned piece of soul! . . . But why have you not sent me your dissertation? Don't you know that I would be one among 1½ fellows who would read it through with interest and pleasure, you wretch? In exchange, I promise you four papers, of which I could soon send the first, because I will soonest receive the free reprints. It treats radiation and the energetic properties of light and is very revolutionary, as you will see if you first send me your paper. The

second paper is a determination of the actual atomic size from diffusion and internal friction of diluted solutions of neutral substances. The third proves that under the assumption of the molecular theory of heat, particles of a size of the order of 1/1000 millimeter suspended in liquids must perform a perceptible, random motion, which is generated by thermal motion. These are motions of inanimate, small, suspended bodies which in fact have been observed by physiologists, which motions were called by them "Brownian motion." The fourth paper exists in concept, and deals with the electrodynamics of moving bodies, by means of a modification of the doctrine of space and time; the purely kinematic part of this paper will surely interest you.[3]

What is missing in this summary of his achievements of the year 1905 is any mention of the famous equation $E = mc^2$. Einstein contrived this equation sometime after writing the above letter to Habicht. In a later letter he added:

One more consequence of the electrodynamical paper has come to my mind. The principle of relativity in conjunction with Maxwell's equations demands that the mass is a direct measure of the energy contained in a body; light transfers mass. A detectable reduction of mass should occur in radium. The argument is jolly and beguiling; but I can't tell whether the Lord isn't laughing about it and leading me by the nose.[4]

Einstein had recognized, in another of his bursts of inspiration, that the formulas for the energy of light included in his paper on relativity held an implication for the energy and mass changes that occur when a physical body—such as a flashlight, or a firefly, or an atom—emits a pulse of light. He had recognized that when light removes energy from the body, it must also remove mass, and he thought he had found a way to calculate how these energy and mass changes are related. This led him to the conclusion that if the body loses an amount of energy E, it must lose an amount of mass m, and the two losses are related by $E = mc^2$, where c^2 is the square of the speed of light. Mass is a measure of the inertia of a body, or the resistance it opposes to changes of motion.

Accordingly, Einstein entitled his paper "Does the Inertia of a Body Depend upon its Energy-content?" He published it in *Annalen der Physik* in September.

What Einstein did not realize was that the Lord had indeed led him by the nose. The Lord had played him for a sucker, revealing to him a part of the puzzle, and fooling him into thinking that the puzzle was complete. The argument in Einstein's paper had a gap—almost big enough for a truck to drive through—and it did not prove what it claimed to prove.

The hidden mistake in Einstein's paper arose from an unwarranted extrapolation. Einstein had proved that some sheep are black, and he rashly extrapolated from this that all sheep are black. He had proved $E = mc^2$ for the simple special case of slow-moving bodies, and he blithely extrapolated this to fast-moving bodies. He had no right to make that extrapolation. Einstein's argument hinges on the formula for kinetic energy, and to prove his result for fast-moving bodies Einstein needed the formula for the kinetic energy of such fast-moving bodies. In 1905, the formula for the kinetic energy of slow-moving bodies was known, but not the formula for fast-moving bodies (the formula for fast-moving *particles* was known, but particles are not sufficiently complex for a test of $E = mc^2$; they have no internal parts—like the internal parts of a flashlight, firefly, or atom—that can emit light). Thus, Einstein was not entitled to draw conclusions about fast-moving bodies,[5] and it remained unclear whether $E = mc^2$ was anything more than a rough approximation that is valid only for low speeds. As Planck wrote some years later in a paper dealing with his own investigation of the mass-energy relation, Einstein's result was based on an "assumption permissible only in the first approximation."[6]

The mistake is the sort of thing every amateur mathematician knows to watch out for. In proving a theorem, it is not sufficient to prove a special case; the proof has to be valid for all cases. Einstein may have been a poor mathematician, but a mistake of this magnitude suggests something far worse than poor mathematics. It suggests that, after the intense creative effort of the preceding four grand papers, Einstein had reached a state of mental exhaustion—he was operating in a daze and not thinking very straight. We know that after finishing his paper on relativity at the end of June, he had collapsed into bed for two weeks,[7] and afterward he had gone on a long vacation to Serbia, visiting his in-laws. He wrote his paper on

$E = mc^2$ in September, after his return to Berne, but maybe he had still not fully recovered?

ALTHOUGH EINSTEIN'S ARGUMENT for $E = mc^2$ was incomplete, Einstein's conjecture that this equation should apply to every kind of body and every kind of energy ultimately proved correct and of great consequence. Einstein had pulled another rabbit out of a hat: his intuition had led him to the correct equation, even though the mathematical reasoning in his argument was botched.

The equation $E = mc^2$, better expressed as $m = E/c^2$, tells us that mass is a form of energy. Mass is a congealed form of energy, or an inactive form of energy. Whereas the other, ordinary, energies are freely exchanged between bodies or particles that participate in collisions or reactions, the congealed energies locked in the masses of the bodies are not altered, or they are altered only by insignificant amounts, in ordinary collisions and reactions. For example, the "congealed" energies, or masses, of two automobiles colliding at an intersection remain unchanged; only their kinetic energies change.

We can think of the congealed energy hidden in the mass of a body as analogous to the congealed water locked in the Antarctic ice sheet, and we can think of the liquid water on the Earth as analogous to the ordinary energy. The Antarctic ice sheet remains unchanged except for minor seasonal fluctuations, while the water level of the oceans and of lakes and rivers rises and falls abruptly with tides, storm surges, floods, droughts, etc. In these events, the water merely shifts from one location to another. In an analogous way, the "congealed" energy locked in the mass of a body or particle remains unchanged in ordinary collisions and reactions, but the ordinary energy shifts from one body or one particle to another.

However, if the congealed water in the Antarctic were to be released by global warming, it would have a dramatic effect on the worldwide water level, with disastrous flooding of all the coasts. Likewise, if the congealed energy in the mass of a body is released and becomes ordinary energy that actively participates in collisions and reactions, its effects become dramatic and violent. And this is exactly what happens in nuclear fission and nuclear fusion.

But the relation $E = mc^2$ tells us more than that. It not only implies that mass is a congealed form of energy, but also that all kinds of energy have mass. Thus, energy and mass are equivalent: wherever there is mass, there is energy; and wherever there is energy, there is mass. Mass can be regarded as a form of congealed energy, and ordinary energy can be regarded as a form of uncongealed mass. Mass and energy are two facets of the same thing—in modern terminology, they are two facets of a generalized concept of energy.

In terms of our water-ice analogy, this simply corresponds to an equivalence of water and ice. We can equally well regard ice as congealed water or water as uncongealed ice. Ice is sold by the kilogram, and water is sold by the liter. It is easy to convert between these units: 1 kilogram of ice is equivalent to 1.0001 liters of water—that is, they are almost exactly the same. But in the case of mass units and energy units, the conversion involves the factor of c^2, which is very large, about 90 million billion. Thus, 1 kilogram of mass is equivalent to a very large amount of energy, about 90 million billion units of energy, or joules. Expressed in kilowatt-hours, this is 25 billion kilowatt-hours. To put this in perspective, this is the total amount of energy used in 1 year by a large city, such as, say, New York.

Conversely, one unit of energy is equivalent to a very small amount of mass. For instance, if you leave the lights of your car on all night, and completely run down the battery, the energy lost from the battery is about 1 kilowatt-hour, but the change of mass of the battery is only a few tenths of a millionth of a kilogram—about equal to the mass of a speck of dust, and undetectable even with the best balance.

In prerelativistic physics, mass and energy were thought to be separately conserved, that is, the sum of masses was thought to be constant and the sum of energies was also thought to be constant. But in relativistic physics, it is possible for congealed energy to become uncongealed, active energy— and vice versa. Only the sum of congealed and uncongealed energies is constant. Einstein saw this joining of the two conservation laws as a historical progression: "We might say that the principle of conservation of energy, having previously swallowed up that of the conservation of heat, now proceeded to swallow that of the conservation of mass—and holds the field alone."[8]

In his paper, Einstein suggested that to detect the small mass changes associated with a release of energy, it would be best to perform experiments with radioactive substances, such as the recently discovered radium, that emit flashes of light that carry away exceptionally large amounts of energy, that is, large relative to the small mass of the emitting atom. This was a prophetic comment on future applications of $E = mc^2$ in the study of subatomic processes.

AFTER FINISHING THE PAPER on the equivalence of energy and mass, there was a hiatus in Einstein's research activity. He published a few very short book reviews in *Annalen* during the fall of 1905, but no other papers until winter, when he revisited the theory of Brownian motion and filled in some more details. He was treading water, eagerly waiting for reaction to his publications of 1905. His sister Maja reports, "The young scholar imagined that his publication in the renowned and much-read journal would draw immediate attention. He expected sharp opposition and the severest criticism. But he was very disappointed. His publication was followed by an icy silence. The next few issues of the journal did not mention his paper at all. The professional circles took an attitude of wait and see."[9] This is somewhat of an exaggeration; reaction to Einstein's papers was actually fairly swift.

The physicist-historian Arthur Miller (not to be confused with the playwright of the same name) has said that Planck made two great discoveries in his lifetime: the energy quantum and Einstein.[10] As editor of *Annalen*, Planck had access to all of Einstein's 1905 papers as soon as they were submitted for publication. He did not approve of Einstein's quantization of the energy of light, but he was very taken with Einstein's relativity. In fact, it was Planck who gave Einstein's theory its name of *Relativitätstheorie*, or relativity theory. Einstein himself continued to call it *Relativitätsprinzip*, or relativity principle, for several years; but when everybody else insisted on calling it the relativity theory, he finally adopted this name. Thus it is fair to say that Planck, the father of quantum theory, was the godfather of the *Relativitätstheorie*.

It is an indication of Planck's intense interest in relativity that in the first week of the fall semester of 1905, he scheduled a colloquium on Einstein's theory and introduced the new ideas about relativity to his students in Berlin. Soon thereafter he wrote to Einstein about some technical details of the theory. This attention from the most famous professor at Germany's leading center for physics research delighted Einstein.

Planck was attracted to relativity because he saw that the methodology of Einstein's approach had some similarities to his own work on the quantization of energy. For Planck, physics was a search for absolutes, and he thought that this was something quantum theory and relativity had in common: "Like the quantum of action in quantum theory, so the velocity of light is the absolute, central point of the theory of relativity."[11]

With his deep perceptiveness, Planck immediately grasped the full extent of Einstein's theory of relativity. And, as his paper on the relativistic equation of motion shows (Chapter 4), in some details he achieved a deeper understanding than Einstein himself. This should not be viewed as a denigration of Einstein. Today, almost any physicist understands relativity better than Einstein did in 1905, which is hardly surprising after 100 years of discussion and dissection of relativity. Planck initiated this discussion and dissection. Einstein, in turn, achieved a deep understanding of Planck's theory of quantization; his paper on the photoelectric effect shows that, in some details, he achieved a deeper understanding than Planck.

BETWEEN 1905 AND 1909, Einstein and Planck danced around each other in a synergistic pas de deux, exhibiting their insights and their creative talents. First Einstein adapted Planck's quantization of energy to light (1905). Then Planck corrected Einstein's mistake in the transverse relativistic mass (1906). Then Einstein made an improvement in Planck's theory of energy quantization of oscillators and applied this to the latent heat of solids, thereby providing the first direct observational evidence for quantized oscillators (1906).[12] Then Planck made an improvement in Einstein's derivation of the $E = mc^2$ equation (1907). Finally, Einstein examined Planck's radiation formula in depth, to extract various subtle implications about the behavior of thermal radiation (1909).

Einstein no doubt felt flattered to have as famous a physicist as Planck as a dance partner. It also must have been highly gratifying to him that Planck was the first convert to the theory of relativity, expressing his solidarity in a letter: "As long as the defenders of relativity form such a small flock, it is doubly important that they agree with one another."[13] And Planck came to appreciate more and more Einstein's perceptive insights into quantum physics. He did not agree with Einstein's interpretation of blackbody radiation as a gas of light quanta, but he expressed his profound admiration for all of Einstein's other contributions. Einstein was much pleased that from the very beginning Planck addressed him as a colleague and an equal, and the special feelings that Einstein thereafter harbored for Planck were to survive all their professional disagreements, their political disagreements in World War I, and even Einstein's hatred of Germans after World War II. When Planck died in 1948, Einstein mourned: "It was a lovely and fruitful time that I was permitted to experience in his vicinity."[14]

Another physicist who took an early interest in Einstein's relativity was Walther Kaufmann, who had been doing experiments on the motion of high-speed electrons at the University of Bonn. In November 1905, he mentioned Einstein's theory, albeit neutrally, in a short presentation of experimental data on the motion of electrons in electric fields. Soon thereafter, Kaufmann pointed out Einstein's mistake in the transverse mass—since Einstein's theory involved the same formulas as Lorentz's theory, Kaufmann concluded that Einstein's transverse mass should be the same as Lorentz's.

Planck reached the same conclusion, and he promptly published his elegant theoretical treatment of the relativistic equations of motion of a particle in which he obtained the correct expression for the transverse mass. Years later, when Einstein was preparing a review article on relativity, he copied Planck's treatment in full (without any acknowledgment).

Planck's favorable reaction to Einstein's work had a decisive influence on other physicists. Planck always spoke in a polite and kind manner, but he had clout. He was the leading physics professor at a university distinguished for its research in physics and chemistry, and whatever he said commanded respect, not only in Germany but also throughout the entire world. The University of Berlin could not boast of the age and renown of Göttingen, Heidelberg, Oxford, or Cambridge. It had been founded quite recently, in 1810, by Alexander von Humboldt, with the express intention of creating a "mother of all modern universities," and by the end of the nineteenth century, the University of Berlin had indeed attained a preeminent position in science. Today it is called Humboldt University, and if you visit its main building, at Under den Linden right across from the Staatsoper, you will find a gallery of its twenty-nine Nobel Prize winners displayed in the entrance hall.

In the hierarchichal structure of the German academic establishment, Planck's imprimatur gave instant respectability to Einstein's relativity and, later, launched Einstein's stellar academic career. Arnold Sommerfeld, a

Humboldt University at Unter den Linden, with a statue of Alexander von Humboldt.

renowned professor at Munich, and also Lorentz and Poincaré, the grandfathers of relativity, soon became interested in relativity. Within a short time, Einstein's theory became a leading contender for the place of the "correct" theory of relative motion.

In those early years, it was often confused with the theory of relativity formulated by Lorentz and Poincaré, and it was usually called the Lorentz-Einstein theory of relativity. Most physicists failed to appreciate the subtle differences between these theories. Lorentz made the same predictions as Einstein, but he stubbornly adhered to the concept of an ether and insisted that the time measured in the reference frame of the ether was the only true time, the "absolute, true and mathematical time" of Newton, although he agreed that observationally there was no way of detecting the ether or finding out which reference frame is the ether frame.

THE FLAW IN EINSTEIN'S 1905 "proof" of the relation of mass and energy remained buried in the pages of *Annalen*. Only Planck recognized the mistake in Einstein's proof of $E = mc^2$, and he made a brief, polite criticism of the mistake in his paper on his own investigation of the mass-energy relation. But this criticism was restricted to a footnote, which did not attract much attention. It seems that the scores of other physicists who read Einstein's paper, discoursed about it, and quoted it never noticed that anything was amiss. Probably they did not notice because Einstein's result was not at all surprising. The proportionality between energy and mass had already been discovered years before, in the study of electrodynamics. It was known that the presence of electric energy in a body contributed to the mass of the body, and in his theoretical treatment of electrons, Lorentz had assumed that the entire mass of the electron arises from the electric energy stored in the electron. Physicists who skimmed through Einstein's paper probably thought that his calculation was merely another alternative way of deriving a result they were already familiar with, except that the previous arguments for $E = mc^2$ focused on electric energy, whereas Einstein's argument was intended for any kind of energy contained in a body.

Einstein probably had an inkling that something about the 1905 proof of $E = mc^2$ was not quite kosher. Over the next two years he published three more proofs of $E = mc^2$, and later, between 1914 and 1945, he published three more (some of these repeatedly). One good proof should have been sufficient, and Einstein's serial production of proofs suggests that he suspected that none of his proofs was a good proof. But if Einstein had doubts about these proofs, he never admitted to that, and he never managed to put his finger on the trouble spots in these proofs.[15]

Some of Einstein's attempted proofs of $E = mc^2$ contain highly original and ingenious arguments—for instance, in one of them Einstein establishes that the stress in a body contributes to its mass.[16] But almost all his proofs were blighted by mistakes, in most cases a recurrence of the mistake of his first proof: the unwarranted extrapolation that a result valid for slow-moving bodies would also be valid for fast-moving bodies. This mistake recurs in his 1906 proof and also in his last two proofs, in 1934 and 1946.

The presentation of the 1934 proof was a media event, held in Pittsburgh, a year after Einstein became a permanent US resident. Einstein was by then world-famous, and an audience of four hundred eager curiosity seekers packed the auditorium of the Carnegie Institute of Technology where he gave his lecture. Thousands more had applied for tickets and had been turned away. *The New York Times* printed a large headline on its front page, "Einstein Offers New View Of Mass-Energy Theorem—EINSTEIN

'REPAIRS' MASS-ENERGY IDEA," and reported, "He talked to the hushed and fascinated group of mathematicians, physicists and cosmologists, partly in English but largely in mathematical symbols. The scene was like watching a Beethoven making the final draft of his Ninth symphony."[17] But it was much ado about nothing. Einstein's new proof bristled with lengthy equations, but it was no better than his first proof—he again failed to recognize that the formulas for the kinetic energy of a high-speed body cannot be constructed by the simple arguments applicable to a high-speed particle. Furthermore, the basic method of his proof was not quite original—it was similar to the method used many years earlier by the American physicists G. N. Lewis and R. C. Tolman, who had devised a simple, clever construction of the relativistic formula for the momentum of a particle by contemplating what happens when two such particles collide (but they did not claim that the same could be done for extended bodies).

Only in one extremely idealized case did Einstein succeed in avoiding his mistake. In a paper published in 1907, he included a special example of the mass-energy relation for electromagnetic radiation or electric fields confined in a hypothetical massless container, and he showed, quite correctly, that the mass of this system is then related to its energy by $E = mc^2$.[18] But this was a highly artificial and totally unrealistic case. In practice, whenever we confine radiation or fields in a container—for instance, thermal radiation confined in an ordinary oven or radio waves confined in a microwave oven—the mass of the container is much, much larger than the mass of the trapped radiation or fields. Einstein's hypothetical assumption of zero mass for the container was as if cows could fly.

The somewhat similar case of the mass-energy relation for thermal radiation was examined by Planck in a paper presented at the Prussian Academy in the same month as Einstein's publication. He showed that the mass of the thermal radiation confined in a blackbody cavity is related to its energy by $E = mc^2$, and he furthermore showed that the pressure that the walls of the cavity exert on the radiation (to keep it confined) contributes an extra amount to the mass.[19] Thus, in some sense, pressure has mass. Pressure is a form of stress, and Planck's result complements and confirms Einstein's result that stress has mass.

Planck's work had an unintended sequel. Johannes Stark, a professor at the University of Hannover who later won the Nobel Prize for his discovery that electric fields alter the emission of light by atoms (the "Stark effect"), read Planck's paper and apparently overlooked the reference to Einstein's work that Planck had included, quite appropriately, in the introductory section. Accordingly, Stark ascribed the discovery of $E = mc^2$ to Planck, and this infuriated Einstein. In an angry letter to Stark, Einstein complained, "I find it rather strange that you would not recognize my priority."[20] Stark replied in a conciliatory and apologetic manner, and Einstein declared himself satisfied. In those early years of relativity, Stark admired Einstein. But later he fell under the spell of the Nazis and became a virulent opponent of Einstein's. He must then have regretted his conciliatory words, and he would have been delighted to know that Einstein's proofs of $E = mc^2$ contained a mistake.

IN 1906, MAX VON LAUE, then an assistant to Planck, traveled from Berlin to Berne, to visit Einstein. Von Laue had been a student at Munich, and he had come to Planck with high recommendations from Arnold Sommerfeld. Planck had recognized von Laue's talent, and appointed him as his assistant. Later, in 1913, von Laue was to move to the front of the queue by winning the Nobel Prize, ahead of Planck and Einstein, the two physicists he admired most.

From the Berne railroad station, von Laue walked a couple of blocks to the nearby patent office where Einstein worked in a small room on the second floor. When the porter announced that an assistant of the famous Professor Planck was waiting for him, Einstein struggled into his jacket and rushed downstairs. Von Laue described their first encounter:

> In the reception room, an official told me to go again into the corridor, and Einstein would meet me there. I did so, but the young man who came toward me had such an unprepossessing appearance that I could not believe he could be the father of the theory of relativity. So I let him pass, and only when he returned from the reception room, did we make acquaintance with each other. What we talked about I recall only in fragments. But I do remember that the cheap cigar that he offered me was so little to my liking, that I "accidentally" dropped it from the bridge down into the Aare.[21]

Von Laue was not alone in his surprise at Einstein's sloppy appearance. In those days, Einstein dressed rather carelessly, and a neighbor in Berne recalled seeing him on the way to his office wearing a pair of multicolored slippers. This sloppiness continued into the early years of his academic career, as is evident from an anecdote told by his sister Maja who attended the University at Berne for a while, as a student of romance languages. One morning, she wanted to sit in on one of her brother's lectures. When the nattily dressed Maja asked the porter at the university for directions to her brother's lecture room, the porter exclaimed, "That slob is your brother? I'd never have thought that."[22] Einstein's sloppy manner also lasted through his tenure in Prague in 1912, where a visiting physicist found him sitting at his desk in his office ". . . without a jacket, without a tie, in a shirt like what the Italian streetworkers wore, and at the back a large triangle was ripped out."[23] Only, later, in Berlin did Einstein begin to dress in the manner expected of a German professor, although with some reluctance: ". . . a certain drill in regard to dress etc., which I am subjected to on the orders of some older fellows, so I won't be counted among the rejects of the local population, disturbs the peace of mind."[24]

Von Laue came away from his visit in Berne with a very poor opinion of Einstein's sartorial standards, but with a very high opinion of Einstein's grasp of physics. In a letter to a colleague he said of Einstein, "He is a revolutionary. In the first two hours of the conversation, he overturned all of mechanics and electrodynamics."[25] And in a letter to another colleague he added, "You should be careful that Einstein doesn't talk you to death. He loves to do that, you know."[26]

VON LAUE LATER MADE two significant contributions to relativity. In 1911 he published the first textbook on relativity and also the first generally valid proof of the $E = mc^2$ relation. The results obtained by Einstein and by Planck had gradually led to the conviction that $E = mc^2$ is indeed the correct equation relating the energy of any physical system to its mass. But until 1911, a general, legitimate proof was lacking, and this was finally supplied by von Laue. His proof was valid for any kind of physical system with a static distribution of energy. His proof was concise and elegant; it took advantage of new mathematical methods introduced by Minkowski and extracted the relation $E = mc^2$ directly from the mathematical properties of the energy-momentum tensor. This permitted von Laue to bypass all the complications that had bedeviled Einstein's various attempts at a proof.

Why did Einstein fail to find the proof that von Laue assembled with such (apparent) ease? The answer is that Einstein had initially formulated his theory of relativity by rather primitive mathematical methods. As Minkowski said, "Einstein's representation of his profound theory is mathematically clumsy—I may say so, because he received his mathematical training from me in Zurich."[27] With his characteristic stubbornness Einstein stuck to these clumsy methods, and he was slow to appreciate the power of the new mathematical methods of the four-dimensional tensor calculus that Minkowski introduced in 1908.[28] His reaction to Minkowski's tensor calculus was at first dismissive; he characterized it as "unnecessary erudition," and he complained that "ever since the mathematicians have thrown themselves on the theory of relativity, I can't understand it any more."[29] Only in 1912 did Einstein finally acquaint himself with tensor methods, in part, by reading von Laue's textbook, of which he said, "His book on the theory of relativity is a little masterwork, and some of what is in it is his intellectual property."[30] He soon thereafter wrote an exposition of special relativity in tensor language.[31] Later, he was to use tensor calculus extensively in his work on general relativity, and he admitted that he could not have developed this theory without it.

In 1911, von Laue was better acquainted with tensor calculus than was Einstein, and thus von Laue found the proof that had eluded Einstein. In his incomplete proofs, Einstein had failed to see the forest for the trees. Von Laue ignored the trees and dealt directly with relevant aspects of the landscape of the forest. His proof illustrates that, with the right mathematical methods, it is sometimes easier to construct a general proof than it is to work out the messy details of special cases. Oddly, even today von Laue's general proof of the $E = mc^2$ relation is not widely known, and it is rarely mentioned in physics textbooks, which mostly content themselves with some handwaving arguments about $E = mc^2$ and some version of Einstein's defective proofs.

IN 1912, EINSTEIN adopted a substantial part of von Laue's proof in a manuscript[32] intended for an omnibus volume on recent developments in physics. The war delayed the publication, and the volume was not published until 1924, without Einstein's contribution, because he refused the publisher's request to update his manuscript. However, Einstein salvaged his version of von Laue's proof from the unpublished manuscript and inserted it into two papers[33] published in 1914 and into his Princeton lectures in 1921, which were printed and published later that same year as a book, *The Meaning of Relativity*. In his usual cavalier manner, Einstein did not acknowledge von Laue as a source, but the time line and the similarities between Einstein's and von Laue's versions of the proof make the relationship quite clear.

Einstein's version of von Laue's proof contains the zaniest mistake in his entire oeuvre, an absolutely nonsensical mistake. The core of von Laue's proof consists of a demonstration that the mathematical properties of the energy and momentum of a system are like those of a particle.[34] Einstein attempted to construct a simplified version of von Laue's proof, with fewer tensor manipulations. He showed, quite correctly, that when a force acts on a system, the resulting *increments* in energy and momentum are particle-like. And he then made a big, unjustifiable, illogical jump: he claimed that the energy and momentum of the system "may be presumed" to behave in the same way as these increments, and that therefore the energy and momentum of the system must be particle-like. This claim is plain nonsense. It is like the fraud perpetrated by some sleazy Italian purveyors of olive oil, who pour a bottleful of genuine olive oil into a barrelful of vegetable oil of unknown provenance and then sell this mix as pure olive oil, *extra vergine*. Einstein was adding an increment of particle-like energy-momentum to the system and claiming that thereby all the energy-momentum of the system became particle-like.

What makes this zany is that Einstein was aware of the mistake, but refused to recognize it for what it was. In a footnote in the 1912 manuscript he said, "To be sure, this is not rigorous," but he dismisses the possibility that the energy-momentum might not behave in the same way as the increments, because "this seems so artificial that we will not dwell on this possibility at all."[35] And what makes it zanier is that after publishing this nonsensical argument in his two 1914 papers, in his Princeton lectures, and in his book *The Meaning of Relativity*, he published a further four revised editions of this book, and never corrected the mistake. The more than 100,000 copies of this book that were printed serve as a monument to his zaniest mistake (the book was recently reissued, without Einstein's intervention; you can find it in your bookstore, and you can find the zany mistake on page 44).

In contrast to many of Einstein's other mistakes, the mistake in his version of von Laue's proof was simply stupid—it was not subtle, and it had no redeeming, fruitful consequences. It was an instance of what Cesare Lombroso, the nineteenth-century Italian anthropologist and forensic psychiatrist, has called the stupidity of men of genius: occasionally, geniuses perpetrate surprisingly stupid mistakes, of the kind that we expect to be the monopoly of nongeniuses.[36] Fortunately, Einstein's mistake did little harm, except perhaps to some students misled by it. No other writers on relativity imitated this mistake—it remains buried and dormant within the pages of Einstein's book.

In 1918, an extension of von Laue's proof was formulated by Felix Klein, one of the great Göttingen mathematicians, perhaps best known for his invention of a mathematical surface called a "Klein bottle," which is a closed surface, but without a distinct inside or outside. Klein had noticed the zany mistake in Einstein's proof, and—after some polite correspondence with Einstein—he contrived a clever modification of von Laue's proof, showing that $E = mc^2$ remains valid even when the energy distribution in the physical system is not static. Thus an energy flow from one part of the system to another does not alter the relation between the total energy of the system and the total mass. The only crucial constraint is that the total energy of the system must be conserved.[37]

Einstein expressed his "sincere admiration"[38] for Klein's ingenious mathematical methods, but he seems to have regarded Klein's proof merely as an alternative to his own—and with his characteristic obstinacy, he merrily continued to do things in his own, mistaken way.

CONTROVERSY BEGAN TO SWIRL around Einstein's first proof of $E = mc^2$ in 1952, when the American physicist Herbert Ives announced that he had spotted a mistake in the logic of Einstein's argument.[39] Ives was an expert in optics; he was interested in testing the time dilation predicted by Einstein by means of optical experiments with light emitted by fast-moving atoms, and he apparently came across this mistake quite by accident, while rereading some of Einstein's early papers. Ives claimed that Einstein had relied on circular reasoning in his argument, that is, he had implicitly assumed what he was trying to prove.

In 1952, this announcement of a mistake by the venerable Professor Einstein did not create much of a stir, because by then other physicists had published other demonstrations of $E = mc^2$, and there were no doubts that the result was correct. But in the technical literature, Ives's claims about Einstein's supposedly circular argument were accepted widely (and somewhat uncritically) and republished in several books and articles. Finally, in 1988, John Stachel, one of the editors of the Einstein papers, showed that the claim was false. There was nothing circular in Einstein's argument—in claiming that Einstein had made a mistake of circular reasoning Ives had himself fallen into the trap of circular reasoning. (So far, so good—but then Stachel overreached by declaring that Einstein's proof was perfectly legitimate, which it was not—the proof was blighted by the mistake that Planck had noticed.)

More controversy erupted in 1953, when Edmund Whittaker, a mathematician at the University of Edinburgh, published a second volume of his *History of the Theories of Aether and Electricity*. In the first volume of this book he had given an encyclopedic survey of eighteenth- and nineteenth-century physics, which had won much praise for its thorough scholarship and critical insight. In the second volume, he continued this history into the twentieth century. Whittaker awarded most of the credit for the theory of relativity to Poincaré and to Lorentz, calling it the "The Relativity Theory of Poincaré and Lorentz," and he also awarded most of the credit for the discovery of $E = mc^2$ to several of Einstein's predecessors, among them J. J. Thomson and Poincaré.[40] Whittaker admitted that although Poincaré had introduced the idea, he "had given practically no proof," and he downplayed Einstein's contribution by saying, correctly, that his proof was valid only for a particular case, and even then only approximately.

Einstein treated the book with contempt, and his friend and biographer Pais accused Whittaker of being ignorant of the literature.[41] In this, Pais's temper got the better of his judgment. Whittaker had a very extensive knowledge of the literature, as is demonstrated by his discussion of von Laue's derivation of $E = mc^2$, which is missing in Pais's writings.

Whittaker was right in questioning the significance of Einstein's contributions to the proof of $E = mc^2$, but he was wrong in dismissing his contributions to the theory of relativity. In 1905, Einstein gave us a more general, more fundamental way to look at relativity than had previously been proposed by Lorentz and Poincaré. His new view of relativity transcended that of Lorentz and Poincaré, by showing that some of the features of the older view—such as the ether and the preferred reference frame of the ether—were superfluous.

ALTHOUGH IN FAIRNESS the credit for the final proof of $E = mc^2$ belongs to von Laue and Klein, Einstein's repeated attempts to find a proof had linked his name to this formula, and it came to be regarded as a quintessential Einstein contribution, not only in the eyes of the public, but even in the eyes of many physicists, who should have known better. As the physicist-historian Jagdish Mehra explained, the reason why Einstein was bound to get credit for anything connected with relativity "is the sociology of science, the question of the cat and the cream. Einstein was the big cat of relativity, and the whole saucer of its cream belonged to him by right and by legend, or so most people assume!"[42]

Einstein himself always regarded $E = mc^2$ as one of his proprietary contributions. After the bombings of Hiroshima and Nagasaki, he exclaimed, "If I had only known, I should have become a watchmaker."[43] He felt that his work on the relation between energy and mass was partly responsible for the deaths of the 200,000 Japanese civilians incinerated by the two atomic bombs. And these feelings of responsibility must have been enhanced by the letter he had written to President Roosevelt in 1939, warning of the German research in fission, and urging the president to initiate a research program in the United States, to explore the feasibility of atomic bombs.

He need not have worried. As Oppenheimer said, "He did write a letter to Roosevelt about atomic energy. I think this was in part his agony at the evil of the Nazis, in part not wanting to harm any one in any way; but I ought to report that the letter had very little effect, and that Einstein himself is really not answerable for all that came later."[44] Fermi had already informed the US government about the power and the dangers of atomic bombs years earlier, as soon as he heard of the discovery of nuclear fission. And furthermore, Einstein's formula $E = mc^2$ was of only marginal significance for the discovery of nuclear fission and for the development of the atomic bomb. True, this formula gave a convenient way to determine the energy released in a nuclear reaction, but there were other, direct ways to measure this energy experimentally, and ultimately the *exact* amount of energy released in the explosion is of little importance. It was known from direct measurements that nuclear reactions release about a million times as much energy as chemical reactions. Hence, weight for weight, a nuclear "explosive" will release about a million times as much energy as a chemical explosive—if you explode a kilogram of uranium, you will get the same result as from the explosion of about a million kilograms of TNT, that is, 1000 tons, or a kiloton of TNT. The 20 or so kilograms of uranium exploded over Hiroshima and over Nagasaki yielded the equivalent of 20 or so kilotons of TNT. And $E = mc^2$ is not needed in this calculation.

The difficulty in designing and building the bomb lies not with $E = mc^2$ or with relativity, but with the complicated calculations of how a chain reaction can be initiated and maintained. The scientists who did the calculations of the critical mass of uranium needed for the explosive release of energy in a bomb had to understand the quantum mechanics of the nucleus and, in particular, the efficiency of the chain reaction in uranium.

Both the American and the German physicists felt that, as a preliminary step toward building a bomb, they needed to study the chain reactions in uranium under controlled conditions in a nuclear reactor fueled with natural uranium U-238, instead of the rare and precious weapons-grade uranium U-235. Chain reactions in such a nuclear reactor require a moderator, either graphite or heavy water, to slow the neutrons released by the fission of uranium and make them more efficient for triggering further fissions. The Americans decided to use graphite as moderator, and they quickly built and tested a reactor. But the Germans—misled by faulty measurements of absorption of neutrons in graphite—concluded that graphite would quench the chain reaction, and they decided to use heavy water as moderator. Heavy water can only be extracted from ordinary water by a laborious process; it was in short supply and, besides, the Allies sabotaged the German efforts to obtain more of it. At the end of the war, the Germans still did not have an operating nuclear reactor, much less any clear idea on how to build an atomic bomb.[45]

This failure of the German physicists to test a nuclear reactor and build an atomic bomb was one of the most fortunate accidents of World War II. The failure was unexpected—before the war, the Germans had been the leaders in fission research, and fission was first discovered in Berlin. The tale of the discovery of fission is one of those convoluted comedies of errors that often precede the birth of a new scientific idea. The comedy that led to the discovery of fission had an international cast of actors with a generous scattering of Nobel Prizes, and it illustrates that sometimes even Nobelists can be pretty dense. "When the mind is not prepared, the eye does not recognize" said Emilio Segrè, one of the Nobelists who missed discovering fission by a hair.[46]

THE TALE BEGINS in Paris in 1934, where Irène and Frédéric Joliot-Curie discovered artificially induced radioactivity. Irène was the daughter of the celebrated Marie Curie, and like her mother, she was to win a Nobel Prize for her work on radioactive elements; and, sadly, like her mother, she was to die of cancer induced by exposure to radioactive contaminants. When Irène married Frédéric they agreed to adopt the hyphenated name Joliot-Curie in a spirit of *egalité, fraternité,* and *cachet*— Irène did not want to lose her world-famous Curie name, and Frédéric was glad to gain it.

The Joliot-Curies bombarded samples of various elements with alpha particles from a radioactive source, and they found that the irradiated samples often themselves became radioactive, which indicated that when an alpha particle penetrates into the nucleus of an atom in the sample it produces drastic modifications in the structure of this nucleus, transmuting it into a new nucleus of a different element. The work of the Joliot-Curies on transmutation of elements by artificially induced radioactivity was an extension of the work of Marie and Pierre Curie on transmutation by natural radioactivity, and for this work the Joliot-Curies received the Nobel Prize in Chemistry in 1935.

In his Nobel lecture, Frédéric Joliot-Curie anticipated that the large energy released in nuclear reactions might be utilized in an explosive device: ". . . we are entitled to think that scientists, building up or shattering elements at will, might be able to bring about transmutations of an explosive type, true chemical chain reactions."[47] He also worried that such a chain reaction might accidentally spread to the entire Earth: "If such transmutations do succeed in spreading in matter, the enormous liberation of usable energy can be imagined. But, unfortunately, if the contagion spreads to all the elements of our planet, the consequences of unloosing such a cataclysm can only be viewed with apprehension." Joliot-Curie's warning did not go entirely unheeded. Years later, while engaged in the design of the H-bomb, or the "Super," Hans Bethe worried that the explosion of such a bomb might trigger a worldwide chain reaction in the atmosphere and incinerate the entire Earth. After some calculations, he dismissed this as highly improbable, but not absolutely impossible. How probable or improbable it is remains classified.

IN ROME, THE GIFTED Italian physicist Enrico Fermi heard about the Joliot-Curie experiments. He had just launched a research program in nuclear physics, and with his characteristic acuteness he immediately recognized that instead of bombarding nuclei with alpha particles, it would be much more efficacious to bombard them with neutrons. Both the alpha particle and the nucleus carry positive electric charge, and the resulting electric repulsion makes it difficult for the alpha particle to penetrate into the nucleus; in contrast, the neutron has no electric charge, and it penetrates the nucleus without difficulty. Fermi began a systematic program to bombard all chemical elements with neutrons. In these experiments Fermi and his nuclear physics group produced radioactive versions of most of the known chemical elements. Many of these artificial radioactive elements were later to prove useful as tracers in chemical reactions, especially biological reactions, and also in medical applications.

But their most dramatic results emerged from the neutron bombardment of uranium, the heaviest of the naturally occurring elements. The chemical analysis of the reaction products suggested that repetitive impacts of neutrons on the uranium nuclei formed new, artificial elements with nuclei larger and heavier than uranium. These putative new elements came to be called "transuranes," that is, beyond uranium. With Italian brio, Fermi's colleagues urged him to bestow names on the new elements. Fermi at first resisted, but, in 1938, when he was awarded the Nobel Prize for his production of radioactive elements by neutron bombardment, he yielded to temptation, and in his Nobel lecture he named two of his transuranes *ausenium* and *hesperium.* These names had to be withdrawn almost immediately, when the transuranes were shown to be an illusion—they were elements "composed of poor chemistry."[48] (Some of Fermi's fascist colleagues had pressured him to use names that reflect the glories of Mussolini. It is too bad that Fermi refused; if he had agreed, the entire nonfascist world would have been treated to the hilarious spectacle of the withdrawal of *Mussolinoleum* or some other such glorious name.)

The affair of the transuranes was one of the few missteps in Fermi's brilliant career. After accepting his Nobel Prize in Stockholm, Fermi and his wife, Laura, who was Jewish, escaped to the United States, where a few years later he succeeded in initiating the first controlled nuclear chain reaction in a reactor built at the University of Chicago. Arthur Compton was on

Fermi's team, and he reported Fermi's success to government officials by phone, in code: "The Italian navigator has landed in the New World." He was asked, "How were the natives?" and he replied, "Very friendly."[49] This chain reaction was the prelude to the development of the atomic bomb. A small memorial on the campus of the University of Chicago marks the site of Fermi's first reactor, and Fermilab, the high-energy laboratory near Chicago, containing what until recently was the world's largest particle accelerator, is named after him.

IN BERLIN IN 1934, Otto Hahn and Lise Meitner, later assisted by Fritz Strassmann, decided to repeat Fermi's experiments. Hahn was the director of the Chemical Section of the Kaiser-Wilhelm Institute. He was a brilliant chemist, who had shown his abilities at an early age, when, as a student, he had discovered a radioactive version of thorium, called meso-thorium. At first, this met with a skeptical reception from his elders—one Yale chemist said that mesothorium was "a compound of thorium and stupidity"[50]—but Hahn soon convinced the skeptics. Hahn began his work on radiochemistry in 1905, in a small laboratory in the basement of the Chemistry Institute in Berlin, and he later moved to the new Kaiser-Wilhelm Institute in Dahlem, on the outskirts of Berlin. His collaboration with Meitner began in those early years. She was a physicist who had come to Berlin from Vienna, where she had studied under Boltzmann. Planck had immediately recognized Meitner's talent, as he had recognized von Laue's, and he had appointed her to a position of assistant. With this appointment, Planck broke new ground: Meitner became the first woman assistant at a Prussian university. Her appointment as an assistant launched her academic career, and she became the first woman physics professor at a German university.

Hahn and Meitner formed a "dream team"; he provided the expertise in analytical chemistry, and she provided the expertise in physics. She was the intellectual leader of the team, and a colleague remembers that she would sometimes point to the stairs leading to the administrative offices on the floor above and admonish Hahn: "Hähnchen, go upstairs and do some chemistry, you understand nothing of physics."[51] Their collaboration lasted for more than thirty years and led to the elucidation of a great many nuclear reactions and decay sequences leading from one radioactive element to another.

The uranium experiments of Hahn, Meitner, and Strassmann continued for several years, from 1934 to 1938. At first these experiments confirmed Fermi's results, and the Berlin team, like the Rome team, thought that the neutron bombardment had indeed produced trans-uranes. But the experimental results had some oddities, and Meitner had to contrive some rather artificial reaction schemes to account for these results. In 1938 she was in the midst of this work when Adolf Hitler, born in Austria as Adolf Schicklgruber (father unknown), decided to invade and annex the land of his birth. The annexation led to rowdy celebrations in Vienna, with robberies and vicious beatings of Jews, and not a few murders.

Meitner was Jewish, but, as an Austrian, she had survived the wholesale dismissal of German Jewish university professors in 1933 with only minor damage—she had lost her position on the faculty of the University of Berlin, but she had retained her professorship at the Kaiser-Wilhelm Institute and her salary. Now, with the Austrian *Anschluss*, she had suddenly become a German Jew, and she was told she must leave the institute. She fled to Sweden, via Holland, one step ahead of the police agents sent to intercept her.

DEPRIVED OF THEIR COLLABORATOR, Hahn and Strassmann continued the uranium experiments on their own. By deft chemical analysis, Hahn established that one of the products from the neutron bombardment of uranium, which he and Fermi had previously taken to be radium, was actually barium. But the weight of the barium nucleus is not much more than half that of a uranium nucleus, and this led Hahn to imagine that the impact of the neutron on the uranium nucleus had split it into two (or more) large fragments, rather like a piece of log split by the blow of an axe. It was a startling discovery. Hahn sent a letter to Meitner asking her to think about such a splitting of the nucleus: "So please think about whether there is any possibility—perhaps a Barium isotope with much higher atomic weight than [the usual] 137?" Meitner was at first doubtful and replied "At the moment the assumption of such an extensive bursting seems very difficult to me," but she admitted "in nuclear physics we have experienced so many surprises, that one cannot unconditionally say: it is impossible."[52]

Although the credit for discovering fission belongs to Hahn, the credit for confirming this discovery and for finding a physical explanation for the fission process belongs to Meitner and her nephew Otto Frisch, a physicist working at Niels Bohr's institute in Copenhagen. He came to visit her in

Sweden for the Christmas holidays in December 1938, a few days after she
received the letters from Hahn. Frisch later described how his aunt began
to understand the mechanism of fission:

> We walked up and down in the snow, I on skis and she on foot (she
> said and proved that she could get along just as fast that way) and
> gradually the idea took shape that this was no chipping or cracking
> of the nucleus but rather a process to be explained by Bohr's idea
> that the nucleus is like a liquid drop; such a drop might elongate and
> divide itself . . .[53]
>
> At this point we both sat down on a tree trunk, and started to
> calculate on scraps of paper. The charge on the uranium nucleus, we
> found, was indeed large enough to destroy the [restraining] effects
> of surface tension almost completely; so the uranium nucleus might
> indeed be a very wobbly, unstable drop, ready to divide itself at the
> slightest provocation.
>
> But there was another problem. When the two drops separated
> they would be driven apart by their mutual electric repulsion, and
> would acquire a very large energy . . . where could that energy come
> from? Fortunately, Lise Meitner remembered how to compute the
> masses of nuclei . . . and in that way she worked out that the two
> nuclei formed by the division of a uranium nucleus would be lighter
> than the original uranium nucleus by about one fifth of the mass of a
> proton. Now whenever mass disappears energy is created, according
> to Einstein's formula $E = mc^2$, and one fifth of a proton mass was just
> equivalent to . . . the source of energy; it all fitted.[54]

From this it is clear that Meitner first calculated the energy released in
fission directly from the electric repulsive forces between the fission fragments, and only afterward checked the calculation by examining the mass
changes. The mass calculation was not really necessary, but because her
first calculation had given her such a surprisingly large number, she felt it
advisable to fall back on the mass calculation for insurance.

Frisch's description of these first calculations of fission energy also illustrates a common misinterpretation of the mass-energy equivalence. Frisch
thought that the mass change was the source of the energy, and that it was
the explanation of these energy changes. Here Frisch is putting the cart
before the horse. The mass change is merely a symptom of the energy

change. The source of the fission energy is not the mass, but the electric energy initially contained in the nucleus, and this energy is fed into the fission fragments by the electric repulsive forces. The decrease of mass is merely a symptom of the release of this electric energy. Energy changes cause the mass changes, not vice versa. In physics, it is always the energy that plays the primary role—it is the "primum mobile"[55] of all of physics. Frisch was thirty years younger than Meitner, but he was old enough to know better.

WHEN FRISCH RETURNED to Copenhagen, he described Meitner's theory of the mechanism of fission to Bohr: "I had hardly begun to tell him, when he struck his forehead with his hand and exclaimed: 'Oh, what idiots we all have been! Oh, but this is wonderful! This is just as it must be! Have you and Lise Meitner written a paper about it?' "[56]

Frisch and Meitner followed Bohr's advice and immediately wrote a short paper about the theory of the splitting of the nucleus. In this paper the word *fission* made its first appearance; Frisch adapted this word from biology, where it is used for cell division. And Frisch went a step further: he set up a simple experiment with particle detectors placed near a sample of uranium. When he bombarded the sample with neutrons, he was able to detect the fission fragments themselves, as they were ejected from the sample. This was direct confirmation of the fission phenomenon discovered by Hahn.

A similar experiment had been performed four years before, by Fermi's nuclear physics group in Rome. Segrè, who was a member of this group, recalls in his memoirs that, by bad luck, they carefully covered their particle detectors with aluminum foil, because they wanted to block alpha particles ejected by the radioactive decay of uranium, and this also blocked the fission fragments. If they had removed the aluminum foil, they would have seen the fission fragments—they missed this discovery by a hair, or by a hair-thin layer of aluminum foil.[57]

The news about fission spread like wildfire, by word of mouth, ahead of the printed reports of Hahn, Meitner, and Frisch. Within less than a month, American physicists heard about it. Fermi, who had arrived in Washington, gave a talk about fission on the radio, and headlines about fission appeared in *The New York Times*. "VAST ENERGY FREED BY URANIUM ATOM" was the headline of an article that reported an experiment performed by Fermi at Columbia University on "the splitting of a uranium atom into two parts, each consisting of a gigantic atomic 'cannonball' of 100,000,000 electron-volts, the greatest amount of atomic energy so far liberated by man on earth."[58]

In 1944, Hahn was awarded the Nobel Prize in Chemistry for his discovery of fission. He was fifty-nine when he made this discovery, exceptionally old for a Nobel Prize winner. Meitner's supporters felt aggrieved that she received no share of the Prize. But it would have been a stretch for the Nobel committee to award her a share on the grounds that she would have anticipated Hahn's discovery of fission if she had not had to flee from Germany. Nobel Prizes are not awarded on hypotheticals.

The participants in the comedy of errors that led to the discovery of fission—except, oddly, Hahn—have had their names attached to various transuranes, when these were later produced in experiments with nuclear reactors and particle accelerators. Thus we have curium (element 96), fermium (100), meitnerium (109), and even nobelium (102). And, to round this out, element 99 was named einsteinium. They all are rather nasty, highly radioactive, short-lived substances, not the kind of thing anybody in his or her right mind would want to give his or her name to—but better not to look a gift horse in the mouth.

ALTHOUGH $E = mc^2$ PLAYED only a marginal role in the discovery of atomic fission and the later development of the A-bomb, it played a crucial role in the discovery of nuclear fusion and the development of the H-bomb. Large, heavy nuclei (such as uranium) release energy when they split into two, or more, fragments. Small nuclei (such as hydrogen) release energy when they merge to form a larger nucleus. The energy released in the fission of a large nucleus can be calculated quite accurately from the electric repulsive force of the two fission fragments; but the energy released in the fusion of two small nuclei hinges on the details of the non-electric nuclear force, and these details were not known during the early years of the study of nuclear physics. Thus, for the calculation of the energy released in fusion reactions, nuclear physicists had to rely on $E = mc^2$. Although for Meitner $E = mc^2$ was merely an alternative way to calculate the energy released in fission, for the nuclear physicists who wanted to calculate the energy released in fusion, $E = mc^2$ was the only game in town.

In 1938, Hans Bethe cleverly exploited $E = mc^2$ for calculations of the energy released by fusion reactions in the center of the Sun. Bethe was another Jewish physicist who had had to flee Germany, first to Britain and then to the United States, where he became a professor at Cornell University. He found that all of the enormous amount of energy radiated by the Sun could be explained by a sequence of fusion reactions that convert hydrogen into helium. Effectively, the Sun "burns" hydrogen and produces helium. This solved the long-standing mystery about the source of energy of the Sun. The Sun has been shining for several billion years (except in the Bible Belt, where, according to zealous preachers, it has only been shining for 10,000 years), and chemical reactions cannot supply enough energy to keep the Sun shining for that long. Even if the Sun were made entirely of carbon, or coal, there would not be enough coal to keep it burning. But nuclear fusion reactions typically release a million times as much energy as chemical reactions, and they will be able to keep the Sun burning for another 10 billion years.

Bethe also investigated the fusion reactions in stars larger than the Sun, and he found that in such stars the sequence of fusion reactions is somewhat different, but the net result is still the burning of hydrogen into helium. For his brilliant solution of the mystery of the energy source in stars, Bethe received the Nobel Prize in 1967. During the war, Bethe participated in the Manhattan Project, as director of the theoretical division. In contrast to Einstein, who expressed regrets about the development of the atomic bomb, Bethe always remained unapologetic about his wartime work on the bomb. However, he later became a strong advocate of nuclear disarmament and of the social responsibilities of scientists.

WHEN EINSTEIN WROTE his scientific autobiography in 1949, he discussed the genesis of his fundamental contributions to physics: the photons, or quanta of light, the theory of Brownian motion, special relativity, general relativity, and even the unsuccessful and chimerical unified theory on which he wasted his final years. But the mass-energy relation is conspicuous by its absence—nowhere in the autobiography is the celebrated equation $E = mc^2$ to be found.[59] Maybe Einstein had finally recognized that his claims for priority in the discovery of the mass-energy relation were shaky, and that his various "proofs" of this relation were equally shaky. Or maybe he had just become tired of seeing himself identified with $E = mc^2$, and he wanted to draw more attention to his other accomplishments.

8

"Suddenly, I had an idea"

Einstein's recollection of his discovery of the Principle of Equivalence in 1907, according to which a freely falling observer feels weightless. He first conceived this idea in Berne, and he developed its consequences during his stay in Prague.

n 1907, Einstein was still stuck in the patent office at Berne. The job that he had initially been happy to find was becoming tedious. He complained, "Every day 8 hours of strenuous work at the patent office, and added to that much correspondence and studies . . . Several papers are incomplete, because I cannot find the time for their composition."[1] He wrote to Grossmann of his plans for a new, academic career, explaining: "Don't imagine that I am driven to strive on this path by delusions of grandeur or some other questionable passion. Rather, I have this craving only because of a fervent desire to continue my private scientific occupation under less unfavorable conditions."[2]

As a first step toward an academic career, Einstein sought to obtain a certificate of *Privatdozent* at the University of Berne. At Swiss and German universities, the *Privatdozent* is a kind of combination of private tutor and instructor. The *Privatdozent* has the right to give lectures, but he receives no pay from the university; his only remuneration is a small fee that the students pay him directly, a throwback to medieval times. Einstein applied for the certificate, but in lieu of the customary qualifying dissertation, he submitted copies of some of his recent, published papers. The Berne professors were unimpressed by Einstein's papers—they could understand little of them, and they refused the application on the technicality that no certificate had ever been awarded without a dissertation. Many years later Einstein dismissed the matter with the comment that in a small faculty such as Berne, a few old nags were bound to stick together and insist on making everybody march to their tune; but at the time he was seriously disappointed.[3]

For a second application, he cobbled together a dissertation by lifting bits and pieces from his published works, and he was granted his certificate. He offered lectures on the theory of heat and of radiation, but the only students who attended his lectures were three of his friends. After two terms, even his friends had had enough, and when only one other student showed up, Einstein canceled his lectures, which placed his academic career on hold.[4]

IT TOOK ANOTHER TWO YEARS for the stolid Swiss academic establishment to wake up to the fact that one of their patent clerks, recently promoted from 3rd class to 2nd class, was a leader in physics, with international renown. In 1909, Alfred Kleiner, then the one and only physics professor at the University of Zurich, felt compelled to propose Einstein for a faculty appointment. Friedrich Adler, who had studied physics at the Zurich Polytechnic during some of the same years as Einstein and who was a competing candidate for this same appointment, wrote in a letter to his father, ". . . for the faculty the situation is that on the one hand they have pangs of conscience over how they treated him earlier, and on the other hand the scandal is felt not only here, but also in Germany, that such a man should be employed at the patent office . . . and that everywhere there is astonishment that Einstein still holds no appointment."[5] Adler had good political connections, and he could have exploited these to gain an advantage over Einstein, but he did exactly the opposite: he very honestly told everyone who would listen that Einstein was the better physicist and deserved the appointment. And in the end the University of Zurich offered Einstein an appointment as an associate professor, which he accepted, after some wrangling over the salary. To one of his colleagues he wrote, "So now I too am an official member of the guild of whores."[6]

Adler's role in promoting Einstein's appointment was not entirely self-less. He was more interested in politics and in philosophy than in physics, and he had become a candidate for the university appointment only under pressure from his father. Adler was a somewhat flaky character, and he later became seriously deranged. In 1916, in the middle of World War I, he walked into a fashionable hotel in Vienna, revolver in hand. In the dining room, he sought out Count Stürgkh, the Austrian chancellor, and he killed him with several shots through the head, yelling, "We want peace!" The Austrians treated Adler with surprising leniency; he was imprisoned for some years, but then pardoned shortly after the end of the war. During his imprisonment he wrote an absurd treatise on a new interpretation of relativity, on the basis of which Einstein offered to testify to his mental derangement, to secure his release. But Adler would have none of that—he believed he was perfectly sane, and he insisted that only he understood the true meaning of relativity.[7]

AT FIRST, EINSTEIN ENJOYED the teaching at Zurich. Students liked his informal manner, as he conducted a seminar with animated discussions at the Café Terrasse, on the right bank of the Limmat River, near the lake.[8] Einstein's lecture notes for his courses are available in the *Collected Papers*.[9] They are often quite messy and sketchy, and their content is unremarkable. They have none of the sharp insights and clever and elegant arguments found in the lecture notes of some of the truly great physics teachers of the twentieth century, such as Arnold Sommerfeld, Enrico Fermi, John Wheeler, or Richard Feynman. In fact, Einstein had little talent and less patience for teaching. He usually put only minimal effort into the preparation of his lectures, and one of his Zurich students recalled that he showed up with lecture notes that "consisted of a scrap of paper, about the size of a calling card, on which was sketched what he intended to cover with us."[10] Good lecturing requires good preparation and often involves some showmanship. Feynman's lively and inspiring lectures were not thrown together on a scrap of paper. He prepared his lectures with great care, even rehearsing them ahead of time in an empty lecture room, sentence by sentence, gesture by gesture, and equation by equation on the chalkboard, like a full-dress rehearsal for a theatrical performance. The charming spontaneity and enthusiasm he displayed in his lectures were carefully choreographed.[11]

Einstein and Mileva lived on the third floor of this house on Moussonstrasse, on the hill behind the Polytechnic, when he became a professor at Zurich University.

Einstein's failure to prepare his lectures had almost scotched his appointment at Zurich, because Kleiner, before offering the appointment to Einstein, had paid a surprise visit to Berne, to attend a lecture. He had come away with a very poor impression. As he had reported to Adler, Einstein was "nowhere near a lecturer, he holds monologues."[12] Einstein himself admitted his performance was below par: "At that time, I really did not give a stellar lecture—in part, because I was not well prepared, and in part because the condition of being-under-investigation made me somewhat nervous."[13] Kleiner had felt obligated to subject Einstein to a further test, by inviting him to present a lecture at the Physical Society of Zurich. "With that I was lucky," said Einstein, "In contrast to my usual practice, I lectured well at that time."[14]

IN ZURICH, THE EINSTEINS lived in a comfortable apartment on Moussonstrasse, on the hill behind the university and the Polytechnic. By coincidence, the Adlers had an apartment in the same house. With the exception of Adler and Kleiner, hardly anyone at the university knew of Einstein's growing preeminence in physics. Einstein felt unappreciated by his other academic colleagues, and, years later, in a letter to his friend Heinrich Zangger, a professor of pathology in Zurich, he said:

> I find the Swiss particularly small-minded. I shall never forget, how the rector . . . reproached me that the heating of my lecture hall was costing so much money! No Frenchman would deign to say that. God only knows, from where you [Zangger] have your blood, so that you have such a completely different attitude than the German Swiss in general, whose brain turns steadily around his little *Fränkli* [as Swiss francs are called in the local dialect].[15]

But the eyes of the Zurich professors were opened by a visit from Walther Nernst, a colleague of Planck's in Berlin. Nernst was not only the preeminent German chemist and director of the Reichsanstalt (the German Institute of Standards), but, more imposingly, he was also a millionaire—he had invented an electric lamp, and he had sold the patent to the German company AEG for the nice round sum of a million Reichsmark, a staggering amount in those days.

Nernst had become very interested in Einstein's paper on the latent heats of solids, because the low-temperature behavior of latent heats predicted by Einstein agreed with measurements recently done at the Reichsanstalt. And, much to Nernst's satisfaction, this low-temperature behavior also agreed with the Third Law of Thermodynamics he had recently formulated, and for which he was to receive the Nobel Prize in 1920. He came to Zurich to discuss thermodynamic matters with Einstein (and, might we guess, to discuss some private financial matters with the gnomes at his bank in Zurich?). The arrival of the famous millionaire professor impressed Einstein's colleagues in Zurich to no end, and made them sit up and take note. They murmured, "This Einstein must be a clever fellow, if the great Nernst travels all the way from Berlin to Zurich to talk to him . . ."[16]

This awareness of Einstein was reinforced when Arnold Sommerfeld arrived for a visit shortly thereafter. Einstein had made some unsuccessful attempts to resolve the conflict between the familiar wave properties of light and the new particle properties of light revealed by his theory of light quanta; as he wrote to a colleague, "I want to see whether I can fully incubate this darling quantum egg in spite of everything."[17] Sommerfeld discussed with Einstein his progress with this quantum problem and also some aspects of relativity. Sommerfeld stayed for a week, and Einstein reported that "his presence was truly a celebration for me."[18]

EINSTEIN'S TEACHING AT ZURICH did not last long, for reasons other than his limitations as a lecturer. Within a year, he got an offer of a better appointment from the University of Prague—twice the salary, and a title of full professor. The offer came in consequence of a strong recommendation by Planck, and the authorities in Prague and Vienna were probably also influenced by a recently published book, in which Planck had compared Einstein to Copernicus and had said that his work on relativity "exceeds in its boldness quite everything that up to now has been achieved in speculative natural science or even in the philosophical theory of knowledge . . ."[19]

Einstein accepted the Prague offer with alacrity. Almost simultaneously, he refused an offer from the University of Utrecht, which would have permitted him to be in close proximity to Lorentz, about whom he claimed, "He meant more to me than all others whom I have encountered in the course of my life."[20] The deciding factor seems to have been the larger salary at Prague—his wife wrote to a friend that he accepted "after ample reflection over the material advantages."[21] As a result of the financial straits of his student years, when he had to rely on the charity of relatives, Einstein had learned to look out for his own *Fränkli* in a thoroughly Swiss manner, and throughout his career he often engaged in rather unseemly haggling over fees and salaries for lectures and appointments.

Apart from salary, little could be said in favor of the appointment in Prague. The University of Prague was one of the ancient universities of Europe, dating back to the Middle Ages, and it could boast of a long tradition in the study of literature; but in physics, it was a nonentity. There was only one other physicist in Prague, an experimentalist, so there was nobody with whom Einstein could (or ever did) have worthwhile discussions about physics. Prague had some assets, such as "a nice Institute with a rich library," and Einstein had a spacious office overlooking a park, where he could pursue his "mullings over scientific problems pretty much undisturbed."[22] The park was attached to an insane asylum and, with his customary sense of humor, Einstein would point to the inmates and say to visitors, "There you see that fraction of the insane who are not working on quantum theory."[23]

WHILE IN PRAGUE, Einstein received an invitation to attend the first Solvay Conference in Brussels. The physicists invited to this first conference in 1911, and the later conferences at intervals of roughly three years (when not interrupted by war), were the crème de la crème. Only two dozen were invited, and by placing Einstein in that illustrious company, the organizers of the conference gave recognition to his status as one of the leaders in physics.

Einstein was delighted to participate. He loved the respectful attention he received from the other participants, all of them older than he: Planck, Lorentz, Poincaré, Rutherford, Langevin, and Marie Curie, the only woman invited to the conference. Lorentz presided over the conference, and Einstein was charmed by him. He wrote to his friend Zangger, "H. A. Lorentz is a wonder of intelligence and delicate tact. A living work of art!"[24]

At this first Solvay Conference, the attention of the physicists was devoted to radiation and quantum theory. But the attention of the newspaper reporters attending the proceedings was absorbed by something quite different. Marie Curie, after the death of her husband Pierre in a street accident, had become involved in a torrid love affair with her colleague Paul Langevin. He was a married man, and his wife got hold of some rather intimate letters written by Marie. These letters were made public in the days of the conference. This disclosure of adultery in the highest ranks of French science caused a major scandal, which, in the newspapers, completely overshadowed the scientific proceedings of the congress. In a clever double pun on Marie's Polish origins, the newspapers called Langevin *le chopin de la polonaise* (*chopin* is French slang for "a good bargain," or "a sweet deal").[25]

Marie Curie already had received the 1903 Nobel Prize in Physics, and she had recently been selected for the 1911 Nobel Prize in Chemistry. The scandal almost aborted her participation in the festive award ceremony in Stockholm, because the Swedes had second thoughts about inviting her as a guest of honor while she was immersed in a flood of ugly publicity. The scandal had a lasting effect on Marie Curie's popularity. In spite of her stellar achievements in the physics and chemistry of radioactive substances, she never became a celebrity of the rank of Einstein. Her reputation was tarnished by the disclosure of her affair with Langevin, and the French bourgeoisie condemned her as not quite comme il faut.

The scandal and the consequent breakup of her relationship with Langevin hurt Marie Curie deeply, turning her into an unhappy, bitter woman. Not even the award of the second Nobel Prize—the first-ever award of a second prize—could restore her spirits. Einstein, who was to cause some scandals of his own by his various indiscriminate adulterous liaisons and apparently thought of himself as somewhat of an authority on women, was at first skeptical about the accusations against Curie and Langevin. He cautioned Zangger, "The horror story bruited about in the newspapers is nonsense . . . I did not have the impression that there was something special hovering between them . . . She has a sparkling intelligence, but despite her passionate nature she is not attractive enough to be a danger to anyone."[26] Some years later, he got to know Marie Curie quite well during a long walking tour from the Swiss mountains to Lake Como. He then wrote of her, "Frau Curie is very intelligent, but meager in emotion, that is to say, impoverished in any kind of joy and pain. Just about her only expression of emotion consists in scolding about things she does not like. And she has one daughter who is worse—like a grenadier. That one is also very talented."[27] The talented daughter, Irène, indeed followed her mother's footsteps in radiochemistry and garnered another Nobel Prize for the Curie family.

EINSTEIN PUBLISHED the first steps toward his theory of gravitation, later called general relativity, while in Prague. But he had begun to think about the deeper meaning of gravitation some years earlier, in 1907, while still working as clerk in the patent office. His germinal idea came to him in another of those flashes of inspiration, which he recollected years later in a lecture at the University of Kyoto:

> I was sitting in my chair at the patent office at Berne. Suddenly, I had an idea: when a person is in free fall, he does not feel his own weight. I was amazed. This simple thought experiment made a deep impression on me. It led me to a theory of gravitation.[28]

His sister Maja claimed that what triggered Einstein's mulling about free fall was the fatal accident of a roofer who slipped from one of the roofs near Einstein's apartment. This tale sounds too good to be true, but Einstein himself also mentions a roof in one of his descriptions of the effects of free fall: "For an observer who is in free fall from the roof of his house, there exists no gravitational field, at least in his immediate surroundings," and he adds that this was "the most fortunate thought of my entire life."[29] The roofs of Berne are certainly steep, and old, and slippery, and roofers probably fall from them by the dozens. If the tale is true, history has almost repeated itself—Einstein's epiphany about gravitation is almost the same as Newton's, except that Newton was watching a ripe falling apple, not a hapless falling roofer.

A roofer in free fall experiences weightlessness because all his body parts fall downward at the same rate, with the same acceleration. And if he releases a hammer or some nails from his hand, they will also fall downward with exactly the same acceleration. If the falling roofer focuses on these phenomena in his immediate vicinity (and manages to forget for a few moments about the impending collision with the ground), he can imagine that he and the hammer and the nails are floating freely in interstellar space, in a region free of gravity.

You can briefly experience such weightlessness in less hazardous conditions by jumping off a diving board at a pool. You will then feel weightless both during the upward part of your jump and the downward part. You can test this by trying to dump water out of a paper cup held in your hand while you fall—you won't be able to dump the water except by squeezing it out of the cup. Astronauts in the space shuttle or the space station orbiting the Earth experience this kind of weightlessness for days or months. They, and the space station, are in free fall, and so are any bodies floating in their vicinity.

There was nothing very original about Einstein's "most fortunate" idea. Fifty years earlier, Jules Verne had anticipated weightlessness for a spaceship coasting from the Earth to the Moon. In his science fiction novel about a bulletlike spaceship launched from the Earth by an enormous gun, Verne correctly described that the cadaver of a dog, thrown out of a porthole, would continue to coast alongside the spaceship; but, oddly, he mistakenly believed that the passengers inside the spaceship would not experience such weightlessness and would walk about or sit on their settees just as in a living room at home.[30]

IF ACCELERATION CAN cancel gravity and produce weightlessness, then acceleration on its own, in the absence of gravity, can simulate gravity, that is, it can produce artificial gravity, indistinguishable from real gravity. This is the Principle of Equivalence of acceleration and gravity. Einstein described it in one of his lectures:

> Two physicists, A and B, awake from a drugged sleep and notice that they are located in a closed box with opaque walls, equipped with all their apparatus. They have no knowledge where the box is positioned or whether or how it is in motion. They now confirm that bodies that they place in the middle of the box and release all fall in the same direction—say, downward—with the same joint acceleration g. What can the physicists conclude from this? A concludes from this that the box sits on a planet, and that the downward direction corresponds to that toward the center of the planet, if it has spherical shape. However, B takes the view that the box might be maintained in uniform accelerated "upward" motion with acceleration g by a force acting on its exterior; no planet needs to be in the vicinity. Is there any criterion by which the two physicists can decide who is right? We know of no such criterion . . .[31]

Again, there was nothing very original about this idea. In Einstein's days, the generation of artificial gravity by acceleration was well understood and widely applied in the technology of centrifuges. The ultramodern Swiss dairies, like all other modern dairies, used centrifuges to separate cream from milk. These centrifuges subject the milk to a high-speed rotational motion, with a large centripetal acceleration, which generates strong artificial gravity of the kind envisioned by Einstein. The cream separates from the milk much faster than if you merely leave a pail of milk sitting on a table, where ordinary gravity slowly pulls the water away from the cream.

What was new was that instead of using artificial gravity to extract cream, Einstein used it to gain insights into the relationship between acceleration and gravity. He argued that the indistinguishability of the effects of acceleration and gravity meant that acceleration, like velocity, is relative. The physicist in the closed box cannot tell whether he is experiencing an acceleration or a gravitational force, unless he looks outside the box and checks its motion relative to some reference point. Thus, Einstein thought there must be a generalized form of relativity, in which both velocity and acceleration are relative. He called the theory of relative velocity he had formulated in 1905 "old" relativity, or *special* relativity; and he called the new theory of relative acceleration *general* relativity.

As in special relativity, the speed of light played a central role in Einstein's attempts to achieve a general theory of relativity. He used the Principle of Equivalence to deduce predictions about the behavior of light in a gravitational field. He reasoned that if the effects of acceleration and of gravity are identical, then it must be possible to deduce what happens in a reference frame at rest in a gravitational field by examining what happens in an accelerated reference frame in the absence of gravity. The latter reference frame is a surrogate for the former, and if we can figure out what light does in the latter, we will know what it does in the former.

EINSTEIN'S FIRST TENTATIVE application of the equivalence of gravitation and acceleration dates back to 1907, when he published a calculation of the effect of gravity on clock rates in a short section of a long review paper on relativity in the *Journal of Radioactivity*. In those days, radioactivity was a hot topic, much more so than relativity, and publications in a journal devoted to radioactivity were bound to get extra attention.

Einstein's initial treatment of clock rates in a gravitational field was only approximate, but it yielded an important result on how clocks in a gravitational field suffer a time dilation, that is, the gravitational field makes clocks run slow. To understand the behavior of clocks in a gravitational field, Einstein began by examining the behavior of clocks in an accelerated reference frame, such as a steadily accelerating train or elevator. He tried to discover the correct synchronization of clocks in such an accelerated reference frame. This is a tricky mathematical problem, because of complications introduced by the time dilation and the length contraction.[32] Einstein had the good sense not to try to solve it exactly, but only for short time intervals and short distances between the clocks. But he failed to understand that even this was a fool's errand—in an accelerated reference frame, there is no "correct," preferred synchronization. Whereas in an inertial reference frame there is a preferred synchronization that leads to the laws of physics having their simple, familiar forms (that is, Newton's laws of motion and Maxwell's equations), in an accelerated reference frame, all synchronizations are bad or worse than bad—they all fail to yield familiar forms of the laws of physics. Some synchronizations are a bit better than others, but none is clearly distinguished by self-evident simplicity.[33]

It was all smoke and mirrors. Einstein managed to confuse himself as well as the (few) physicists who read this paper. It is a measure of his success at seeding this confusion that Abraham Pais—a friend of Einstein's and, later, his biographer—characterized the discussion as "less general, more tortured, and yet, oddly, more sophisticated."[34] Today, Einstein's 1907 attempt at extracting information about the effect of gravity on the behavior of clocks is forgotten because he later found a simpler way to extract the same result.

It seems that Einstein himself recognized his treatment of synchronization in an accelerated reference frame was faulty. Four years later, he wrote that this treatment "had not satisfied me."[35] He never returned to this idea, and he never again tried to introduce a special synchronization for accelerated reference frames. In his later discussions of such reference frames, he used the synchronization of unaccelerated reference frames.

EINSTEIN'S NEXT RESULTS about gravitation were obtained in Prague, in 1911. Again, he relied on the Principle of Equivalence for acceleration and gravitation. First he asked what happens to the frequency of a light wave emitted at one place in an accelerated reference system and received at a different place, say, a light wave emitted by a flashlight at the floor of an elevator cage and received at the ceiling, while the elevator cage accelerates upward (in the absence of gravity!).[36] Under these conditions, the upward speed of the elevator cage is increasing, and when the light wave strikes the ceiling, the speed of the ceiling will be larger than the speed the floor had when the wave was emitted. This means that the light wave arriving at the ceiling is subject to a Doppler shift—its frequency when received at the ceiling will be smaller than its frequency was when emitted from the floor (this is analogous to the reduction of frequency of an ambulance siren when the ambulance is moving away from you).

The upward accelerating elevator cage is a surrogate for a cage at rest on the surface of a planet, where the cage experiences the effects of gravity. Einstein therefore concluded that in the gravitational field of a planet, the frequency of a light wave will decrease as it travels upward. A decreasing frequency of light means a shift of color toward the red, hence this is called the gravitational redshift of light. (Conversely, the frequency of a light wave will increase if it travels downward; this results in a gravitational blueshift.)

However, Einstein recognized a problem with this gravitational redshift: if the source of light emits wavecrests at a steady rate, and these wavecrests arrive at the receiver of light at a slower rate—that is, with a reduced frequency—then what is happening to the excess wavecrests? An analogy makes it clear that something doesn't add up here: imagine that cars enter one end of a tunnel at a steady rate of, say, five cars per minute, but only three cars per minute emerge from the other end of the tunnel—then you know that there is some catastrophic problem in the tunnel. Faced with this conundrum, Einstein declared, "The answer is simple . . . Nothing compels us to assume that the clocks U at different [locations] must be regarded as going at the same rate . . . If we measure the time at $S1$ with the clock U, then we must measure the time at $S2$ with a clock that goes . . . more slowly than the clock U." And he concluded, "We must use clocks of unlike constitution, for measuring time at places with differing gravitation."[37]

Einstein was right in saying that clocks in a gravitational field run more slowly. But he was wrong in saying that this is so because we elect to adjust the clocks in such a way as to make them run slowly. This was another of Einstein's mistakes.

The conceptual pathology of this new mistake was the same as in his earlier mistake with the speed of light, the mistake that drove poor Crowhurst into madness. He had then proposed that, by his own free will, he could select the synchronization of clocks in such a way as to achieve a constant speed of light, so as to enforce the principle of relativity of unaccelerated motion. And now he was proposing that, by his own free will, he could select the rates of clocks in a gravitational field in such a way as to enforce the principle of equivalence of acceleration and gravitation. It was the same mistake all over: In 1905, he had ignored the mechanical laws that apply to clocks and determine their synchronization when the clocks are (slowly) transported from one place to another. And now he was again ignoring the mechanical laws—if two clocks are built to exactly the same specifications, they necessarily run in the same way, and any fooling about with their rates does not result in new physics, but merely in confusion.

Only much later did Einstein recognize that the real answer to the conundrum about clock rates in a gravitational field is found in the curvature of spacetime. Time is "slower" in a gravitational field, that is, the gravitational field produces a time dilation, and this automatically makes the clocks go slower—there is no need to fiddle with their adjustment. This slowing of time in a gravitational field is a direct consequence of the curvature of spacetime associated with gravity. In fact, the gravitational redshift is the most direct observational evidence we have for the curvature of spacetime. But in 1911, Einstein's intuition failed him. He was not yet ready to imagine a curved spacetime. Here, again, Segrè's dictum held sway: When the mind is not prepared, the eye does not recognize.

Oddly, Einstein immediately proposed to test this time-dilation effect by using atoms on the Sun and on the Earth as little clocks. In those days, physicists believed that the internal vibrations of the atoms result in the emission of light of a frequency that corresponds to this internal vibration. This is not exactly right—quantum mechanics has given us a new, somewhat different picture of the emission of light—but it is close enough for present purposes. Hence an astronomer on Earth can compare the frequency of vibration of atoms on the surface of the Sun with that of atoms on the Earth by simply comparing the light arriving from the atoms on the Sun with the light emitted by similar atoms here on Earth. However, in proposing this test, Einstein forgot that according to his own, mistaken, idea he was supposed to use clocks of "unlike constitution," that is, he was supposed to adjust the rate of vibration of the atoms on the Sun to a slower rate. Since there is no practical way to do such an adjustment, the experimental test he proposed was inconsistent with his own theoretical concepts about the time dilation. But Einstein was lucky. The inconsistency in his proposal compensated the mistake in his theoretical concepts—thus his proposal for the experimental test was correct.

ALTHOUGH EINSTEIN was dead wrong about the reason why clocks in a gravitational field run slow, he correctly concluded that the slowed clock rate held an implication for the speed of light. He assumed that when the speed of light is measured at some location in a gravitational field with the local slowed clock, the speed still has its standard value, 300,000 klicks per second. But since the clock is running slow, the actual speed must be slowed by as much as the clock is slowed—in a gravitational field, light propagates more slowly than outside the field. This means that in regard to the propagation of light, the neighborhood of the Sun behaves like a large glass-filled globe encasing the Sun, so this glass slows the propagation of light. The glass is most dense near the Sun, less dense farther out, and it gradually fades away into empty space at large distances (where the speed of light resumes its standard value).

Einstein recognized that such a slowing of the speed of light would bend rays of light in the same way that a globe of glass bends rays of light—it bends rays of light that strike the right half of the globe toward the left, and it bends rays of light that strike the left half toward the right, that is, it always bends the rays of light toward the centerline.

Alternatively, this bending of rays of light can be understood by considering the propagation of a ray of light across an accelerating elevator cage. In the absence of gravitation, the ray propagates horizontally on a straight line, but because the elevator is accelerating upward, the ray will appear to sag downward relative to the elevator—if it is emitted from, say, the midpoint of one wall of the elevator cage, it will strike below the midpoint of the opposite wall. According to the Principle of Equivalence, we then expect a corresponding downward sag, or bending, of a ray of light propagating across the gravitational field of the Sun.[38]

Einstein calculated the numerical value of this bending for a ray of light passing by the Sun at a distance equal to the Sun's radius, that is, a "grazing" ray of light that just barely avoids hitting the Sun. He found that for such a ray of light the deflection is 0.8 seconds of arc. This is a very small angle; it is roughly the width of a thumb when seen from a distance of 5 klicks (or 3 miles) from your eye.

Einstein immediately understood that his prediction of the bending of light could serve as a crucial test of his theory, and he contacted several astronomers with requests to investigate the matter observationally. Because the glare of the Sun hides stars from view, observation of the bending of starlight can be made only during an eclipse of the Sun, when the sunlight is blocked out, and stars near the Sun become visible. He urged astronomers to make preparations for observations during the eclipse of August 1914. But confirmation of the bending of light was actually not obtained until the eclipse of May 1919, after World War I (see Chapter 10).

Continuing his work in Prague, Einstein tried to construct a theory of gravitation based on a slowed speed of light. He thought that since the speed of light appears in the relativistic equations for the momentum and the energy of a particle, it should be possible to formulate an equation of motion in which the changing speed of light alters the momentum and the energy of a particle and therefore brings about the orbital motion of planets, satellites, and comets. Newton's law of universal gravitation is replaced by a law for the speed of light, and everything supposedly follows from that. This was a mistaken idea, but it led to a new expression for spacetime distances, and it later brought Einstein to the recognition that spacetime is curved.

EINSTEIN SOON FELT unhappy in his isolated position in Prague. Within half a year of his arrival, he longed to leave: ". . . it is certain that I would depart from the half barbaric Prague with an easy heart."[39] In 1912, after a stay of only one and a half years, he turned his back on Prague and returned to Zurich. The Polytechnic had just been reorganized and expanded, with new privileges of a federal Swiss university, and it needed a new professor of theoretical physics. On the instigation of Einstein's friends Zangger and Grossmann, the position was offered to Einstein, who gladly accepted.

Sometime just before his move from Prague to Zurich Einstein made the revolutionary discovery that curved spacetime is the real explanation of gravity. Einstein arrived at the profound insight that any gravitating mass curves, or warps, the spacetime geometry in its vicinity, and that this curved spacetime geometry then acts on other nearby masses and affects their motion. As John Wheeler later expressed it succinctly, Mass grips spacetime, telling it how to curve, and spacetime grips mass, telling it how to move.[40]

In his last Prague paper on the variable speed of light, Einstein had added an appendix in the printer's proofs, in which he showed that the effect of the variable speed of light on the equation of motion of a particle could be interpreted as a demand that the particle follow a spacetime path with a shortest effective length between the starting point and the end point of its motion. The "effective length" he adopted for this purpose was a length modified in a suitable way by the variable speed of light.[41] This gave him the clue that the spacetime geometry is affected by the variable speed of light, so distances and times in the vicinity of a gravitating body are different from distances and times in empty space. Distances and times in empty space correspond to a flat spacetime geometry, familiar to our intuition; but distances and times in the vicinity of a gravitating body correspond to a new, counterintuitive curved spacetime geometry. Einstein found this key to the theory of gravitation by remembering some lectures on curved surfaces he had attended while a student at the Polytechnic in which he had become acquainted with the formulas for distances on curved surfaces, and he suddenly saw that the formulas in the Prague paper were of the same kind.[42]

When Einstein added the appendix in the proofs of the last Prague paper, he had not yet quite recognized that his formulas were those for a curved spacetime geometry—if he had, he would surely have said so in this appendix. But it seems he recognized this during his final days in Prague. In a letter to Besso, he wrote, "Recently, I have been working furiously on the gravitation problem. It has reached a stage where I am ready with the statics. I know nothing as yet about the dynamic field . . . Every step is devilishly difficult."[43]

Upon arrival in Zurich, he enthusiastically rushed ahead with this work, and at first he thought that he had reached his goal. He announced, "Gravitation is coming along splendidly. Unless I am much mistaken, I have now found the most general equations."[44] But the jubilation was premature. He ran into new difficulties with the complicated mathematics in his equations for curved spacetime, and in desperation he turned to his friend Grossmann, exclaiming, "Grossmann, you must help me, or I'll go crazy!"[45]

CURVED THREE-DIMENSIONAL SPACE—or, even worse,

curved four-dimensional spacetime—is impossible to visualize. If our three-dimensional space is curved, it must be curved into some dimension beyond three dimensions. Our mind is attuned to three dimensions, and it does not permit us to visualize anything with more than three dimensions. Some mathematicians claim they can visualize a curved three-dimensional space, but if so, they are crazy, that is, crazy in the sense of abnormal. The best a normal person can do is to visualize a curved surface, such as the surface of an apple or the surface of the Earth. Such a surface is a two-dimensional curved space, which curves into the visualizable third dimension.

The curved four-dimensional spacetime of general relativity curves into a fifth, sixth, . . . or even a tenth dimension. But since we can't step out of our four-dimensional spacetime to contemplate its curvature from "outside," we will have to focus on those features of the curved geometry that we can measure within the four-dimensional space, without stepping out into any extra dimensions.

In the study of a two-dimensional curved surface, this corresponds to focusing on what we can measure on the surface, without stepping into the third dimension. For instance, we can lay out triangles and circles on the surface of the Earth (at sea level, so we don't have to worry about the irregularities that mountains and valleys add to the curvature of the Earth's surface), and we can check the geometry of such triangles and circles. When we do this, we find that triangles and circles laid out along the surface of the Earth do not obey the theorems of ordinary, flat, Euclidean geometry. Thus, on the curved surface of the Earth, the square of the hypotenuse of a right triangle is *smaller* than the sum of the squares of the two sides. The Modern Major General in Gilbert and Sullivan's *Pirates of Penzance*, who claims to know "many cheerful facts about the square of the hypotenuse," would have been shocked to discover this (but the Captain of the *Pinafore* would have known all about it from his training in celestial navigation and spherical trigonometry).

The Major General would have been even more shocked to discover that on the Earth's surface, the sum of the interior angles of a triangle is *more* than 180 degrees, and the circumference of a circle of radius r is *less* than $2\pi r$. By measuring these deviations, we can detect the curvature of the Earth's surface without stepping out of this surface, into the extra, vertical dimension. Likewise, in a curved three-dimensional space or in a curved four-dimensional spacetime, we can detect the curvature by measuring deviations in the geometry of triangles and circles, even though we cannot step outside the space and contemplate the curvature from the outside.

Einstein quickly came to the realization that for the mathematical treatment of the curved four-dimensional spacetime he needed to rely on a generalization of the mathematical methods that Minkowski had introduced into relativity years before. He later admitted that without Minkowski's tensor calculus "the theory of general relativity . . . would perhaps have got no farther than its swaddling clothes."[46] Minkowski had also introduced a geometrical, graphical way to look at relativity, by spacetime diagrams in which the motion of particles can be represented by lines, called *worldlines*, giving the position of particles as a function of time. These graphical methods are today a commonplace tool in the study of relativity, but it took many years for them to gain popularity. Einstein rarely drew any spacetime diagrams (there are a couple in his 1921 Princeton lectures), and as late as 1960, J. L. Synge, an expert on relativity, complained, "When, in a relativistic discussion, I try to make things clearer by a spacetime diagram, the other participants look at it with polite detachment and, after a pause of embarrassment as if some childish indecency had been exhibited, resume the debate in their own terms."[47]

Grossmann acquainted Einstein with the metric tensor, or the "fundamental" tensor, devised by the great German mathematician Bernhard Riemann to establish the connection between the coordinates and the distances in a curved space. This tensor can be regarded as a calculating device, or what John Wheeler likes to call a "machine," for extracting the real, measurable distance between any two nearby points (it is called the *metric* tensor because it yields the distances that you will find when you measure the curved space with a meterstick). You feed into the metric-tensor machine the difference between the locations of the first point and the second point, and it instantly spits out the distance. The metric-tensor machine performs this trick by means of ten code numbers stored in its memory. These ten numbers encode all the information about the geometry of spacetime.

You can think of the ten code numbers in the metric tensor as a "team," analogous to a soccer team consisting of ten players, not counting the goalkeeper (if the players wear jerseys with numbers, you can improve on this analogy by regarding the players as numbers, so the soccer team is a team of ten numbers). The ten code numbers in the metric tensor are called the components of the tensor. These components are different at different locations, because spacetime might have a different geometry at different locations. This is analogous to having different players on the soccer teams in different towns. For a complete description of the geometry of the curved space, we need to know the ten components of the metric tensor at each location. This seems complicated, but it is no more so than giving a complete description of the status of soccer in the world by listing the ten players on the "home team" in each town.

GROSSMANN WAS WILLING to help Einstein with the mathematics, but he washed his hands of the physics. He insisted that Einstein would have to take full responsibility for the physics. The paper that emerged in 1913 from the Einstein-Grossmann collaboration was called an *Entwurf* of a generalized theory of relativity, that is, an outline or a sketch. Because of the sharp division of labor, this *Entwurf* is a rather odd thing, part fish, part fowl: it consists of a first "physical" part under the name of Einstein and a second "mathematical" part under the name of Grossmann.[48] The connection between these parts is somewhat vague and gives the impression that Einstein did not fully understand the mathematics, and Grossmann did not fully understand the physics.

The first part of the Einstein-Grossmann paper holds much promise. It starts out with a sound perception of what was required by the physics. Einstein gives an incisive argument for why the spacetime geometry near a gravitating mass must be curved, an argument that is in fact much better than the argument he adopted later in 1915 in his final formulation of general relativity.[49] In essence, he relies on the appendix of his Prague paper to establish that in the presence of a gravitating mass—and a variable speed of light—the distances and times are of the mathematical form expected in a curved spacetime.

Einstein proposed that the metric tensor should play the role of the gravitational field and determine all the gravitational effects—the behavior of clocks and measuring rods as well as the motion of particles and of light signals, such as the motion of planets around the Sun and the bending of light rays by the Sun. Einstein understood he needed to discover the exact relationship between the magnitude of the metric tensor and the amount of mass in the Sun, or, more precisely, the amount of energy and momentum in the Sun.

Mathematically, the energy and momentum of the matter in the Sun or in any other gravitating body is described by the energy-momentum tensor, or, as John Wheeler likes to call it, the "momenergy" tensor. This is another ten-component tensor whose components give the density of energy, the density of momentum, and the flow of energy and momentum. Thus, to complete his program, Einstein needed to find the right equation linking the metric tensor to the momenergy tensor, that is, he needed to find the equation that governs the dynamics of the metric tensor.

This looks to be an awesome problem, but it actually is much simpler than it looks. In physics, all the equations that govern the dynamics of anything—a particle, or a cluster of particles, or an electric field, or a gravitational field—are pretty much the same in their broad features. In essence, they are all variants of Newton's law of motion, $ma = F$. On the right side of this equation, we have the force, or the cause, of the motion, and on the left side we have the resulting acceleration. Einstein expected that the field equation that governs the dynamics of the gravitational field would be of the same form. On the right side of the equation he expected to find the momenergy tensor (playing the role of the "force" that causes the curvature of spacetime), and on the left side he expected to find the "acceleration" of the metric of spacetime. Acceleration is really a rate of change of a rate of change (for instance, for the motion of a car, the acceleration is the rate of change of the speed, and the speed is the rate of change of the distance). Hence Einstein expected to find a rate of change of the rate of change of the metric tensor on the left side of his equation.

This was a good start, but because the metric tensor has ten components and spacetime has four distinct directions, there are very many ways to construct rates of change of rates of change of the metric tensor. Einstein correctly surmised that the rate of change of the rate of change he was looking for should be some kind of a tensor. This simplified the problem considerably, because the mathematicians who had investigated curved geometries had established there are three, and only three, ways of constructing a tensor from the various rates of change of rates of change of the metric tensor. The first is the Riemann tensor (with twenty components), the second is the Ricci tensor (with ten components), and the third is the curvature invariant (with one component). The first two of these tensors are named after Bernhard Riemann and Gregorio Ricci, the German and Italian mathematicians, respectively, who had investigated the mathematics of curved geometries in the nineteenth century. The Riemann and Ricci tensors and the curvature invariant directly characterize the curvature of the geometry—angles of small triangles and circumferences of circles can be expressed in terms of these tensors. Thus, these tensors are directly related to measurable aspects of the curved geometry, and it seemed clear to Einstein that his equation for gravitation should contain some or all of these three curvature tensors.

BUT ALTHOUGH EINSTEIN'S intuition about the physics was clear and incisive, his mathematical implementation of the physics was a disaster, and it prevented him from constructing a consistent theory. Einstein contributed one gross mistake and Grossmann the other. Both of these mistakes arose from a poor grasp of the mathematics of curved spacetime. Einstein and Grossmann were novices at this game. They had only a superficial knowledge of the available mathematical literature on curved spaces (especially the publications of the Italian mathematicians Ricci, Tullio Levi-Civita, and Luigi Bianchi), and they did not fully understand what the mathematical equations in curved spacetime meant and what they entailed.[50]

Einstein's mistake was a misconception about the meaning of the terms in the equation of motion for a particle in curved spacetime.[51] He thought that the terms in this equation implied that particles with different velocities would fall with different accelerations unless the curvature of spacetime is restricted to the time part of the spacetime geometry (with no curvature in the space part). To make all particles fall with the same acceleration, he therefore assumed that the curving of the spacetime geometry in the vicinity of, say, the Sun existed only in the time part of the geometry. This was a disastrous mistake—in fact, the spacetime geometry around the Sun is curved in time *and* in space in roughly the same way. With his wrong assumption Einstein got the wrong results for the gravitational field, and he therefore thought that his theory was totally inconsistent with Newton's law of universal gravitation.[52] Of course, he expected some deviations from Newton's theory, because his new theory of gravitation was supposed to be an improvement on Newton's. But for the outermost planets of the Solar System, he expected the deviations between the two theories to disappear—and they did not.

Grossman's mistake was even worse and casts serious doubts on his competence as a mathematician. Following Einstein's program, he wanted to construct an equation relating the Riemann or Ricci tensors to the momenergy tensor. He knew that the Ricci tensor has ten components, and that the momenergy tensor also has ten components. Because of this neat match of the number of components, Grossmann proposed as the fundamental field equation of general relativity that the Ricci tensor should equal the momenergy tensor (except for a constant of proportionality, which we will ignore, to simplify the discussion). In a simplified, semimathematical notation, Grossmann's equation was

$$\text{Ricci tensor} = \text{momenergy tensor}$$

Big mistake! The trouble with this equation is that it is mathematically inconsistent. It is an impossible equation, because it conflicts with the conservation of energy and momentum.[53]

Grossmann never noticed this inconsistency. He also overlooked an obvious alternative choice for the fundamental equation of general relativity. This alternative uses a combination of the Ricci tensor and the curvature invariant to contrive a consistent equation. If Grossmann had not overlooked this alternative, he might have saved Einstein several years in the lengthy quest for the right equation.

In a performance worthy of Elmer Fudd marching off to hunt "wabbits" and failing to notice that Bugs Bunny is sitting on top of his hunting cap, Einstein failed to recognize the mistake in Grossmann's equation, and he proceeded to dismiss the equation for entirely different and entirely spurious reasons. Einstein had assumed that the curvature exists only in the time part of the spacetime geometry, and he therefore thought that his equations were in conflict with Newton's law of universal gravitation. And he also thought they had another defect: the equations seemed to have what he called a "hole," that is, they seemed incapable of producing a unique solution for the curved spacetime geometry generated by a gravitating body. Einstein failed to see that this "hole" was a figment of his imagination—the spacetime geometry is actually unique, but it does not *look* unique, because it can take on different appearances depending on the choice of coordinates (for instance, in ordinary flat three-dimensional space, the geometrical formulas for distances or areas in rectangular and in spherical coordinates *look* different, even though the geometry described by these coordinates is really the same).

Thus, Einstein discarded his initial, nearly correct equation for his theory of general relativity, and in desperate attempts to repair the imagined defects in this equation, he descended into a quagmire of alternative mathematical constructions involving nontensor quantities, that is, quantities other than the Riemann tensor, the Ricci tensor, and the curvature invariant. This futile pursuit of various mistaken versions of the theory of general relativity was to cost him three years of tiresome work. As he himself admitted later, "The series of my publications on gravity is a chain of wrong turns."[54]

EINSTEIN THOUGHT HE MIGHT FIND some confirmation for his theory by examining fine details in the motion of planets, especially Mercury. Toward the middle of nineteenth century, the French astronomer Urbain Leverrier had noticed a deviation in the motion of

Mercury that could not be explained by Newton's law of gravitation. Mercury's elliptical orbit around the Sun precesses, that is, it gradually rotates around the Sun, so Mercury completes each of its revolutions about the Sun on an orbit that is slightly displaced relative to the preceding orbit. Most of this precession is explained by the attractions that the other planets exert on Mercury, attractions which perturb the ideal, fixed elliptical orbit that would exist in their absence.

By calculations based on the data accumulated over many years by various observatories, Leverrier found that the precession of Mercury's ellipse was larger than it should be—the excess precession amounted to one extra rotation of the ellipse around the Sun in 3 million years. Leverrier had earlier found something similar in the case of Uranus, which had led him to predict the existence of an extra planet whose attraction perturbs the motion of Uranus. The subsequent discovery of Neptune was a spectacular triumph for celestial mechanics. With thoroughly French logic, Leverrier therefore made a similar prediction in the case of Mercury, attributing the excess precession to another extra planet, Vulcan, supposedly in an orbit very near the Sun and hidden in its glare. But nature can be fickle, and on this occasion she chose to disappoint Leverrier—no sign of Vulcan was ever found.

Einstein hoped that the excess precession of Mercury's ellipse might be explained by his new theory of gravitation. To calculate the precession from his theory, he enlisted the help of his friend Michele Besso, who had been a better student at the Zurich Polytechnic and was better trained in mathematics than Einstein. The calculation was lengthy and messy, but straightforward. And in the end it brought Einstein no joy—the calculated precession accounted only for about half of the actual excess precession of Mercury. Einstein thought it best to keep quiet about this disappointing result. He filed the calculation away, until 1915, when he revised it by means of an improved theory of gravitation—and this was to bring him a spectacular success, fulfilling all his hopes.

WHILE EINSTEIN WAS STRUGGLING with the problems of his gravitational theory in Zurich, serious maneuvers to entice him to Berlin were in progress at the Prussian Academy of Sciences. For some years, Nernst had planned to bring Einstein to Berlin, and his plans were supported by Planck and other leaders of the German scientific establishment. Among theses supporters was Fritz Haber, the influential director of the Kaiser-Wilhelm Institute for Physical Chemistry and winner of the 1911 Nobel Prize in Chemistry for his discovery of a process for the synthesis of ammonia from the nitrogen in air, a process that is a first step in the production of chemical fertilizers (and also a first step in the production of high explosives, which was to prove crucial for the German munitions industry in World War I). Berlin was then the world capital of physics, and it seemed appropriate that the man Planck had called the "new Copernicus" should settle there among other distinguished physicists, instead of remaining in Zurich, a place better than Prague, but then without much distinction in the study of physics.

Besides offering Einstein the favor and prestige of their company, the Berlin physicists thought to entice him with an extraordinarily generous salary and a handful of titles and perquisites: member of the Prussian Academy of Sciences, professor at the University of Berlin, director of the soon-to-be founded Kaiser-Wilhelm Institute for Theoretical Physics, and, last but not least, freedom from any teaching duties whatsoever. In short, Einstein would be paid a royal salary for merely being present in Berlin and doing as much or as little as he pleased.

In a meeting of the academy in the summer of 1913, Planck and Nernst proposed Einstein for membership. With such influential support, it was no surprise that the academicians voted overwhelmingly in favor of Einstein. The vote was secret, by means of white or black balls deposited in a bag, and of the twenty-two balls cast by the academicians, only one was black.

A few weeks later, Planck and Nernst traveled to Zurich to surprise and overwhelm Einstein with their offer. They wanted an immediate answer, but Einstein asked for a day's delay to make up his mind. He promised he would meet them at the railroad station and wave a handkerchief, if he agreed to their terms. The next day he waved a white handkerchief, in sign of surrender.[55]

In Berlin, the bureaucratic process for completing Einstein's appointment ground forward slowly, and it took several more months for Einstein to receive the official notification that his election to the academy "had been confirmed by an All-highest decree of His Majesty the Kaiser and King."[56] With that, Einstein unwittingly became a subject of Kaiser Wilhelm II and a German citizen, a detail that he (and everybody else) overlooked at the time and that was to lead to complications when he later won the Nobel Prize.

WHY EINSTEIN ACCEPTED the offer from Berlin was somewhat of a puzzle to some of his friends and colleagues in Zurich. He had no obvious need for the higher salary offered by the Berliners—he had a good appointment in Zurich, and could live in complete Swiss bourgeois comfort on his professorial salary. Einstein was fond of the Zurich milieu. In spite of its wealth and its role as an international banking center, Zurich was and is a somewhat sleepy and provincial city, really a provincial town in sextuplicate. As Alain de Botton put it, "Zurich's distinctive lesson to the world lies in its ability to remind us how truly imaginative and human it can be to ask of a city that it be nothing other than boring and bourgeois."[57] In contrast, Berlin was lively, thoroughly cosmopolitan, and it encompassed a wide social range, from aristocratic to bourgeois to bohemian. It had become a great center of political power in 1871, when Bismarck had promoted King Wilhelm I of Prussia to emperor of Germany and founded the German Reich, with Berlin as its capital.

But Einstein had no interest in the hustle and bustle of the great German metropolis. The friendly and informal attitude of the Swiss suited Einstein's temperament much better than the disciplined and formal attitude of the Germans, and especially the Prussians. Of the Berliners he said, "Unfortunately, this lack of personal refinement is typical of Berliners . . . How crude and primitive they are! Vanity without an inner self! Civilization (well-cleaned teeth, elegant cravat, manicured mustache, immaculate suit), but no personal culture (crudeness in speech, gestures, voice, emotions)."[58] Nevertheless, he was to spend some twenty years, spanning the peak of his scientific career, in Berlin.

He could not help being impressed by Berlin's success in surpassing Paris and London as a center of the sciences, especially by the founding of the Kaiser-Wilhelm Institute in Dahlem, just outside Berlin. Kaiser Wilhelm donated the land for the institute, and wealthy Germans, among them many Jews, volunteered contributions that paid for the buildings and for generous salaries for scientists. Besides, Einstein had deep feelings of gratitude and intellectual kinship for the Germans, because they had been the first to offer recognition for his achievement in relativity. These feelings were especially strong toward Planck, who had treated him as an equal from their first contact in 1905, when Einstein was merely a miserable 3rd Class patent clerk, who might have expected to be far below the notice of the topmost professor of the German hierarchy.

Einstein was aware that Planck had done much to promote relativity and also to promote his academic career with glowing recommendations for his first appointments in Zurich and in Prague. His feelings toward Planck lacked some of the warmth of his feelings toward Lorentz, whom he venerated as a father figure. But he held Planck in deep respect and later said of him, "His gaze was aimed at eternal things, but he nevertheless took an active interest in everything belonging to the human and temporal sphere . . . The hours that I was permitted to spend in [his] house, and the many private conversations that I held with him, will remain among my most beautiful recollections for the rest of my life."[59]

And membership in the Prussian Academy of Sciences was a feather for Einstein's cap. The academy had an illustrious history; it had been founded by Leibniz 200 years before, and in the late nineteenth century it had risen to the status of the premier scientific society of all of Europe. But many of its members were elderly or overelderly statesmen of science, liable to drift off into a nap whenever a lecture lasted for more than a few minutes. Einstein said, "The academy reminds me in its demeanor of some kind of faculty. It seems that most members limit themselves to display a certain peacock-like grandiosity in their writing, otherwise they are quite human."[60] One clear advantage of his new position was the freedom from all teaching duties. When he was about to leave for Berlin, he wrote to a friend, ". . . by Easter I go to Berlin as an academy man without any duties, almost as a living mummy. I look forward to such a difficult profession!"[61]

But it seems that the main attraction that Berlin held for Einstein was not the generous salary, nor the prestige of the position and the association with the best and brightest physicists of Germany, nor the liberal working conditions, but a flouncing bit of skirt and the prospect of an adulterous liaison. During a previous visit to Berlin, he had there met his cousin Elsa Löwenthal, a divorceé with two grown daughters. She was his double first cousin (that is, a cousin both on his father's side and his mother's side), and a German proverb warns that *Heiraten ins Blut tut selten gut* (Marrying into blood is rarely good). But this had not deterred Einstein. He and Elsa had engaged in some mutually agreeable, suggestive flirtation. He wrote to her, "I consider myself a man in every which way. Perhaps sometime there will be an opportunity to prove that to you. Kisses from *your* Albert."[62] During a second visit they engaged in a bit more than flirtation, with expectations of further fun to come. Elsa thought that Einstein was negligent in matters of personal hygiene, and she gave him a *necessaire* with a hairbrush and a toothbrush. Einstein used the former, but not the latter. As he informed her, "Hairbrush is in regular use, and also in other regards relatively thorough cleaning. Otherwise way of life is so-so la-la. Toothbrush has again been put into retirement for purely scientific considerations regarding the dangers of pig's bristle: pig's bristle can drill through diamonds; how could my teeth resist it?"[63]

In the invitation from Planck and Nernst, Einstein saw above all the opportunity to enjoy himself, intimately, with his cousin. Years later he admitted to Zangger that the charming cousin was "The main reason for which I went to Berlin."[64]

"The theory is of incomparable beauty"

*Einstein's announcement to his friend Heinrich Zangger
of his discovery of the theory of general relativity,
according to which all gravitational effects arise from a
curvature of space and time.*

On the eve of his departure from Zurich in the spring of 1914, Einstein's friends and colleagues gave him a dinner at the Kronenhalle, his favorite restaurant, renowned then as now for its hefty Swiss adaptation of French cuisine. Leaving the restaurant, after much food and good cheer, Einstein remarked to an acquaintance, "In Berlin they speculate on me like on a prize hen; but I don't know whether I can still lay eggs."[1] And, indeed, his egg-laying abilities were not much in evidence during his first year in Berlin. He had the whole year to do research, unencumbered by any teaching duties, but he produced only a handful of papers, and they were all unremarkable—short in length and short in substance.

Among the productions of his first year in Berlin, there was only one paper that looked long and substantial, on the formalism of general relativity; but this was really no more than a rehash of the Einstein-Grossmann *Entwurf* paper of the preceding year. Einstein merged the two parts of this previous paper into a single piece, placing the physics in the correct mathematical context. While doing so, he mastered the mathematical tensor formalism that Grossmann had supplied in the previous paper, which he had initially found difficult to grasp, and about which he had said, "I have become imbued with high respect for mathematics, which in its subtler part I had regarded in my simple-mindedness as pure luxury until now. Compared with this problem, the original relativity is child's play."[2] But now he felt sufficiently confident of his mastery of the math to do without Grossmann, and he deleted his name from the new paper (later, he even declared on one occasion that Grossmann had not contributed anything).

In his inaugural address at the Prussian Academy, Einstein apologized for his lackluster performance: "I beg you to remain convinced of my sense of gratitude and of the fervor of my striving even when the fruits of my efforts appear meager."[3] Planck, as permanent secretary of the academy, replied soothingly, ". . . as your works have shown us, you understand not only how to formulate the program of theoretical physics, but also how to execute it"[4] but with Prussian forthrightness he then commented unfavorably on Einstein's notion of photons, and he added some criticism of Einstein's efforts on general relativity, saying that the theory (as formulated at that time) did not succeed in making gravitation equivalent to acceleration, because it failed to treat all coordinate systems in the same way, "as you yourself have proved recently."[5] Privately, Planck had advised Einstein against pursuing his attempts on general relativity: "As an older friend I must advise against it . . . In the first place you won't succeed; and even if you succeed, no one will believe you."[6]

EINSTEIN'S POOR PERFORMANCE may have been the result of fatigue. He had spent months with Grossmann and with Besso in his frantic search for the correct equations for general relativity. "I work like a horse," he wrote to a friend, "even if the cart does not always move very far from the spot."[7] Again and again the answer had seemed within his grasp, only to elude him at the last moment. When he moved from Prague to Zurich and started to work on his new theory, he had believed that "Gravitation is coming along splendidly. Unless I am much mistaken, I have now found the most general equations."[8] But a year later, he still had not obtained the result he had hoped for, and he had admitted, "The thing still has big snags, so my confidence in the validity of the theory still wavers . . . The theory contradicts its starting point; it floats on air."[9] This was shortly followed by, "The gravitational affair has been resolved to my complete satisfaction."[10] And so it went, successes alternating with failures. For Einstein, the lengthy quest for his revolutionary theory of general relativity was the best of times and the worst of times. As he described it later, "The years of searching in the dark for a truth that one feels but cannot express, the intense desire and the alternations of confidence and misgivings until one breaks through to clarity and understanding, are known only to him who has himself experienced them."[11]

The intense effort on general relativity with Grossmann and Besso may have left him in a state of exhaustion, similar to his exhaustion at the end of his miracle year, 1905. Besides, he was in poor health, suffering from some kind of stomach ailment, possibly ulcers. He was put on a diet of half a liter of raw milk, twice per day, and only one meal.

To this were added the distractions from the marital and extramarital complications of his life. Mileva had been deeply unhappy in Prague. She had no friends there, she disliked the place, and she had been relieved to return to Zurich. She was strongly opposed to the move to Berlin, wailing "about Berlin and her fear of the relatives." Above all she feared Einstein's formidable mother, now also moving to Berlin, who had fought long and hard against Einstein's marriage and had never become reconciled to the daughter-in-law. Einstein wrote of her, "My mother is usually kindhearted, but as mother-in-law she is a veritable devil. When she is with us, then everything is like filled with explosives."[12]

Mileva may also have had some suspicions about the developing relationship between Einstein and Elsa. In a letter to Elsa, Einstein reported, "She did not ask about you, but I think she does not underestimate the significance that you have for me."[13] Mileva was given to intense jealousy, and she knew from some previous incidents that Einstein was liable to stray. She had a needy personality, with bouts of depression and deep feelings of inferiority, arising in part from her physical deformity (she had a severe limp), in part from her failure at the Zurich Polytechnic, and in part from unfair comparisons with Einstein's genius. Her needy character had suited Einstein very well when he had seduced her in their student days, but now her pathetic attempts to cling to him were not welcome. He wrote to Elsa, "At home it is more awful than ever before: icy silence."[14]

Elsa prodded Einstein to divorce Mileva. But he replied that there were difficulties: "Do you think it is so easy to get a divorce, when one does not have any proof of the guilt of the other party, when the latter is cunning and—must I say it—mendacious? And I really don't even have any proof that convinces me of the existence of facts that a court would regard as 'adultery' . . . On the other hand, I treat my wife like an employee which I cannot dismiss. I have my own bedroom, and I avoid being alone with her."[15] And in a later letter he badmouthed Mileva some more, declaring she had the evil eye: "She is an unfriendly, humorless creature, who has nothing of life, and undermines everybody's joy of life by her mere presence (malocchio)!"[16]

Upon receiving the offer from Planck and Nernst, Einstein had immediately informed Elsa. "No later than next spring, I will come to Berlin permanently . . . I very much look forward to the lovely times that we will spend with each other! Don't tell anybody about this matter. A resolution in the plenary session of the academy is still needed, and it would make a bad impression if the news got around."[17] Elsa promptly marched off to Haber's office to urge him to accelerate the processing of the bureaucratic paperwork for the appointment. Einstein thought this was a lark: "The impudence with which you went to pester Haber is thoroughly Elsa . . . did your black soul decide on that on its own? I wish I had been there."[18]

MILEVA AND THE BOYS arrived in Berlin at the end of April, one month after Einstein. They moved into an apartment in Dahlem, a suburb southwest of Berlin, where the new Kaiser-Wilhelm Institute was located. By a recently installed suburban train line, Einstein could quickly reach the center of Berlin, where the academy had its headquarters next to the university.

Einstein was delighted with Berlin. "I have to confess that I like the life here very well," he wrote to his friend Zangger, "How much of proficiency and ardent interest in science are to be found here! I am always bewitched by the colloquium and the Physical Society. And you want to know about the people? Basically they are the same everywhere. In Zurich they act conventionally republican, here militarily stiff and disciplined, but they are governed by the same basic drives here and there."[19]

Einstein did not spend much time at home. Sometimes he would disappear for a week, leaving Mileva ignorant of his whereabouts.[20] Mileva suspected he was spending days and nights in the arms of the plump and eager Elsa. We can date the start of Einstein's adultery with Elsa quite precisely to June or July 1914 from a deposition he made in his divorce proceedings in December 1918: "It is true that I committed adultery. I am

The entrance to the Prussian
Academy of Sciences, next to
Humboldt University.

living since about 4½ years ago with my cousin, the widow Elsa Einstein,
divorced Löwenthal, and have been in intimate relations with her continuously since then. My wife, the complainant, has known since summer 1914
that I am in intimate relations with my cousin. She has made me aware of
her indignation about that."[21]

We do not know how Mileva found out about the adulterous liaison.
But we do know that in July, after a violent domestic quarrel, she suddenly
moved out of their apartment and, with the boys, went to live in the home
of the Haber family. The July date is consistent with the "summer 1914"
date in Einstein's deposition, so it's a good guess that Mileva did what the
injured wife usually does—she moved out because she found out.

For a while, the only communication between Albert and Mileva was by
written notes carried back and forth by Haber. Mileva could not bear to
let Albert go, whereas Albert was more than willing to let her go, but did
not want to lose his sons. In a memo to Mileva, he laid down a long list of
conditions under which he would permit Mileva to come back to live with
him. This document is without doubt the oddest and most obnoxious item
in the large collection of Einstein's papers and other memorabilia. It reads
like a set of orders issued by a Prussian officer to his batman:

Conditions

A. You will see to it

1. that my clothes and my wash are kept in order

2. that the three meals are regularly served to me in *my room*.

3. that my bedroom and my workroom are always kept in good order, and particularly that my desk is *exclusively* at my disposal.

B. You renounce any personal relationship with me, except when absolutely required for social reasons. In particular, you renounce

1. that I sit with you at home

2. that I go out with you or travel with you

C. You agree explicitly to observe the following point in your relationship with me

1. You are not to expect tenderness from me nor are you to make any accusations.

2. When you direct any talk at me, you are to stop immediately if I demand it.

3. You are to leave my bedroom or workroom immediately without contradiction if I demand it.

D. You agree not to denigrate me by word or by deed in the eyes of the children.[22]

What makes this memo even uglier is the undercurrent of hypocrisy that runs through it. Einstein acts as though *he* were the offended party, and that Mileva was guilty of adultery or some worse offense.

Maybe he expected and hoped that Mileva would never accept the humiliating conditions imposed by this memo, and he would then be rid of her. But Mileva disappointed his hopes—she accepted. So he wrote another letter grinding her deeper into the dust: "I have to write you again to make your situation perfectly clear. I am prepared to return to our apartment because I do not want to lose the children and because I do not want that they lose me, and *only* for that reason . . . After what has happened, there can be no possibility of a companionable relationship with you." And he adds another choice bit of hypocrisy: "If a joint life on this basis is not possible for you, I will resign myself to the necessity of a separation."[23] This letter apparently did the trick: Mileva decided to return to Zurich, with the children.

ELSA AND HER TEENAGE DAUGHTERS had left town in a hurry, for discreet rustication in Bavaria. She probably had thought it advisable to drop out of sight temporarily, as was befitting for a bourgeois

widow discovered in intimate relations with a married man. After signing
the financial arrangements for his separation, Einstein went directly to
Elsa's empty apartment from where he wrote, "Now you have the proof
that I am capable of making a sacrifice for your sake! What you have suffered in the past days has made such an impression on me that I could not
do otherwise, regardless of the children . . . Tonight I sleep in your bed!
It is odd how confusing one's feelings are. It is just a bed like any other, as
though you had never slept in it. And yet I feel it is a delight that I can lie
down, it is like tender intimacy."[24]

Elsa returned to Berlin within a week, highly satisfied with the new
developments. Mileva left for Zurich a few days later. Einstein saw her off
at the train station. He described the scene to Elsa: "The last battle has
been fought. Yesterday my wife and the children left forever. I went to the
train and gave them the last kiss. I cried yesterday, I howled like a little boy,
yesterday afternoon and yesterday evening, after they left."[25] On the advice
of Haber, Einstein refrained from meeting with Elsa for a while: "Haber
emphasized to me that we have to take dreadful precautions to avoid that
we, i.e. you, do not get into the gossip mill. Do not go about town unac-
companied! Haber will inform Planck, so my nearest colleagues do not
hear about the matter by way of rumors. You will have to do wonders in tact
and restraint so you are not regarded as a kind of murderess; the appearances are strongly against us."[26]

In the normal course of things, the Einstein-Elsa scandal would have
become the talk of the town. The Berlin newspapers had their gossip col-
umns, and Berliners were very fond of satirical cartoons. When *l'affaire*
Curie-Langevin came to light in 1911 during the Solvay Conference, the
famous Berlin satirical magazine *Kladderadatsch* (Berlin dialect for "fra-
cas") had been merciless:

> Mme. Curie and her assistant Professor Langevin are being insulted in
> a most extraordinary way. Mme. Curie supposedly abducted Professor
> Langevin; whereas they merely traveled to a scientific congress in
> Brussels. It was maliciously said of them that there is an intimate
> relation between them. This is defamation; Mme. Curie merely
> performs scientific experiments, in which Professor Langevin assists
> her. At this moment, they are busy with experimental investigations
> on the creation of life.[27]

It is easy to imagine what *Kladderadatsch* would have made of the Einstein-Elsa affair, involving extramarital adventures in the highest ranks of the staid Royal Prussian Academy of Sciences. Einstein would have weathered the storm of scandalous publicity somehow, but Elsa probably would have had to flee Berlin. Einstein and Elsa were saved from ignominy by World War I, which had begun on July 28, with the declaration of war of Austria-Hungary against Serbia, just two days before Mileva's departure for Zurich. The Berliners suddenly had more momentous matters to worry about than a juicy scandal in academe. Regiments paraded along Unter den Linden, jingoistic poems, reports of battles and acts of heroism, and lists of casualties filled the newspapers . . . and nobody paid any attention to Einstein and Elsa.

THE BEGINNING OF THE WAR spoiled a German effort to take advantage of the solar eclipse of August 21, 1914, for an observational test of the bending of starlight by the Sun. Already in 1911, during his first draft of his theory of gravitation, Einstein had been eager to obtain observational confirmation for his prediction. He had encouraged the Berlin astronomer Erwin Freundlich to organize an expedition to south Russia for the eclipse of August 1914. He had told Freundlich not to worry about the funding: "If the [Prussian] Academy will not touch this, we will get the bit of moolah from private contributions . . . If all else fails, I will pay for the thing out of my own little savings, at least for the first 2000 M."[28] It may be that the "little savings" that Einstein had in mind were the 15,000 Reichsmark he had recently obtained as a gift from a generous anonymous admirer. But in the end, the 2000 Reichsmark were paid by the academy, and an extra 3000 Reichsmark for travel and transportation expenses were contributed by the Krupp armaments conglomerate. Krupp could well afford it—the German-British arms race preceding World War I had proved immensely profitable for all manufacturers of armaments, especially Krupp, which had fully exploited the opportunity to sell cannons and armor plate in escalating alternation.

In the summer of 1914, Freundlich and two companions, three cameras, and other equipment, traveled to the Crimea, where they arrived at the end of July, in good time for the eclipse of August 21. But on August 1, less than a week after their arrival, while they were engaged in preparations for the eclipse, Russia mobilized and Germany declared war. The Russians promptly interned Freundlich and confiscated all his equipment. Einstein complained to Zangger, "The observation of the solar eclipse has surely been suppressed by the Russian knout, so the decision about the most important result of my scientific struggles will not come in my lifetime."[29] In the end, the Russians did little harm to science or to Freundlich—the day of the eclipse was cloudy, unsuitable for good photography, and Freundlich was soon released in an exchange of prisoners of war.

An American team of astronomers led by William Campbell, the director of the Lick Observatory in California, was also in the Crimea, with plans to measure the bending of starlight. The Russians did not interfere with them, but the weather did—they were rained out.

The confirmation of the bending of starlight would have to wait until the eclipse of 1918, after the end the war. This delay was a lucky break for Einstein, because in 1914 his prediction for the bending of light was 0.9 seconds of arc, which was the wrong value. A year later, Einstein revised his theory, and he increased his prediction to 1.8 seconds of arc, which is the correct value. If Freundlich had completed the measurements of the bending of light in August 1914, his result would have contradicted Einstein's prediction, such as it was at that time, and Einstein would have found himself in the embarrassing position of having to revise and correct his theory in response to observational evidence. The failure of Freundlich's expedition gave him the extra time he needed to make the correction on his own initiative and to harvest the full glory of an accurate prediction at the next eclipse of the Sun.

THE SOUND AND FURY of the war did not interrupt Einstein's scientific work in Berlin. He locked himself in his study, focused on his work, and ignored the disturbances outside his window, even when these disturbances involved mayhem and death for millions. Later in the war, he was to experience some deprivation, when the blockade of German ports took effect, and Germans suffered severe shortages of food. For many Berliners, this meant starvation—it is estimated that more than 200,000 residents of Berlin died of starvation during the war (in all of Germany, about 760,000). But for Einstein, the shortages were merely a minor inconvenience; he was kept fairly well supplied by packages of foodstuffs mailed to him from Switzerland and from Swabia by friends and relatives.

In the first months of 1915 Einstein still was not making any progress with general relativity. For the sake of variety he turned to an entirely different problem, a problem having nothing to do with gravitation, a problem that was experimental rather than theoretical. Although Einstein was the quintessential theoretician, he occasionally liked to get his hands dirty and tinker with experimental apparatus, for which he had developed a taste in his childhood, when he often visited his father's factory of electromechanical devices.

During his years in Berne, he had tried to perform an experiment to confirm the existence of electric currents within the atoms of permanent magnets. The existence of such currents had been suspected since the early nineteenth century, when the French physicist André Marie Ampère had discovered that electric currents flowing in coils of wire produced the same magnetic effects as permanent magnets. This led him to the conjecture that permanent magnets, also, must have currents flowing within them, and these hypothetical electric currents came to be called Ampèrian currents.

Einstein thought that these currents might consist of electrons performing a circulatory, orbital motion within the interior of atoms. He proposed an experimental test of this idea by measuring the "amount of magnetism" and the "amount of circulatory motion" in a magnetized piece of iron. If the magnetism of iron is the result of a circulatory motion of the electrons, then any change of magnetism must produce a corresponding change of circulatory motion. The ratio of the amount of magnetism to the amount of circulatory motion is called the gyromagnetic ratio, and for magnetism produced by orbiting electrons, Einstein calculated that the gyromagnetic ratio should equal 1, in suitable units.

In his experiment, Einstein suspended an iron rod from an untwisted, fine thread, and he used a magnet to change the magnetism of the iron, thus changing the circulatory motion of the electrons in the iron atoms. But the total amount of circulatory motion cannot change (in the absence of a twist, the total circulatory motion must remain constant, that is, the total angular momentum must remain constant), and therefore the iron rod has to compensate for the change of circulatory motion of the electrons in its interior by a bodily rotation in the opposite direction. This is rather like what a cat in free fall does to reorient its body, so as to land on its feet. The cat rotates the stretched-out ends of its hind legs very quickly in one direction, and thereby makes its body rotate in the opposite direction. In his experiments, Einstein indeed found that whenever he changed the magnetism of the rod, the rod began to rotate, giving a qualitative confirmation of his idea.

IN BERLIN, EINSTEIN wanted to repeat this experiment quantitatively. He secured the help of Johannes de Haas, Lorentz's son-in-law, who was temporarily working as an assistant at the Reichsanstalt, and they set up the experiment in a laboratory there. They initially had some trouble with the suspension of the iron rod, which "performed extremely adventurous motions," but they fixed this by replacing the thread with a glass fiber and by various other modifications of the experiment. In February, Einstein reported their experimental result to the Physical Society. He claimed that the measured gyromagnetic ratio was approximately 1.02, very close to the theoretically expected value of 1. In fact, Einstein wondered if maybe the experimental result was too good, saying, "The agreement with the theory is of course accidental; but enough of it is real, so as to silence any doubt about the correctness of the theory."[30] He wrote to Besso, "A wonderful experiment! A pity you didn't see it." In a reference to the experimental difficulties, he added, "And how treacherous is nature when one wants to snare her experimentally."[31]

But that was not the end of the story. Other physicists at other laboratories repeated the Einstein–de Haas experiment, with more patience and more care. And instead of a gyromagnetic ratio equal to 1, they found a gyromagnetic ratio equal to 2. What had gone wrong with Einstein's "wonderful" experiment? It seems that Einstein had committed a classic error of cherry-picking. He had a theory for the gyromagnetic ratio that said it should equal 1, and he picked the data to match his theory. Years later, de Haas admitted that in the experimental measurements they had actually found the value 1.02 on one occasion and 1.45 on another occasion. These discrepant results should have sent up a warning flag, and the experiment should have been repeated, to resolve the discrepancy. But Einstein, convinced that the "right" value had to be 1, decided that 1.02 was the "best" value and he discarded the other value—and only the "best" value was reported to the Physical Society. Even six years later, after several repetitions of the experiment had established the value 2 beyond reasonable doubt, Einstein stubbornly persisted in his belief that the gyromagnetic ratio was 1.

IN THE AUTUMN of 1915, while the German army launched futile assaults against the French trenches in the Champagne, Einstein launched his own final assault on his theory of general relativity. He later described his struggle to Lorentz: ". . . last autumn the gradually dawning realization of the incorrectness of the old gravitational equations gave me nasty times,"[32] and to Sommerfeld: ". . . I had one of the most exciting and strenuous times of my life; but also one of the most successful ones."[33] In November of 1915, this long struggle with the theory of general relativity finally bore fruit. In a series of four communications to the academy at weekly intervals, Einstein gave birth to the fully formed theory. All of these communications contain mistakes, except the last one. And all begin in more or less the same manner: "In a recently published communication I have shown. . . ,"[34] but now "I find that the necessity of adopting this path . . . rests on a mistake."[35] What the members of the academy thought about these once-a-week communiqués followed by once-a-week retractions is not known. Maybe most of them thought of nothing at all and simply dozed off, because the papers were all quite heavy on tensor mathematics, which only the experts of the mathematical section of the academy would have been able to understand.

Einstein would have done better to stay silent until he had completed all the details of his theory. He should have remembered a rule well-known to carpenters: Only a fool shows unfinished work. He admitted as much later, when he said, "Unfortunately, I have immortalized the last errors in this struggle in academy publications."[36] But at the beginning of November, he apparently could not contain his excitement about the progress he was making on his pet theory. In the first communication he gushed, "Hardly anyone who has truly grasped this theory will be able to escape its magic; it is a real triumph of the methods . . . of the absolute differential calculus."[37]

EINSTEIN'S FEVERISH EFFORTS of that November illustrate his mystical, intuitive approach to physics. His intuition told him he finally had found the right answer, but he still couldn't quite discern all the details. Over a period of three weeks he immersed himself in hectic labors, until the fog gradually lifted, and then the complete theory stood revealed before him. The starting point of his labors was the recognition that his first attempt at a theory of general relativity, the *Entwurf* of 1913 with Marcel Grossmann, was *almost* right. All that was lacking was the correct equation relating the curvature of the spacetime geometry to the distribution of gravitating mass that causes this curvature.

At first, Einstein repeated Grossman's gross mistake, adopting the mathematically inconsistent equation relating the Ricci tensor to the energy-momentum tensor:

$$\text{Ricci tensor} = \text{momenergy tensor}$$

But then, with more experience in tensor calculus, he recognized the inconsistency, and he tried a patchwork repair by assuming that all the gravitating mass behaves like electric and magnetic energy.[38] He quickly gave up on this cheap fix, and in the last of the four communications, on November 25, 1915, he finally found the correct equation.

On the left side of his new equation he again included the Ricci tensor, but now he added an extra term involving the curvature invariant. The final equation looks like so:[39]

$$\text{Ricci tensor} - \tfrac{1}{2} \times \text{curvature invariant} = \text{momenergy tensor}$$

This is Einstein's equation for the gravitational field. (Because the Ricci tensor on the left side and the energy-momentum tensor on the right side of this equation each have ten components, it is actually an equation with ten components, and it might be more accurate to call it the Einstein equation*s*, in plural; some physicists use the plural, some the singular . . . whatever.) Many years later, this equation, written in the concise form $R_{\alpha\beta} - \tfrac{1}{2} R g_{\alpha\beta} = \kappa T_{\alpha\beta}$, was engraved on the Einstein memorial at the Academy of Sciences in Washington, DC, and it has also appeared on postage stamps and sweatshirts, although not as often as $E = mc^2$.

From his new equation for the gravitational field, Einstein immediately calculated the details of the curved spacetime geometry surrounding the Sun, and from that he calculated how the curved geometry affects the motion of planets and the propagation of light rays. Upon repeating the calculation of the perihelion precession of Mercury he had first done with Besso's assistance three years before, he was delighted to find that his new theory gave a precession in exact agreement with the available astronomical data. He wrote to Sommerfeld, ". . . this is the most valuable discovery I have made in my life . . . The result on the motion of perihelion of Mercury gives me great satisfaction."[40] And he later told Ehrenfest, "For several days I was beside myself with joyful excitement."[41]

Einstein also recalculated the bending of light by the Sun, and found a

result twice as large as what he had predicted in 1911. And he also calculated the bending of light by Jupiter, finding it to be about 1/100 of that by
the Sun. His astronomer friend Freundlich thought this might be measurable, but all his attempts proved unsuccessful, and the bending of light by
Jupiter was not confirmed in Einstein's lifetime.

THERE IS A LENGTHY CONTROVERSY connected with
Einstein's equation for the gravitational field. Einstein first presented the
equation in the session of the academy on November 25. But five days
before, on November 20, the celebrated Göttingen mathematician David
Hilbert had already presented this selfsame equation at a session of the
Royal Society in Göttingen, stealing a march on Einstein. This has led to
an earnest dispute over priority: Does Einstein deserve the credit for discovering this equation or does Hilbert? Needless to say, each of the principals claimed that the equation was his, and each trumpeted its beauty.

In the printed version of his paper, Hilbert suggested that with his equation he had completed the theory that Einstein had only sketched: "The
differential equations for gravitation thus arrived at are, it seems to me, in
accord with the theory of general relativity established in broad outlines by
Einstein in his later publications."[42] And in his letter to Einstein he wrote
of his theory, somewhat incoherently: "I consider [it] math[ematically]
ideally beautiful and also inasmuch as it contains no calculations that
are not completely transparent and absolutely compelled by axiom[atic]
meth[ods], and trust therefore in its reality."[43]

Einstein complained to Zangger that Hilbert was trying to expropriate,
or "nostrify," his theory: "The theory is of incomparable beauty. But only
one of my colleagues [Hilbert] has understood it thoroughly and that one
seeks to 'nostrify' it in a clever way . . . Never in my personal experience
have I come to know human wretchedness better than in regard to this
theory and its connections."[44]

Among physicists and historians of science, it became widely accepted
that Hilbert discovered the equation first, and that Einstein discovered it
independently, a few days later. With Solomonic wisdom, even Einstein's
friend and biographer Abraham Pais proposed that the credit be shared,
so ". . . both he and Hilbert should be credited with the discovery of the
fundamental equation."[45]

The dispute became acrimonious when some of Einstein's supporters accused Hilbert of plagiarism, and Einstein's opponents returned the favor, accusing *him* of plagiarism, sometimes in violent and intemperate language.[46] It is not possible to decide what each of the principals knew and when he knew it by merely examining the printed reports of the sessions of November 20 in Göttingen and November 25 in Berlin. These final printed versions of Hilbert's and Einstein's papers do not necessarily coincide with what was presented orally on those days. It was, and is, not at all unusual for the lecturers at the sessions of scientific societies to hand the printer somewhat revised, improved versions of their speeches.

To settle the matter of priority and plagiarism, we need to examine the circumstances leading up to the sessions of November 20 and November 25. Hilbert's involvement with general relativity began in June 1915, when Einstein visited Göttingen and gave a series of colloquia on his theory of gravitation, with generous financial support from the Wolfskehl foundation. He and Hilbert discussed the remaining troubles of his theory at length. Einstein came away from this visit with a highly favorable impression of Hilbert: "I was in Göttingen for a week, and there got to know and love him. I held six two-hour long lectures on the now well-developed theory of gravitation and, to my joy, I completely convinced the mathematicians there."[47]

During the next months, Hilbert thought some more about Einstein's theory, and he soon found an elegant mathematical formulation for it. He wrote to Einstein, who asked for a copy of Hilbert's notes with his calculations. Einstein apparently received these notes sometime before November 18, because on that date he replied to Hilbert, saying, "The system [of equations] established by you is in complete agreement—as far as I can see—with what I have found in recent weeks and presented to the academy."[48] Presumably, this "system of equations" included the equation that is at the center of the controversy. And if Einstein received this equation from Hilbert before November 18—that is, before his own presentation of this equation on November 25—he is indeed under vehement suspicion of plagiarism.

In the spirit of tit for tat, Einstein's supporters made the contrary accusation. They conjectured that neither the notes that Hilbert sent to Einstein nor his presentation on November 20 actually contained the equations, and that he inserted these equations into the proofs at a later time, after reading Einstein's printed paper. They base this imaginative conjecture on a set of printer's proofs found in 1997 in the Göttingen library, with handwritten alterations by Hilbert. These proofs differ somewhat from the final published paper, and Einstein's supporters were delighted to announce that ". . . the first set of proofs of Hilbert's paper . . . does not include the explicit form of the field equations of general relativity."[49] But this is disingenuous: the reason why the proofs do not contain the equations is that half a page is torn off from the set of proofs, exactly at the place where the equations would have been. This, of course, suggests conspiracy theories aplenty: Did an Einstein supporter mutilate the proof to hide the presence of the equation? Or did a Hilbert supporter do the dirty deed to hide the *absence* of the equation?

IN FACT, THE MISSING HALF PAGE is a red herring. In the proofs, Hilbert states the equation in an implicit, abstract form in the discussion of his action principle at the beginning of this paper, and any moderately competent mathematician can deduce the explicit form of the equation from that.[50] The claim that Hilbert's proofs do not include the equation is misleading, and it is also irrelevant.

In their eagerness to award or deny priority, the champions of each party overlooked one crucial detail in Hilbert's papers (both the papers that exist as proof and the final, printed paper): Hilbert's equation *looks* like Einstein's final equation, but this is merely a matter of appearances. In its substance, Hilbert's equation actually is exactly Einstein's *previous* equation of early November. Hilbert obtains an equation that looks like Einstein's new equation,

Ricci tensor − ½ × curvature invariant = momenergy tensor

But in Hilbert's scheme of things, the curvature invariant is zero, and therefore his equation really is

Ricci tensor − 0 = momenergy tensor

which is exactly the equation that Einstein had attempted to use in early November and then discarded. Thus, Hilbert gives us what looks like a new dog-and-pony show, but in reality the new pony is merely the old nag that Einstein had ridden into exhaustion several weeks before, wrapped in an elegantly embroidered caparison.[51]

So Einstein was absolutely right in replying to Hilbert that his equations were ". . . in complete agreement—as far as I can see—with what I have found in recent weeks and presented to the academy." It is of course possible that Hilbert's notes inspired Einstein to think some more about the field equation and to make the required final modification. But that does not affect the issue of priority: Einstein got to the final equation first, whereas Hilbert never got there at all—he remained stuck on Einstein's previous equation. Hilbert lost the race for the same reason that Moitessier lost the 1969 *Sunday Times* sailboat race. Both Hilbert and Moitessier were ahead of the competition, but, just before the finish line, they decided to go off on tangents, pursuing dreams of their own—a misguided unified theory of gravitation and electricity for Hilbert, and a cruise to Tahiti for Moitessier. Hilbert was fixated on an ambitious attempt to explain the foundations of all of physics by electromagnetic and gravitational fields. He could have been the first to obtain the full, final Einstein equation, but he decided to try something more audacious, at which he failed. *Fortuna audentes juvat . . .* but not always.

This is not to say that Hilbert's work was worthless. It reformulated Einstein's theory in the elegant and powerful mathematical language of the calculus of variations, and it established that Einstein's equations could be derived from an extremum principle: Einstein's equation corresponds to an extremum (that is, a minimum or a maximum) of the curvature invariant, which is, in essence, an average of all the curvatures contained in the Riemann tensor. Accordingly, the English mathematician Edmund Whittaker later was to say, "Gravitation simply represents a continual effort of the universe to straighten itself out."[52] Lorentz soon applied Hilbert's method to forms of matter other than electromagnetic, and he thereby derived Einstein's final equation in an elegant way. But this did not happen until several months after Einstein's publication of his final equation, and does not affect the priority debate.

THE CLAIMS AND COUNTERCLAIMS in Einstein's and
Hilbert's publications left some bad feeling between them, and a year later Einstein made a peace offering in a letter to Hilbert: "There has been between us somewhat of an ill mood, the cause of which I do not wish to analyze. I have fought against the related feeling of bitterness, and in this I have achieved complete success. I now think of you again in unclouded friendship, and I beg you to attempt the same in regard to me. Objectively,

it is a pity if two good fellows, who have achieved something significant in this shabby world, do not achieve joy in each other."[53]

Apparently Hilbert accepted this offer of continuing friendship, and he did not push his claims of priority. According to one of Einstein's assistants, Hilbert even apologized in writing to Einstein, but the letter appears to be lost.[54] Maybe he recognized that his treatment of the gravitational problem had not gone beyond Einstein's previous attempt. But maybe he just did not attach all that much importance to the gravitational field equations. For Hilbert, finding the gravitational field equations was a trivial matter—he had found the equations almost immediately, and so did not consider this much of an achievement.

In attempting to formulate a unified theory of gravitation and electromagnetism he was after bigger game, and that was the real objective of his November 25 paper, which he had grandiosely entitled "The Foundations of Physics." He thought he was creating a complete theory of all of physics, in which gravitation was supposed to explain electromagnetism and that, in turn, was supposed to explain ordinary matter, with all particles treated as entities consisting of gravitational and electromagnetic fields. In Hilbert's paper, the field equation for gravitation appears only in passing, while Hilbert goes on to this more ambitious target of a unified theory of all of physics.

However, Hilbert's mathematical virtuosity was not matched by a clear grasp of physics. His attempt to extract electromagnetism from gravitation arose from a misconception about the physical role of the energy-momentum tensor, and his grandiose scheme was a delusion. He continued to publish papers on this subject for several years, and those finally sank into obscurity. Today nobody remembers his work in general relativity, except for his introduction of the elegant technique of variational calculus.

Hilbert praised Einstein, "Every boy in the streets of Göttingen understands more about four-dimensional geometry than Einstein. Yet, in spite of that, Einstein did the work and not the mathematicians."[55] Nevertheless, Hilbert did not trust physicists. "Actually," he said, "physics is far too difficult for physicists."[56] This may be true, but the history of physics shows it is even more difficult for mathematicians.

ALL'S WELL THAT ENDS WELL. For three years Einstein had bumbled through several wrong versions of his theory of general relativity, each announced with great enthusiasm and naïveté as the answer to

his quest to formulate a geometric interpretation of gravitation. In the end he stumbled on the right one. Maybe he did not fully understand the conceptual foundations and mathematical complexities of his theory, but neither did anybody else at that time—some crucial aspects of general relativity were not understood until half a century later, and the study of all aspects of the theory continues to this day. But in November 1915 Einstein had found a pearl, and he knew it, and so did the most discerning of his contemporaries.

Einstein was very aware that his theory of general relativity implied the complete overthrow of Newton's ideas about mechanics. Special relativity had modified Newton's ideas about space and time, but it had kept intact his basic scheme of mechanics, with force playing a fundamental role as the cause of acceleration. In special relativity, Newton's Second Law is not discarded, it is merely modified. But in general relativity, the cause of gravitational accelerations is not a force, but the curvature of spacetime—mass tells spacetime how to curve, and curved spacetime tells mass how to move. In his autobiography, Einstein included an apology to Newton: "Newton, forgive me; you found the only way that, in your times, was just about possible for a man with the highest powers of thought and creativity. The concepts you created still guide us today in our thinking in physics, although we now know that they have to be replaced by others, more remote from the realm of immediate experience."[57]

Arthur Eddington, who was widely known as the leading expert on general relativity and who understood some of it better than Einstein himself, declared, "By his theory of relativity, Albert Einstein has provoked a revolution of thought in physical science," and he added, "Einstein stands above his contemporaries even as Newton did."[58]

Ernest Rutherford, the discoverer of the atomic nucleus, chimed in, "The theory of relativity of Einstein, quite apart from is validity, cannot but be regarded as a magnificient work of art."[59] Although this comment is nicely put, it is somewhat doubtful that Rutherford knew exactly what he was talking about. He was a brilliant experimenter, but the mathematics of general relativity was probably beyond his reach. In contrast to Eddington, Rutherford never published anything having to do with relativity, not even special relativity. When a German physicist later told Rutherford that "no Anglo-Saxon could understand relativity," he concurred and replied, "No, they have too much sense."[60]

And Hermann Weyl described general relativity as "one of the greatest examples of the power of speculative thought."[61] His emphasis on the speculative side of Einstein's theory was on target. Whereas special relativity is firmly rooted in the experimental investigations of the speed of light by Michelson-Morley, general relativity emerged from Einstein's purely theoretical and somewhat misguided speculations about a possible relativity of acceleration.

That Einstein's notion of a relativity of acceleration and his Principle of Equivalence ultimately proved will-o'-the-wisps does not detract from Einstein's achievement, but, in a way, enhances it. Only a great genius (or a madman) could succeed in the construction of a correct theory on such a misbegotten basis. Einstein's profound intuition permitted him to extract from his defective notions the building blocks he needed to proceed to the completion of his greatest work of art. It was sleepwalking, à la Koestler, but sleepwalking of the highest order. But the murky basis of general relativity was to prove a persistent source of confusion for students of relativity. As the highly respected English theoretical physicist James Jeans put it, "Einstein's theory may meet with an unfavorable reception on account of the somewhat metaphysical—one might almost say mystical—form in which its results have been expressed," and he emphasized that "the more concrete part of Einstein's work is quite independent of the metaphysical garment in which it has been clad."[62]

The early encomiums of relativity were confirmed by later writers. Some of the highest praise comes from three winners of the Nobel Prize in Quantum Theory: Paul Dirac, ". . . probably the greatest scientific discovery ever made"; Max Born, "The foundation of general relativity appeared to me then, and still does as the greatest feat of human thinking about Nature, the most amazing combination of philosophical penetration, physical intuition, and mathematical skill"; and Lev Landau, ". . . represents probably the most beautiful of all existing physical theories."[63]

But beauty is in the eye of the beholder, and Einstein's theory was beautiful only to those who understood it. To many outsiders, it looked like a mess, an orgy of mathematical formalism. The equations on page 216 seem deceptively simple when written in terms of the Ricci tensor; but for actual calculations, they have to be rewritten in terms of the metric tensor, and then the equations become much more messy. For instance, even for the relatively simple case of a metric tensor with only four nonzero components (that is, a metric tensor in which six of the usual ten components are zero), the Einstein equations fill three large book pages, in small, tight print.[64] Today, physicists often use computer programs to write the Einstein equations in terms of the metric tensor, and they also use computer programs to solve the equations.

If Einstein's theory is more beautiful than Newton's theory, it is also more complicated. A witty wordsmith felt compelled to add two lines to Alexander Pope's epitaph in praise of Newton:

> *Nature and Nature's law hid in night.*
> *God said "Let Newton be" and all was light,*
> *It did not last: the Devil howling "Ho!*
> *Let Einstein be!" restored the status quo.*[65]

IN MAY OF 1916, Einstein published a long explanatory paper in *Annalen* on "The Foundations of General Relativity Theory."[66] This was intended as a complete, coherent introduction to the theory that he had presented in bits and pieces in several papers in the preceding year. It followed the pattern of the Einstein-Grossmann *Entwurf* paper, in that it contains separate physical and mathematical parts. The mathematical parts are obviously largely lifted from Grossmann, but Grossmann's name is not mentioned. The only references mentioned in the paper are one brief reference to Hilbert and an equally brief reference to Einstein's calculation of the perihelion precession. The mathematical part is a self-contained, concise primer on tensor calculus, which even today is helpful for beginning students. Einstein initially had much trouble with tensor calculus, and he anticipated that other physicists might have the same trouble. The mathematical part is Einstein's expository writing at its best.

The same is not true of the physical part. It begins with a long, muddled discussion of why the relativity of velocity, which is at the core of special relativity, should be extended to a more general relativity that embraces the relativity of acceleration and why this general relativity should involve a curved spacetime geometry. Einstein gives a long list of arguments for all of this, reminiscent of Cyrano de Bergerac's long list of methods for traveling to the Moon. Cyrano's list was inane, and Einstein's is not; but most of his arguments suffer from mistakes of one sort or another.

HIS FIRST ARGUMENT draws on an idea of Ernst Mach, the Viennese professor famous for his critical exposé of the logical foundations of Newtonian mechanics, who had died a few months earlier. Einstein had written his obituary, praising Mach's insights into the defects of Newton's concepts of space and time and claiming that Mach "was not far from demanding a theory of general relativity [of acceleration], as long as half a century ago!"[67] Mach had faulted Newton's theory for its failure to explain the inertial effects that we feel when we are subjected to an acceleration, for instance, while sitting in a train that suddenly speeds up or slows down or goes around a sharp curve. According to Newton, such inertial effects are associated with our acceleration relative to "absolute" space, but Mach objected to this, because space is merely nothingness, and it seems nonsensical to reckon our acceleration relative to nothingness; he thought that inertial effects would disappear if the universe were empty. Mach proposed that we should reckon the acceleration relative to the average distribution of all the other masses that make up the universe— planets, stars, galaxies, clouds of interstellar gas, etc.—and that inertial effects should arise from our acceleration relative to these distant masses. Einstein adopted Mach's idea and argued that if acceleration is reckoned relative to masses rather than relative to space, then we need a theory of relativity of acceleration, that is, a theory of general relativity.

The trouble with Mach's idea is that it is purely metaphysical—we cannot remove the masses in the universe to test whether this eliminates inertial effects. Furthermore, although Einstein used Mach's idea to motivate his theory of general relativity, his theory doesn't actually exhibit an explicit dependence of inertia on the mass distribution in the universe. And according to Einstein's own Principle of Equivalence, the inertial effects in a freely falling box are completely independent of any surrounding masses, near or distant.[68]

EINSTEIN'S SECOND ARGUMENT relies on his Principle of Equivalence, according to which all experimental results obtained by a physicist enclosed in an accelerated box in empty space are identical to the results obtained in a box that sits at rest in the gravitational field of some planet. Because of this interchangeability of the effects of acceleration and gravitation, argued Einstein, the physicist in the box cannot identify the acceleration absolutely, but only relative to some external landmark. And "the implementation of the general theory of relativity [for velocity and for acceleration] leads directly to a theory of gravitation; because we can 'produce' a gravitational field by a mere change of the coordinate system."[69]

It should have given Einstein pause that his 1911 calculation of the bending of rays of light, which was based on the Principle of Equivalence, yielded a result *half* as large as the new calculation based on his new theory of gravitation. Einstein understood that the reason for this discrepancy was that the new calculation included an extra deflection coming from the warping of space, whereas the 1911 calculation had effectively included only the warping of time. But he gave no thought to the implications of this discrepancy for the Principle of Equivalence. For the physicist enclosed in a box, the implication is clear: the bending of a ray of light in an accelerated box is half as large as the bending in a box at rest in a gravitational field. This means that acceleration and gravitation are not interchangeable—and the Principle of Equivalence fails!

Admittedly, the bending of a light ray that travels from one side of a small box to the other is quite small and practically undetectable. But in principle it is possible to bounce a light ray back and forth between the sides of the box many times, so it accumulates a detectable amount of bending.

What is more, a criticism of the Principle of Equivalence had already been published three years earlier, in 1913. Einstein's friend and colleague Ehrenfest, who had just succeeded Lorentz as a professor at Leiden and who had an exceptional talent for spotting the mistakes in the theories of others, had published a short paper presenting a general proof about the failure of the Principle of Equivalence for the propagation of light.[70] Ehrenfest relied on the obvious fact that in a steady gravitational field, such as the gravitational field of a planet, light rays can be sent repeatedly from one point to another by the same path, and they can also be sent back from the second point to the first along that same path (that is, light signals are repeatable and reversible). He then asked the question, Is this also true in an accelerated box? By a brief and elegant argument, he proved that this is almost never true in an accelerated box, not even if we permit different parts of the box to have different accelerations (it is true only if the original gravitational field is artificially tuned to make it true). Einstein choose to ignore Ehrenfest's result.

And there are other ways to defeat the Principle of Equivalence. One consequence of the Principle of Equivalence is that a physicist in a box in free fall in a gravitational field can't sense the gravitational field—supposedly the effects of the downward acceleration exactly cancel the effects of gravitation. But in fact, the physicist can sense the presence of the gravitational field by performing careful measurements on the motions of small masses in the box.[71] One simple experiment that the physicist might perform is to observe the relative motion of several small masses floating freely in the box. For instance, while the box is in free fall in the gravitational field of the Earth, the physicist might release a coin near the bottom of the box and another coin near the top. The coin at the bottom is nearest the Earth, and it therefore experiences a larger gravitational attraction and a slightly larger downward acceleration than the coin at the top, whereas the freely falling box experiences an intermediate acceleration, slightly smaller than the acceleration of the bottom coin, but slightly larger than the top coin. The net result is that relative to the box, the bottom coin accelerates downward, whereas the top coin accelerates *upward*. By observing this motion of recession of the coins, the physicist can tell that she is indeed in a gravitational field—if she were floating in empty interstellar space, the two floating coins would simply remain at rest.

The relative acceleration of two neighboring particles gives us a quantitative measure of the strength of gravitation, and this relative acceleration can be regarded playing the same role in Einstein's gravitational dynamics as the electric field plays in Maxwell's electrodynamics. In the words of the English physicist and cosmologist Hermann Bondi: "Thus everywhere an observable of gravitation may be measured through the fact that, although different particles fall equally fast if they are at the same place, there is a difference in their accelerations if they are in different places, even if these are close by. *So the universal observable of gravitation is the relative acceleration of neighbouring particles . . . A gravitational field is a relative acceleration of neighbouring particles.*"[72]

Experiments on such a relative acceleration have been performed in artificial satellites orbiting the Earth (and therefore in free fall). Instead of depositing coins in the satellite, the experimenters installed sensitive electronic accelerometers at opposite ends of the satellites, and they found that these accelerometers did indeed indicate accelerations in opposite directions, whereas in a truly inertial reference frame in interstellar space, they would have indicated zero acceleration. The excess accelerations observed in such satellite tests were about a millionth of the normal acceleration of gravity, small but easily detectable (for instance, the accelerometers installed on the GOCE satellite of the European Space Agency can detect accelerations 10 million times *smaller* than that).[73]

The relative motion of the two coins in the box is called a tidal effect, because the tides of the oceans of the Earth arise from a mechanism of this kind. The Earth can be regarded as a "box" in free fall in the gravitational field of the Sun, Moon, and other planets. Because of this free-fall motion, we do not experience the gravitational pulls of these celestial bodies in any obvious way. (The pull of the Sun on your body is about as much as the weight of an egg. This is small, but it would be detectable with an accurate balance, if it were not hidden by the free fall of the Earth.)

However, the free fall of the Earth does not completely hide the gravitational pulls of the Sun and the Moon. A small mass on the side of the Earth nearest the Moon has a slightly larger free-fall acceleration toward the Moon than a small mass on the side of the Earth farthest from the Moon, while the body of the Earth has an intermediate acceleration. If the small masses are coins lying on the ground, they will not move relative to the Earth, because the gravity of the Earth holds them firmly against the ground. But if the small masses are parcels of ocean water, they will move and pile up above the normal sea level. The water on the side of the Earth nearest the Moon has an excessive free-fall acceleration; it gets ahead of the Earth and rises toward the Moon. Whereas the water on the side of the Earth farthest from the Moon has an insufficient free-fall acceleration; it lags behind and rises away from the Moon. This gives the ocean water tidal bulges on opposite sides of the Earth. These bulges stay in a fixed orientation relative to the Moon while the Earth rotates. Hence the tidal bulges travel over the surface of the rotating Earth, and when they intercept the seashores, they produce two high tides per day.

Although Einstein was an avid sailor and sailboats were his favorite hobby, he never recognized that the ocean tides are an obvious counterexample to his Principle of Equivalence. Of course, when sailing on the Swiss lakes and on the lakes near Berlin he would not have seen any tides. But he certainly saw them later, when he sailed on the North Sea, at Kiel, and on the coastal waters of New Jersey and on Long Island Sound. The eye sees what the mind has prepared it to see (Segré)—and Einstein's faith in the Principle of Equivalence made him blind to the tides.

There is an odd coincidence here: Galileo, Newton, and Einstein all came to grief on the tides. Galileo contrived a wrong theory of the tides; Newton had the right theory, but botched the quantitative calculation of the height of the tides; and Einstein failed to recognize that the tides undermine the Principle of Equivalence that he adopted as the cornerstone of his theory of general relativity. The three greatest physicists ever known all suffered from tidal dysfunction.

EINSTEIN'S THEORY OF GENERAL RELATIVITY not only includes the tidal effects of Newton's theory of gravitation but also extra, relativistic tidal effects. In fact, the curvatures contained in the various components of the Riemann curvature tensor correspond to various tidal effects. We can detect these curvatures by measurement of tidal effects or, alternatively, by measurements of lengths or areas or volumes in space-time. The existence of these curvatures is in clear conflict with the relativity of acceleration and the Principle of Equivalence. In essence, general relativity is a theory of the curvature of spacetime—and because of that it cannot accommodate the Principle of Equivalence, except in an approximate way.

That the Principle of Equivalence is, at best, an approximation, was recognized not only by Ehrenfest, but also by other physicists. Thus, August Kopff, an astronomer at Heidelberg, mentioned it in a lecture in 1920.[74] Two years later, the defects of this principle were fully dissected by Arthur Eddington, the Cambridge astronomer who had also recognized the mistake in Einstein's claim that the constant speed of light in special relativity is merely a stipulation (see Chapter 4). In 1922, Eddington published one of the first textbooks of general relativity, with a systematic exposition of the mathematics and also of the conceptual foundations of the theory.[75] This book included an incisive critical assessment of the Principle of Equivalence:

It is essentially a hypothesis to be tested by experiment as opportunity offers. Moreover, it is to be regarded as a suggestion, rather than a dogma admitting of no exceptions. It is likely that some of the phenomena will be determined by comparatively simple equations in which the components of the curvature of the world do not appear; such equations will be the same for a curved region as for a flat region. It is to these that the Principle of Equivalence applies. It is a plausible suggestion that the undisturbed motion of a [single] particle and the

propagation of light are governed by laws of this specially simple type
. . . But there are more complex phenomena governed by equations in
which the curvatures of the world are involved . . . Clearly there must
be some phenomena of this kind which discriminate between a flat
world and a curved world; otherwise we would have no knowledge of
world-curvature. For these the Principle of Equivalence breaks down
. . . There can be no infallible rule for generalising experimental
laws; but the Principle of Equivalence offers a suggestion for trial,
which may be expected to succeed sometimes, and fail sometimes.

The Principle of Equivalence has played a great part as a guide in
the original building up of the generalized relativity theory; but now
that we have reached the new view of the nature of the world it has
become less necessary.[76]

Eddington's views were echoed in 1960 by J. L. Synge, a respected
expert on relativity and brother of the Irish poet J. M. Synge. Perhaps
inspired by his poetic brother, J. L. expressed his opinion about the Principle of Equivalence in blunt and colorful language rarely found in physics textbooks:

I have never been able to understand this Principle . . . Does it mean
that the effects of a gravitational field are indistinguishable from the
effects of an observer's acceleration? If so, it is false. In Einstein's
theory either there is a gravitational field or there is none, according
as the Riemann tensor does or does not vanish. This is an absolute
property; it has nothing to do with any observer's word-line . . . The
Principle of Equivalence performed the essential role of midwife at
the birth of general relativity, but, as Einstein remarked, the infant
would never have gone beyond its long-clothes had it not been for
Minkowski's concept [of spacetime]. I suggest that the midwife be
now buried with appropriate honours and the facts of absolute spacetime faced.[77]

Equally bluntly but less colorfully, Vladimir Fock, a Soviet-era academi-
cian and professor at the University of Leningrad (now re-renamed St.
Petersburg, *slava bogu*), wrote that "the terms 'general relativity' . . . or 'general principle of relativity' should not be admitted. This usage . . . reflects
an incorrect understanding of the theory itself. However paradoxical it may

seem, Einstein, himself the author of the theory, showed such a lack of understanding when he named his theory." And he continued, "We call the theory of Einstein space the Theory of Gravitation, not the 'general theory of relativity,' because the latter name is nonsensical."[78] Fock wrote this in 1962, when Soviet scientists were required to pay lip service to communist political correctness, so he felt compelled to add, "The teachings of dialectical materialism helped us to approach critically Einstein's point of view concerning the theory created by him and to think it out anew." That bit of bull probably earned him an Order of Lenin—and perhaps a new apartment.

American authors have been somewhat more restrained in criticizing Einstein for the name he gave to his theory. John Wheeler, one of the greatest teachers and exponents of the theory, preferred to call it *geometrodynamics* (in analogy to *electrodynamics*) to indicate that the essence of the theory is in its dynamic geometry, varying over space and time. But, as a thoroughly polite gentleman of the old school, he refrained from criticizing Einstein.

ANOTHER OF EINSTEIN'S arguments was based on an odd fact about length measurements on a merry-go-round, that is, a large rotating turntable. Imagine that such a turntable is rotating very fast (much faster than any speed that has ever been attained in practice), and that you measure the circumference of the turntable by laying a sufficiently large number of short rulers along its circumference, end to end. Because the rulers are all moving at high speed, they will all be contracted relative to rulers at rest on the ground, and therefore it takes more rulers to complete the circumference of the rotating turntable than for a nonrotating turntable, that is, the circumference of the rotating turntable will appear to be *longer* than normal. For a nonrotating turntable, or for a nonrotating circle, the circumference is $2\pi \times$ radius, and for a rotating turntable the circumference will therefore appear to be *more* than $2\pi \times$ radius. This contradicts ordinary geometry, or Euclidean geometry, and Einstein took this to mean that on the rotating turntable, Euclidean geometry is not valid, that is, the geometry is not that of a flat space but that of a curved space.

The rulers on the rotating turntable have a centripetal acceleration because of their circular motion, and Einstein attributed the curved geometry on the turntable to this acceleration. He then argued that if acceleration produces a curvature of the geometry, then gravitation should also produce a curvature of the geometry, because the Principle of Equivalence demands that acceleration and gravitation produce similar effects.

The puzzling features of the geometry measured by rulers on a rotating turntable had been discussed in the physics literature since the early days of special relativity, and the puzzle was called the Ehrenfest paradox. Many physicists besides Einstein had come to believe that the geometry on a rotating turntable is indeed curved. But this was a delusion. The turntable is, in essence, a rotating coordinate system. The space that surrounds this coordinate system is unquestionably flat before the coordinate system begins to rotate, and it cannot become curved merely because we choose to adopt a rotating coordinate system. Space is either flat or it is curved, and this is an objective property of space, which cannot depend on how we select our coordinates.

The basic error of Einstein's and of his contemporaries was that they failed to recognize that the use of accelerated rulers is an improper way to measure length. A "good" measurement of length requires that all the rulers be in the same inertial reference system, and that all have the same velocity. Rulers placed along the circumference of the turntable have different velocities (for instance, two rulers on opposite ends of the turntable move in opposite directions), and it is not legitimate to combine the lengths measured by such incompatible rulers—it is like adding apples and oranges, which gives you neither apples nor oranges but fruit salad.

EINSTEIN ALSO ARGUED that the coordinates in a curved space or a curved spacetime cannot be constructed by the methods that a surveyor would use in a flat spacetime, and that in such a curved spacetime there are no preferred coordinates that lead to a particularly simple formulation of the laws of nature. For instance, on the surface of the Earth, we can use latitude and longitude or some other coordinates to specify the position of, say, Berlin or San Francisco. But regardless of how we select our coordinates, we won't have a simple way to calculate the "straight," or airline, distance between these cities—the Pythagorean theorem is not valid on the curved surface of the Earth, and calculations of distances are much more complicated than on flat surfaces. Anybody with a vague recollection of high school geometry can calculate the distance between the top corner of a flat piece of paper and a point 3 centimeters down and 4 centimenters to the right, but to calculate distances from latitude and longitude on a curved spherical surface requires extra training in math. Because of such complications, Einstein concluded that in a curved space "there is nothing for it but to regard all imaginable systems of coordinates, on principle, as equally suitable for the description of nature" and he therefore laid down the requirement that "The general laws of nature are to be expressed by equations which hold good for all systems of co-ordinates . . ."[79]

Einstein used this requirement as a criterion to decide what equations are and are not acceptable in general relativity. But, as the German physicist Erich Kretschmann pointed out some years later, Einstein did not practice what he preached. The requirement that a law of nature must be expressed by equations that hold good in all systems of coordinates is no more than a bureaucratic formality, devoid of any physical significance. Any equation that pretends to govern physical phenomena will necessarily hold good in all systems of coordinates. Laws of nature relate phenomena—for instance, if you push some numbered buttons on your phone in San Francisco, a phone will ring in Berlin 5 seconds later—and the coordinates used to describe this don't really matter. The coordinates are merely an accounting procedure for describing phenomena, and it is trivially true that any law of nature can be expressed by any accounting procedure whatsoever.

The criterion that Einstein actually used in practice was stronger. He required not only that the laws of nature be expressed by equations that hold good for all systems of coordinates, but that the metric tensor of spacetime, and *only* the metric tensor, was to be used to adapt the equations to different systems of coordinates. Einstein's failure to spell out the exact meaning of his criterion for selecting the equations of general relativity caused a great deal of confusion, and even led to accusations that Einstein himself did not understand this criterion. Even fifty years later, experts were still arguing about the meaning of Einstein's criterion for the role of coordinates in the equations of general relativity.[80]

Einstein must have had some inkling that none of his arguments was compelling. After presenting these arguments, he conceded that he had no simple logical basis for general relativity, and he appealed to the psychology of the reader: "In this treatise it is not my goal to represent the general relativity theory as the simplest possible logical system with a minimum of axioms. Instead, it is my main goal to develop this theory in such a manner that the reader will feel the psychological naturalness of the path that is being followed and that the underlying assumptions will seem supported as much as possible by experience."[81]

He then proceeded to develop his theory on the basis that gravitation is to be interpreted geometrically, so the metric tensor of spacetime represents not only the spacetime geometry but also the gravitational field.

IN OCTOBER 1918, just a few weeks before Germany surrendered to the demands of the Allies, Mileva surrendered to Einstein's demands for divorce. She initiated divorce proceedings in Zurich, on the grounds of adultery. Einstein readily admitted to the adultery in an affidavit filed with a court in Berlin, and the divorce was granted four months later. The divorce decree of the Zurich District Court required Einstein to pay alimony and, besides, to give Mileva the full amount of any Nobel Prize money that might be awarded to him. The divorce decree also forbade him to remarry for two years, a condition routinely imposed on the guilty party in Swiss divorces.

It is not clear who hit upon the slick idea that the divorce should be financed out of the not-yet-awarded Nobel Prize. Probably the idea came from Einstein, who, in financial matters was rather more wily than might be expected of the resident of an ivory tower. We know he was a shrewd negotiator in matters of salaries and fees for lectures and guest professorships. We also know that he had a secret bank account in the Netherlands, where he deposited royalties from his books and his patents, playing a game of hide-and-seek with the German tax collectors. And we know that when he moved to Princeton in 1934, he extracted a princely salary from the Institute for Advanced Study. When he retired from the institute in 1946, he refused to accept the customary (and contractual) pension at half salary, instead demanding, and getting, a pension at full salary by threatening to leave Princeton and to publicize his complaints about how badly he was being treated.

By 1918, Einstein felt sure that his Nobel Prize was just a matter of time. He had received multiple nominations each year, and he must have reckoned that the Nobel committee found itself in the role of Buridan's ass—they had such a long list of Einstein's achievements before them, that they couldn't quite make up their minds where to begin.* Of course the deal with Mileva had to be kept quiet; the Nobel committee would have thought it offensive that Einstein was making deals with the prize money in anticipation, especially under the scandalous circumstances of a divorce being negotiated while the guilty party persisted in the practice of flagrant adultery (in 1911, the committee had almost kept Marie Curie out of the award ceremony in Stockholm for less offensive adulterous behavior than that). Both parties in the Einstein divorce had an interest in keeping the deal quiet, and the details of the arrangement only emerged many years later. An enterprising newspaper reporter could have dug the details out of the records of court proceedings at the District Court in Zurich, where the divorce decree was (and is) available to public inspection; but apparently nobody looked for this juicy bit.[82]

At first sight, Einstein's grant of his Nobel Prize money to Mileva creates an impression of noble generosity. But in reality only 50 percent of this generosity was ever implemented. After he received the prize money, he gave half to Mileva (in two installments, separated by several years), but he never gave her the other half. Under the terms of the divorce decree, he was supposed to deposit that in a Swiss bank. Instead, he transferred it to an investment company in New York, beyond the reach of the Swiss courts. He paid alimony out of that, until he lost the investment in the stock market crash of 1929. Legally, this did not release him from the obligation of restoring the money to Mileva, but he never did, and his alimony payments became intermittent.[83]

During the first years following his divorce, Einstein maintained some contact with his sons, visiting them in Zurich or inviting them to join him for vacations in Germany. But when his oldest son, Hans Albert, vehemently objected to Einstein's refusal to hand over the full amount of the Nobel Prize money, their relationship became bumpy. When Hans decided to get married, Einstein tried to block that, but Hans was just as stubborn in making his own choice of spouse as Einstein had been in his choice of Mileva. Hans and Einstein later became reconciled, after Elsa's death in Princeton in 1936.

The younger son, Eduard, or "Tete," was sickly as a teenager, and he developed schizophrenia as a young man, apparently in consequence of a disturbing love affair with an older woman. Mileva tried to take care of him in her apartment in Zurich, but he had to be confined in the nearby Burghölzli mental institution, first intermittently and then permanently when she became too ill to cope. After Mileva's death in 1948, Tete was placed with foster families and then again confined until his own death in 1965.

Einstein made a quick visit to Tete at the Burghölzli in 1933, before leaving for the United States. After that, he broke off all contact and sent no letters; he claimed that he was under the sway of "an inhibition, which I am incapable of analyzing completely."[84] In his will Einstein left some money for Tete's keep, but is was too little, too late, and Tete was lodged in a dormitory as a "third-class" patient.

Tete had an oversensitive and poetic nature, and he had his father's musical talent. He passionately loved to play the piano, especially Chopin, but instead he was compelled to work in the Burghölzli fields. A Swiss journalist who visited him two years before his death described him: "He was dressed in a blue smock and wooden shoes, because he had worked in the fields. He looked rather solid and pale, with a large nose, and he resembled his gifted father so much that I found it scary. What was most beautiful about him were his eyes, very large, deep, luminous eyes of a child, and he looked at us just like Albert Einstein looks at us from his pictures." Tete meekly told the journalist that "he would have liked to play the piano, but the playing disturbs the other inmates, and he understands that. He did not like working in the fields, but he understands it is good for him. He would like to sleep by himself, but he understands that this is impossible."[85]

AFTER MILEVA'S PRECIPITOUS DEPARTURE in 1914, Einstein performed a gradual advance toward Elsa's apartment. He first moved from the home he had shared with Mileva to the Wilmersdorf district of Berlin, a few blocks from Elsa's apartment. This made it much more convenient for discreet encounters. In the fall of 1917, he moved into an apartment adjoining Elsa's at Haberlandstrasse 5. This apartment was mostly for show—he effectively began to live in Elsa's apartment. And in the spring of 1919, he gave up all pretence and gave up his apartment.

When his divorce from Mileva was finalized, in February 1919, Einstein had been living in Elsa's apartment for a long while. He now came under heavy pressure to marry Elsa, pressure from her, from her parents, and from his and her friends. At first he resisted, replying to the remonstrations about the effect of his domestic arrangements on Elsa's reputation that the attentions of a famous professor, such as he, could not ruin the reputation of Elsa and her daughters, but, on the contrary, only enhance it: "Snobbery is here so highly developed," he claimed in a letter to Zangger, "that by [my attentions], these women do not lose respect, but on the contrary gain it. If I let myself get caught, my life would become complicated."[86]

It seems that what Einstein most loved about Elsa was her cooking. The Germans say *Die Liebe geht durch den Magen* (Love goes through the stomach), and it is certainly true that Germans love food—they love it solid, and they love it solidly. In one of their early letters, Einstein waxes lyrical over the *Schwammerl und Gansgrieben* (mushrooms and goosecrackle, a revolting dish consisting of little bits of carbonized goose skin immersed in congealed goose grease), which Elsa cooked for him and mailed to his office in Zurich.[87] When she mailed him a second package, he cursed the postman for looting the goosecrackle, leaving only a grease-stained piece of wrapping paper.[88]

But in spite of Elsa's lovely cookery, by the time he married her, his roving eye had already wandered elsewhere—he had made a pass at Ilse, Elsa's oldest daughter, a pretty Fräulein of twenty years, whom he had hired as his secretary. This was a major tactical blunder, and it was not well received, by the mother or by the daughter. At first he persisted, indicating his preference for marrying the nubile daughter rather than the aging mother. But after a tearful scene from Elsa, he capitulated, declaring he did not really care whom he married, mother or daughter, and that he would leave the decision to them. Ilse did not want him, Elsa did, . . . and that settled that. They married in June 1919, violating the two-year ban imposed by the Swiss divorce decree. Since Einstein had given up his own apartment some months earlier, he listed a nearby boarding house as his official address on the marriage certificate.

One noticeable effect of his liaison and subsequent marriage to Elsa was sartorial: during his Berlin years, under her tutelage, he would dress quite appropriately to his lofty professorial and directorial status, and on special occasions he would look positively elegant. Later, after her death in Princeton, he would revert to his casual, sloppy manner, and he would acquire some notoriety for not wearing socks.

10

"The world is a madhouse"

Einstein's comment on the storm of publicity about him and his theories triggered by measurements of the light deflection by English astronomers during the solar eclipse of 1919.

The event that elevated Einstein to the status of a worldwide celebrity was a formal joint session of the Royal Philosophical Society and the Royal Astronomical Society in London on November 6, 1919, just a few days before the first anniversary of the armistice that ended World War I. Einstein was not invited, but the crème de la crème of English physicists and astronomers were in attendance: Sir J. J. Thomson, the discoverer of the electron; Sir Ernest Rutherford, the discoverer of the nucleus; Sir Frank Dyson, the Astronomer Royal; and Arthur Eddington, later Sir Arthur, the brilliant Cambridge professor and director of the Cambridge Observatory.

They all understood it was a momentous occasion. Nothing less than the overthrow of Newton's theory of gravitation was at stake. The philosopher Alfred North Whitehead, who also attended, described the staging of the event:

The whole atmosphere of tense interest was exactly that of the Greek drama: we were the chorus commenting on the decree of destiny as disclosed in the development of a supreme incident. There was a dramatic quality in the very staging—the traditional ceremonial, and in the background the picture of Newton to remind us that the greatest of scientific generalizations was now, after more than two centuries, to receive its first modification. Nor was the personal interest wanting: a great adventure in thought had at length come safe to shore . . .[1]

Dyson, as Astronomer Royal, reported the results of two expeditions that the royal societies had sent to Africa and South America, to observe the solar eclipse of May 29 and to obtain photographs of stars near the Sun. He described the careful measurements of the apparent positions of the stars in the photographic plates, and he concluded, "After careful study of the plates, I am prepared to say that they confirm Einstein's prediction. A very definite result has been obtained that light is deflected in accordance with Einstein's law of gravitation."[2]

The astronomer Ludwick Silberstein urged his colleagues to be cautious in rejecting Newton: "We owe it to that great man," he said, "to proceed very carefully in modifying or retouching his law of gravitation."[3] But the majority of the physicists and astronomers attending the session were prepared to accept the observational evidence presented to them, and J. J. Thomson, as president of the Royal Philosophical Society, closed the session, saying, "This is the most important result obtained in connection with the theory of gravitation since Newton's day, and it is fitting that it should be announced at a meeting of the society so closely connected with him." He described Einstein's new theory as "one of the highest achievements of human thought," but then he added, "I have to confess that no one has yet succeeded in stating in clear language what the theory of Einstein really is."[4]

PLANS FOR THE ENGLISH astronomical expeditions that led to this dramatic result had been a long time in the making, and Eddington had been the chief instigator (and participant) in this effort. Eddington had become an early convert to general relativity, and he was influential in securing the acceptance of the theory in England. There are some similarities between the biographies of Newton and Eddington. Like Newton, Eddington was born in a small country town. His father died before his birth, and his mother raised him. At school, he was an outstanding student, with a remarkable talent for mathematics, and he gained a scholarship to attend Trinity College in Cambridge. In his second year at Trinity, he won first place in the celebrated tripos math exams, something never before achieved by a second-year student.

Eddington acquired some early experience in practical astronomy while working as assistant to Dyson at the Greenwich Observatory. He then quickly rose to the top of his profession, and returned to Cambridge as Plumian Professor of astronomy and director of the Cambridge Observatory. There he devoted himself to the application of mathematics and physics to astronomical problems, especially the theory of the internal structure of stars. He recognized that what prevents the crushing of a star under its own weight is not only the pressure of the gas within the star, but also the pressure of the thermal radiation trapped in the star, what is called the "pressure of light." With his theoretical work on the internal equilibrium of stars, he became the founder of modern astrophysics.

Eddington is said to have been arrogant; if so, he had a great many things to be arrogant about. But in the 1930s his arrogance went a tad too far. He became involved in a lengthy and futile dispute with Subrahmanyan Chandrasekhar, a bright student of astrophysics from India, and later a professor at the University of Chicago, who discovered that heavy stars, with a mass of more than 1.4 times the mass of the Sun, cannot remain in equilibrium once they exhaust their internal fuel and cool down. Such a star will be crushed by its own weight, that is, it will collapse on itself, like an overdone soufflé. Eddington stubbornly refused to believe this, despite the evidence of Chandrasekhar's calculations, and he tried to block acceptance of this new idea. Ultimately, Chandrasekhar received a Nobel Prize for his discovery, whereas Eddington never did.

Like Newton, Eddington spent his final years in the pursuit of delusional speculations. For Newton, these speculations involved zany biblical chronologies; and for Eddington, they involved an attempt at a numerological derivation of the constants of physics. He thought that the fine-structure constant (which characterizes the strength of the electric forces at the atomic level) could be constructed by pure mathematics. According to Eddington's wacky numerological scheme, this constant was exactly equal to $1/136$, and the total number of proton particles in the universe was exactly 136×2^{256}. When it was pointed out to him that the experimental value of the fine-structure constant is closer to $1/137$ than to $1/136$ (actually $1/137.035$ . . .), he amended his calculation, adding 1 to the denominator, for which the satirical magazine *Punch* gave him the title of Sir Adding-one.

THE WAR HAD INTERRUPTED the routine shipment of copies of German scientific journals, such as *Annalen der Physik*, to England. Nevertheless, a few copies of Einstein's 1916 paper on general relativity had found their way to England by roundabout routes, via Holland and Switzerland, and by 1917 the English had access to Einstein's work. When Einstein's 1916 paper on general relativity reached Cambridge, Eddington immediately appreciated its profound significance. His mathematical expertise made it easy for him to grasp the mathematical content of Einstein's theory, and he also developed a sharp insight into its conceptual physical basis. As described in Chapter 9, Eddington had a much clearer perception of the true (and limited) role of the Principle of Equivalence than Einstein. Overall, his understanding of the theory was at least as good as Einstein's, maybe better.

Alerted by Eddington, English astronomers took a lively interest in Einstein's new prediction for light deflection by the Sun. Eddington cast the light deflection in the guise of a competition of Newton versus Einstein, which caught the attention of his sporting English colleagues. In his *Opticks*, Newton had speculated that light might be deflected by gravitational attraction: "Do not Bodies act upon Light at a distance, and by their action bend its Rays, and is not this action (other things being equal) strongest at the least distance?"[5] Newton had not attempted to calculate the deflection of light by the Sun or by some other body, and Eddington filled this gap somewhat simplistically by assuming that Newton's law of gravitation applies to light, and that the motion of light under the influence of Newton's gravitational force can be calculated by assuming that a ray of light behaves like a particle moving at the speed of light. Sir Isaac, who had refrained from any such simplistic calculation, even though he believed in a corpuscular theory of light, must have rolled over in his grave. Apparently Eddington was not aware that in 1801, the German astronomer Johann Soldner had done a similar calculation. In those days, the calculation attracted little attention, because astronomical observations were not accurate enough to measure such a small deflection, which required the use of photographic techniques. Soldner's calculation was later rediscovered in the 1920s, and Einstein's opponents then tried to exploit it to discredit general relativity.[6]

Eddington calculated a deflection of 0.9 seconds of arc for a ray grazing the Sun's limb (about the width of a thumb seen at a distance of 5 klicks), exactly the same as what Einstein had predicted from his first, naïve calculation based on the Principle of Equivalence.[7] Eddington drew attention to the difference between this "Newtonian" value for the deflection of light and Einstein's new prediction of 1.8 seconds of arc, twice the Newtonian value. He suggested that a measurement of the deflection would be a definitive test deciding between Newton and Einstein. In an appeal to the gallery, he gave this test the name of "weighing light." The challenge caught the fancy of Frank Dyson, the Astronomer Royal, who proposed the organization of expeditions to measure this deflection during the solar eclipse of May 29, 1919. Two expeditions were planned, one to the small island of Principe, off the west coast of Africa, and one to Sobral, in northeast Brazil.

ALTHOUGH EDDINGTON WAS more of a theoretical than an observational astronomer, he was sent on the expedition for a tangential reason having to do more with religion than with astronomy. Eddington was a devout Quaker, and as such he intended to refuse military service as a conscientious objector. In 1917, the British Army, in need of a fresh supply of cannon fodder for the blood-soaked battlefields of Flanders, had upped the maximum age of draftees to thirty-five years, and Eddington was due for call-up. With the country suffering painful losses, any conscientious objector was considered a social disgrace, and this disgrace extended not only to his family, but also to all of his known associates. The academic establishment at Cambridge sought to save Eddington and themselves from shameful publicity by convincing the Home Office that his services as a scientist were of far greater value for the war effort than his services as a soldier. Eddington received a notice of deferment from the Home Office, and he countersigned it . . . but with true Quaker obstinacy he added a postscript declaring that he would have refused military service anyhow, as a conscientious objector, deferment or not.

This provocation did not amuse the Home Office. Eddington was about to be arrested and shipped off to one of those dreary "camps" in the north of England, where he would have had to peel potatoes for the duration of the war, when Dyson intervened to save him from the consequences of his religious folly. By influence in high places, Dyson achieved a compromise with the British Admiralty: they would excuse Eddington from military duty, under the condition that he serve his country by participating in the solar-eclipse expedition in 1919.

Eddington and his colleague E. T. Cottingham were placed in charge of the African expedition. "As the problem presented itself to us," Eddington later recalled, "there were three possibilities. There might be no deflection at all; that is, light might not be subject to gravitation. There might be a 'half-deflection,' signifying that light was subject to gravitation, as Newton had suggested, and obeyed the simple Newtonian law. Or there might be a 'full deflection,' confirming Einstein's instead of Newton's law. I remember Dyson explaining all of this to my companion Cottingham, who gathered the main idea that the bigger the deflection, the more exciting it would be. 'What will it mean if we get double the deflection?' 'Then,' said Dyson, 'Eddington will go mad, and you will have to come home alone.' "[8]

THE EXPEDITIONS SET OUT in good time, arriving at their destinations about a month before the eclipse, and they completed the installations of all their telescopes and cameras. But, to the dismay of Eddington and his colleagues, the weather on Principe was unfavorable on the day of the eclipse. Sun and Moon were shrouded in a layer of cloud that made it difficult to see the stars whose light the Sun was supposed to deflect. "There was nothing for it but to carry out the arranged programme and hope for the best," wrote Eddington. "There is a marvelous spectacle above, and . . . a wonderful prominence-flame is poised a hundred thousand miles above the surface of the sun. We are conscious only of the weird half-light of the landscape and the hush of nature, broken by the calls of the observers, and beat of the metronome ticking out the 302 seconds of totality."[9]

Of the photographs obtained in Principe, some showed no stars at all, and only two plates were good enough to measure the deflections of starlight. These gave a result of 1.61 seconds of arc for starlight grazing the edge of the Sun, close to the value predicted by Einstein. This sufficed to convince Eddington that he had found confirmation of Einstein's theory: "Although the material was very meagre compared with what had been hoped for, the writer (who it must be admitted was not altogether unbiassed [*sic*]) believed it convincing."[10]

But the photographs obtained at Sobral threw some doubts on Eddington's assessment. The weather in Brazil had been fine on the day of the eclipse, and each of the two telescopes there had yielded good plates. The plates from one of the telescopes gave deflections of 1.98 seconds of arc, and those of the other gave deflections of about 0.9 seconds òf arc. Thus, the results from one telescope were larger than Einstein's prediction (although not much larger, and, given the experimental uncertainties, perhaps within a tolerable range), and the results from the other telescope were close to the Newtonian prediction.

Eddington performed some creative cherry-picking. He argued that the plates from the second Sobral telescope should be disregarded, because the images on them were somewhat fuzzy, probably because a mirror used to aim the starlight into the telescope had become distorted by the hot Brazilian sunlight prior to the eclipse. The argument was somewhat questionable—the fuzziness was no worse than on some of the Principe plates and did not prevent measurements of the images. And there was no direct evidence of a problem in the mirror that would reduce the apparent deflection of the starlight. It seems that Eddington decided to discard the data from the second telescope for the same reason that Einstein–de Haas had decided to discard one of their sets of data—he thought he knew the right answer, and he selected the observational results according to his theoretical prejudices. But there was a crucial difference between the cases of Eddington and Einstein–de Haas: he picked the right cherry, and they picked the wrong one. Eddington somehow convinced his astronomical colleagues to support his shaky position, and this led to the dramatic announcement at the session of the Royal Philosophical Society in November 1919.

Although the English astronomers went along with Eddington's manipulations of the data, other astronomers were more cautious. Campbell, who had been unsuccessful in his attempt to measure the deflection in the Crimea in 1914, had made another attempt during the eclipse of June 8, 1918, sending his assistant Heber Curtis to take photographs at a site in Washington state. The weather was iffy, and the few photographs that were obtained were of poor quality. Curtis felt that, as best he could tell, there was no deflection of light, and Campbell had reported this to the Royal Society. But Eddington chose to ignore that.

Curtis was skeptical not only of Eddington's analysis of the 1919 eclipse data, but he was also skeptical of the data from later eclipses. "There may be a deflection," he wrote, "but I do not feel that I shall be ready to swallow the Einstein theory for a long time to come, if ever. I'm a heretic." And he later described Einstein's theory as "beautiful but bizarre, clever but not a true representation of the physical universe."[11]

WHEN THE SESSION of the Royal Society ended, Silberstein approached Eddington and said, "Professor Eddington, you must be one of the three persons in the world who understands general relativity." When Eddington demurred to this, Silberstein prompted, "Don't be modest, Eddington." And Eddington famously replied: "On the contrary, I am trying to think who the third person is!"[12]

If Eddington had taken the question seriously, he could have mentioned a handful of persons who not only understood relativity but also had made significant contributions to it. For instance, Hilbert had come close to winning the race to find the field equation for general relativity, and he lost to Einstein not by any lack of understanding but by the hubris of his attempt to unify gravitation and electromagnetism.

Eddington could also have mentioned Karl Schwarzschild, a German astronomer, director of the Potsdam Observatory. Schwarzschild had volunteered to serve at the Russian front during the war, where his mathematical expertise was put to practical use in the calculation of trajectories of artillery shells. When off duty, he continued to work on astronomical problems, and while studying Einstein's publications on general relativity, he found an *exact* solution of Einstein's equation for the curved spacetime geometry surrounding a spherical mass, such as the Sun.

Einstein had done his calculations of the light deflection and the perihelion precession with an *approximate* solution—he had not even tried to find an exact solution, because he thought that the complicated form of his equation would make an exact solution extremely difficult or impossible. But Schwarzschild was a better mathematician than Einstein, and he found the exact solution within a few days of reading Einstein's papers. He sent his solution to Einstein, with the comment, "As you can see, the war means to be friendly with me, in that, despite heavy artillery bombardment within a quite terrestrial distance, it permitted me this excursion into your land of ideas."[13] Einstein expressed his astonishment: "I would not have thought that the rigorous treatment of the problem could be so simple,"[14] and he offered to forward Schwarzschild's calculation to the academy.

During his service at the Russian front, Schwarzschild contracted pemphigus, a rare and then-incurable autoimmune skin disease. He died in 1916, soon after he was invalided back to Germany. Einstein wrote to Hilbert, "I also find his death shattering. It can be supposed that there are few among the living who know to use mathematics with the virtuosity that was characteristic of him."[15] Although Schwarzschild did not live to enjoy the triumphal success of Einstein's theory in predicting the deflection of light, he is not forgotten. Today, the Schwarzschild solution is found in all textbooks of relativity; it plays a central role in the study of neutron stars, gravitational collapse, and black holes.

EDDINGTON LATER ADMITTED he first learned relativity from the publications of the Swiss mathematical physicist Hermann Weyl and the Dutch astronomer Wilhelm de Sitter. Weyl had taught a course on general relativity at the Zurich Polytechnic in 1917, and he had published the first textbook on the theory a year later. This book, *Space-Time-Matter*, is still in print; it is a more comprehensive and detailed introduction to the theory than any writings by Einstein. In a review, Einstein himself professed his enthusiastic admiration for Weyl's book: "Again and again I feel compelled to reread the individual parts of this book; because each page reveals the extraordinarily sure hand of a master, who has penetrated the subject from diverse directions."[16] Eddington acknowledged that his own book was partially based on Weyl's (Eddington's book was first published as a mathematical supplement to a French translation of Weyl's).

De Sitter's involvement with general relativity was brought about by a paper on cosmology that Einstein wrote in 1917. This was Einstein's first attempt to apply the equations of general relativity to the overall distribution of stars throughout the universe. Although the application of the laws of terrestrial or Solar System physics on such a grand scale looks like a very daring thing to do, Einstein was following in the footsteps of Newton, who had attempted something similar with his law of gravitation. Newton, like Einstein in 1917, thought that the distribution of stars was static, or nearly static; that is, the stars remained in fixed positions (which is why they are called *fixed* stars), in contrast to the planets (called *the wanderers*). Newton asserted that if an infinite number of stars is uniformly distributed over infinite volume, then each star will remain in equilibrium: "The fixed stars, being equally spread out in all points of heaven, cancel out their mutual pulls by opposite attractions."[17] Although Newton's assertion seems intuitively obvious, nobody ever managed to provide a mathematical proof.[18]

In the late nineteenth century, some astronomers came to believe that Newton's law of gravitation gave ambiguous results when applied to an infinite distribution of stars.[19] They thought that to resolve the ambiguities, a boundary condition needed to be imposed on the infinite distribution of stars, that is, a condition that the stars gradually thin out in the most distant portions of the universe. Years later, it was shown that this boundary-value problem was a red herring created by a misguided insistence on keeping the distribution of stars static. If, instead, the stars are allowed to move outward (or inward), giving a motion of expansion (or contraction) of the entire universe, then Newton's law of gravitation uniquely determines the deceleration (or acceleration) of this motion, and there are no ambiguities whatsoever.[20] But in 1917, it was not yet known that the universe is expanding, and the insistence on a static distribution of stars was creating much confusion.

IN HIS 1917 PAPER on cosmology, Einstein accepted and repeated the confused criticism of Newton's laws, and he tried to solve the problem by applying the equations of general relativity. First he toyed with the red herring of boundary conditions, and he described these unsatisfactory attempts in some detail, saying: "In what follows I lead the reader along the rather indirect and bumpy road that I traveled myself, because only thereby can I hope that he will show some interest in the final result. I have actually reached the opinion that the field equations for gravitation proposed by me still require a small modification."[21]

Then Einstein decided to bypass the problem of boundary conditions altogether, by adopting a closed, uniformly curved universe, that is, a universe without boundaries. The spatial geometry of such a universe is analogous to the curved surface of an ordinary sphere, on which you can walk in any direction without ever reaching any boundary—if you walk straight ahead in any direction, you ultimately come back to your starting point after walking completely around the sphere. Since there are no boundaries in this kind of universe, boundary conditions play no role, and the spacetime geometry is completely determined by the mass distribution and the field equations. Einstein believed that this was a distinctive and highly desirable feature of the closed universe. He interpreted it as an expression of Mach's principle: the spacetime geometry—and therefore the inertial properties of particles moving in this geometry—is completely determined by the mass distribution in the universe, and this geometry depends on nothing else.[22]

However, Einstein found that, according to his field equations, the closed, uniformly curved universe does not remain static. This kind of universe expands with a gradually decreasing speed, then stops, and then recollapses with a gradually increasing speed. Since Einstein, and all his contemporaries, labored under the misconception that our universe is static, he decided that he needed to modify his field equations so they would permit such a static universe. For this purpose, he added one more term on the left side of his original equation (see p. 216), the so-called cosmological term represented by the Greek letter lambda. With this, Einstein's equation became

$$\text{Ricci tensor} - \tfrac{1}{2} \times \text{curvature invariant} + \lambda = \text{momenergy tensor}$$

Einstein was not happy about this complication, but he could see no way of avoiding it. He wrote to Ehrenfest, "I have again committed a crime on the theory of gravitation, which puts me somewhat in danger of being interned in a nuthouse. I hope there is none in Leyden, so I can again visit you without endangering myself."[23]

The cosmological term is quite small, and therefore its effects do not show up within the Solar System. But it modifies the motion of expansion of the universe, because it effectively generates a gravitational repulsion throughout the universe, which, on a large scale, cancels the usual gravitational attraction of the masses in the universe. Einstein selected the magnitude of the cosmological term to be equal to the average mass density of the universe. With this choice for the value of the cosmological term, the gravitational attractions and repulsions are in equilibrium on a large scale, and the universe is static.

HOWEVER, EINSTEIN'S COSMOLOGICAL PROGRAM

contained two mistakes. De Sitter immediately pointed out that it was possible to contemplate a universe that has a zero (or nearly zero) mass density but a nonzero cosmological term, and in such an empty universe a particle will still have inertia. This undermines Einstein's claim that inertia is caused by the mass distribution of the universe. The empty de Sitter universe is not static. It expands with ever increasing speed, getting larger and larger—instead of holding the universe in equilibrium, the cosmological term in the de Sitter universe generates an accelerating expansion. Because of this expansion, the de Sitter universe has been described as "motion without matter," in contrast to the static Einstein universe which is "matter without motion."[24] (The modern view is that the cosmological term represents a mysterious form of matter called "dark energy," uniformly distributed throughout all space. Thus, today the de Sitter universe might be described as motion without *ordinary* matter, but with dark energy.)

In 1917, there was not yet any clear evidence of the expansion of our universe, and de Sitter's model of the universe was thought to be unrealistic. But even regarded as an unrealistic, purely mathematical model, the de Sitter universe demonstrated that Einstein's ideas about Mach's principle were off the mark, and that general relativity did not provide a coherent explanation of the origins of inertia. Toward the end of his life, Einstein gave up Mach's principle entirely, declaring, "Actually, one should no longer speak of Mach's principle at all."[25]

The other mistake in Einstein's cosmological program did not surface until 1930, when Eddington discovered that Einstein's static universe has a catastrophic defect: this universe is unstable. If the mass density and the cosmological term are mismatched by even the tiniest amount, then the large-scale gravitational attractions and repulsions will not balance exactly. The universe will then either expand or contract, and this expansion or contraction will proceed with an ever increasing rate, that is, the universe explodes outward or implodes inward. The universe is like a pencil standing on its point; even if the pencil is initially well balanced, it will soon tumble toward one side or another.

When Eddington discovered this mistake, he did not attract much attention, because by then he was flogging a dead horse. By 1930 it was known that our universe is expanding, and Einstein's static universe was no longer of any interest. By then, de Sitter's empty universe also was no longer of interest; it had been discarded in favor of alternative cosmological models formulated by the Russian mathematician Aleksandr Friedmann and the Belgian astronomer Georges Lemaître, a Jesuit who was following in the footsteps of the Jesuit astronomers of Galileo's days, in the long tradition of astronomical studies within the Society of Jesus (even today, almost all the astronomers of the Vatican Observatory are Jesuits).

BUT BEFORE IT WAS DISCARDED, the unrealistic, empty de Sitter model rendered an important service in cosmology: it lead cosmologists to the discovery of the expansion of our real, not-empty universe. Paraphrasing Synge's comment on the Principle of Equivalence, we can say that the de Sitter model played the role of midwife in the discovery of the expansion of the universe. The first observational clues of this expansion were found in 1912, when the astronomer Vesto Slipher, at the Lowell Observatory in Flagstaff, Arizona, began to measure the redshifts of light reaching us from spiral nebulae. This redshift is analogous to the Doppler shift of sound—it indicates a motion of recession. Slipher did not know that these nebulae were galaxies, similar to our Milky Way Galaxy, but he did measure their speeds of recession, and he found that some were moving away with speeds as large as one-fourth the speed of light. If some handfuls of particles are sprinkled into the empty space of de Sitter's expanding model of the universe, the expansion of this space brings about an expansion of the distance between the particles, that is, a mutual motion of recession of the particles. This recession was called the "de Sitter effect," and astronomers conjectured that it might be the explanation of Slipher's redshift measurements.

The conclusive evidence for the expansion was not obtained until the late 1920s, when the astronomer Edwin Hubble, at the Mt. Wilson Observatory in California, identified the nebulae as separate galaxies, far beyond our Galaxy, and developed methods for measuring the distances to these nebulae. He confirmed that the speeds of recession of galaxies are directly proportional to their distances, a relationship now known as Hubble's law.

By the time the evidence for the expansion was in hand, cosmologists had found other, more realistic theoretical models of the universe. These models, by Friedmann and by Lemaître, retained the expansion of the de Sitter universe, but they avoided the unrealistic emptiness of that universe.[26] From an observational point of view, the essential difference between the Friedmann and the de Sitter models is that in the former the expansion decelerates, and in the latter it accelerates. The deceleration in the Friedmann universe is a direct consequence of the gravitational attraction of galaxies on each other.

Thus, the credit for first inspiring astronomers to think about an expanding universe belongs to de Sitter, not to Einstein. What is more, in recent years, it has been discovered that the expansion of our universe is accelerating—we live in a universe that is somewhat similar to an empty de Sitter universe, with a relatively large cosmological term and a correspondingly large amount of dark energy, endowed with gravitational repulsion.

Our universe contains more repulsive dark energy (71 percent) than attractive mass (29 percent). Although Einstein supplied the theoretical building blocks for relativistic cosmology in 1917, de Sitter assembled these building blocks into a more fruitful model than Einstein.

IN HIS LATE YEARS, Einstein told the physicist George Gamow that the cosmological term was the "biggest blunder" he ever made in his life.[27] But the cosmological term, in itself, was not really a blunder. Today, we know that the cosmological term, or the dark energy, is an essential feature of our universe, needed to produce the accelerated expansion revealed by recent observations. Einstein's mistake was not in proposing the cosmological term, but in selecting the wrong magnitude for this term—he did not anticipate that if he was going to introduce a cosmological term, he needed to make it much larger than the average mass density of the universe. This mistake arose from his misconception that our universe was static. Einstein's blind acceptance of this misconception shows that he had no special insight into the workings of the universe—he was merely taking his cue from his astronomical colleagues.

It would have been a spectacular coup for Einstein to have proposed an expanding universe in 1919, anticipating the discovery of this expansion by ten years; and it would have been even more spectacular for him to have proposed an *accelerating* expanding universe, anticipating the discovery of this acceleration by eighty years. But Einstein missed the target in every which way. First he inserted a cosmological term to stop the expansion; later, he removed the cosmological term entirely, and missed the acceleration. Besides, Einstein's first attempt was intrinsically flawed—he failed to notice the instability of his static model of the universe.

In a recent discussion of Einstein's mistakes in the magazine *Physics*

Today, Steven Weinberg explained, "I don't think that it can count against Einstein that he had assumed the universe is static. With rare exceptions, theorists have to take the world as it is presented to them by observers."[28] This is true enough, and it would serve as a good excuse for most theorists. But Einstein had often insisted on *not* taking the world as it was presented to him. He believed, quite rightly, that he had an exceptional talent to see beyond the horizon and to discover the unexpected. In his treatment of cosmology, this exceptional talent is conspicuous by its absence—his deep intuition about most aspects of physics failed him.

In a series of books and museum exhibits published in Germany in celebration of the "Einstein Year 2005," the editors sought to exalt Einstein by calling him the "engineer of the universe."[29] But to call a physicist an engineer is not a compliment. Einstein's engineering of the universe was clumsy—a plodding, uninspired search for the "right formula," without any real feeling for the deeper implications. His construction of the static model of the universe was like the work of an inept engineer who designs a structure on the principle of a house of cards, in which the cards are not tied, or glued, together. The house of cards is in equilibrium, but it will collapse on the slightest provocation, because the equilibrium is not stable. In cosmology, Einstein did not have the *Fingerspitzengefühl*, or the sixth sense, of a good engineer to spot defects in the structure, and he did not have the mystical insights that had led him to special relativity and general relativity.

THERE WAS, HOWEVER, one detail on which his intuition guided him in the right direction: he recognized that the cosmological term needed some physical explanation, in terms of the physics of elementary particles. The connection between the cosmological term and elementary-particle physics remains one of the most fundamental, unsolved problems in physics to this day, and Einstein deserves credit for first recognizing this problem and for making the first of many failed attempts at its solution.

The solution that Einstein proposed in 1919 was probably inspired by Hilbert's attempt to interpret all of matter as purely electromagnetic. Einstein speculated that maybe elementary particles, such as electrons and protons, consist of nothing but electric charge, held together by gravitation. Without some mechanism to hold it in equilibrium, a cloud of electric charge would immediately spread out and disperse in all directions, because of the mutual electric repulsion of the charge. Hence Einstein proposed that the attractive gravitational force acting on the electric mass-energy of the cloud might hold it together. And he found that to make this work, he needed to alter his field equations once more, changing the coefficient in front of the curvature invariant (see p. 216) from ½ to ¼. This was the fourth change that Einstein made in his equations in four years. He wryly commented on his continual remodeling of his equations: "That fellow Einstein makes things convenient for himself. Each year he retracts what he wrote the year before."[30]

By some mathematical manipulation of his equations and by a rather artificial modification of the momenergy tensor of matter, Einstein was able to show that his new equation could be massaged to produce a cosmological term. However his new equations had a bad defect: they could not produce the observed properties of the electron and the proton. Einstein found that with his new equation a spherical cloud of positive or negative electric charge could be held in equilibrium by its own gravitational attraction, but if squeezed or stretched by an external disturbance, the cloud would change its size and still remain in equilibrium. This meant the size of an electron or a proton would not remain fixed—whenever you push or pull such electrons or protons around, they might shrink or expand like, say, a piece of cotton candy pushed or pulled by your sticky hands. This is not how real electrons and protons behave; each real electron or proton has exactly the same size as every other electron or proton, and its size never changes, no matter what you do to it.

At the end of this speculative paper, Einstein admitted, "Without further details, the problem of the constitution of elementary quanta [of charge] cannot yet be solved on the basis of the given field equations."[31] This was rather an understatement. Einstein soon abandoned this approach and its remodeled field equations; he touched on it briefly some years later, and then never returned to it.[32]

Today, the attempts to relate the cosmological term to the properties of elementary particles are much more sophisticated, involving quantum theory and string theory. But even so, all attempts have been total failures—by comparison with the value of the cosmological term observed in our universe, the calculated theoretical value is too large by a factor of more than one googol, that is, 1 followed by a hundred zeros.[33] The abysmal failure in the calculation of the value of the cosmological term is a painful blow to the pride of theoretical physicists—it is the largest failure in physics, ever. There is a Nobel Prize waiting for whoever solves this problem.

AT A DINNER given by the Royal Astronomical Society to celebrate the return of the 1919 expedition, Eddington read a poem composed in honor of his findings, ending with the verse:

> *Oh leave the Wise our measures to collate,*
> *One thing at least is certain. LIGHT HAS WEIGHT*
> *One thing is certain, and the rest debate—*
> *Light-rays, when near the Sun, DO NOT GO STRAIGHT.* [34]

A somewhat more recent and less pretentious verse by an unknown wordsmith, probably a graduate student, includes relevant mathematical details (the last two lines are to be read *very* fast):

> *Twinkle, twinkle little star,*
> *How I wonder where you are.*
> *1.8 seconds of arc from where I seem to be,*
> *For dl equals 1 minus 2GM over r times dt.* [35]

The 1919 eclipse expedition and Eddington's somewhat slanted data analysis were lucky breaks for Einstein. Subsequent eclipse expeditions to Australia, Sumatra, Siberia, etc., produced erratic results; some yielded numbers in good agreement with Einstein's prediction, but some yielded numbers that were considerably larger. All of the eclipse expeditions suffered from the difficulties inherent in establishing an improvised observation post in a remote, awkward location, often afflicted by nasty weather, awful living conditions, and, sometimes, nefarious natives.

Really good and reliable numbers were not obtained until much later, after 1970, when radio telescopes became available, which permitted precise measurements of the deflection of radio waves emitted by pulsars. The detection of such radio waves is not inhibited by the glare of the Sun, and this means the measurements can be done without waiting for an eclipse and organizing an expedition to the ends of the Earth. The results from radio telescopes have confirmed Einstein's prediction with high precision, and it has even become possible to measure the deflection of radio waves by Jupiter, which produces a deflection a hundred times smaller than the deflection produced by the Sun, because of Jupiter's smaller mass.

Today the deflection of light and of radio waves is used as a tool in astronomy. Astronomers can deduce the masses of distant galaxies by measuring the deflections that these galaxies produce when acting on the light and radio waves from even more distant galaxies in the background. Such measurements have confirmed that galaxies are surrounded by vast clouds of dark matter, that is, nonluminous matter (this dark matter associated with the galaxies is not to be confused with dark energy, which has no association with galaxies and resides in the intergalactic void). The dark matter reveals itself by its gravitational effects in the deflection of light and radio waves, but because we can't see this matter, we can't tell what it is—maybe clouds of dust, or new exotic particles, or maybe swarms of cometlike bodies or Jupiter-like bodies, which remain invisible because of the absence of illumination from a nearby star.

The most remarkable examples of deflections are observed when two galaxies lie on exactly the same line of sight from the Earth, so one galaxy is hidden behind the other (a kind of "eclipse" of one galaxy by the other). The foreground galaxy then bends the rays of light of the background galaxy, and this distorts the apparent image of the background galaxy into a ring, concentric with the foreground galaxy. Such a ring image is called an Einstein ring. Many beautiful instances of Einstein rings have been found with radio telescopes. These rings are a spectacular demonstration of the light deflection by gravitational effects.[36]

THE FIRST NEWS about the result of the English eclipse observations reached Einstein via Lorentz, who telegraphed that the preliminary results were between 0.9 and 1.8 seconds of arc. This wide range of numbers made an objective decision between the general relativistic and the Newtonian predictions impossible. But Einstein interpreted the preliminary results optimistically, and declared that the numbers confirmed his theory, although with "low precision." He reportedly told a visiting student, "I always knew that the theory was right." The student asked him, But what if the measurements had contradicted your theory? and received the grand reply, "Then I would feel sorry for the Dear Lord; the theory is right anyway."[37]

It was not until the end of October 1919, during a visit to the Netherlands, that Einstein obtained more precise information about Eddington's favorable interpretation of the measurements. He wrote to Planck, exaggerating the accuracy of Eddington's results somewhat: ". . . the precise measurements of the plates yielded exactly the theoretical value for the deflection of light. It is a gift of fate that I have lived to experience that."[38] With Einstein on display on the stage, Lorentz gave a lecture at the Royal Academy in Amsterdam, and thousands of students cheered.

The news quickly spread through the international academic community. From Switzerland, Einstein's Swiss ex-colleagues at the Zurich Polytechnic sent him a congratulatory telegram, and his friend Zangger wrote, "So for you everything is getting better, even light goes bent for your sake for several million years, and the heavenly bodies make perihelia, and perhaps you will command them other cabrioles. Under the circumstances I can understand that you would rather write to Isak [sic] Newton than to me." Zangger added a comment that reveals that Einstein already had had an unerring intuition about the bending of light during his days in Zurich: "Your confidence, the confidence of your thinking, that light would have to go bent around the Sun, e.g., at the time when you were with us, is for me a tremendous psychological experience. You were so certain, that your certainty had an overwhelming effect."[39]

BEFORE 1919 EINSTEIN was, of course, well-known among physicists. But with the confirmation of the bending of light, he suddenly acquired worldwide fame among the general population. This meteoric rise to fame can be attributed to several fortuitously coincident factors, two of which were war weariness and sensationalism of the press, as perceptively described by Rutherford in 1933 in an after-dinner conversation at Trinity College at which Eddington and Chandrasekhar were present. One of the diners asked why Einstein was accorded greater public acclaim than Rutherford, the discoverer of the atomic nucleus. In an expansive mood, Rutherford turned to Eddington and began, "You are responsible for Einstein's fame." But then he continued, "The war had just ended; and the complacency of the Victorian and Edwardian times had been shattered. The people felt that all their values and all their ideals had lost their bearings. Now, suddenly, they learned that an astronomical prediction by a German scientist had been confirmed by expeditions to Brazil and West Africa and, indeed, prepared for already during the war, by British astronomers. Astronomy had always appealed to the public imagination; and an astronomical discovery, transcending worldly strife, struck a responsive cord. The meeting of the Royal Society, at which the results of the British expedition were reported, was headlined in all the British papers; and the typhoon of publicity crossed the Atlantic. From that point on, the American press played Einstein to the maximum."[40]

American newspapers were indeed eager to give their readers some uplifting distraction from the depressing horrors and sufferings caused by the war. A new topic was needed, and an outlandish and sensational topic, such as a spectacular wonder in the heavens, admirably suited the purpose of newspaper editors. They could say, with Oscar Wilde, "We are all in the gutter, but some of us are gazing at the stars." Of course it was a bit difficult to make a sensation out of a bending of light rays by 1 second of arc, an angle so small that it can only be seen with a telescope (remember: 1 second of arc is the width a thumb seen at a distance of 5 kicks). The newspaper editors would have preferred the discovery of centaurs on Venus or mermaids on Mars; but, in lack of better, the Einstein story wasn't half bad.

THE NEWSPAPERS HELPED the story along by loud headlines. In *The Times* of London, the headline for the report about the meeting of the Royal Society was "REVOLUTION IN SCIENCE/NEW THEORY OF THE UNIVERSE/NEWTONIAN IDEAS OVERTHROWN," and similar headlines were repeated in the next days.[41] These revolutionary ideas were attributed to the "famous Professor Einstein," but the newspaper did not mention his first name, his affiliation, or his nationality. Across the Atlantic, the story was picked up by *The New York Times*, with the headline "LIGHT'S ALL ASKEW IN THE HEAVENS."[42] For the enlightenment of its readers, the newspaper reported that stars were not where they seemed to be, but that there was no need to panic. And it also reported that only twelve scientists in the whole world could understand Einstein's writings—a journalistic canard that served to impress the public with the depth of the mysteries being announced and also served to excuse the befuddled writing by the newspaper reporter, who evidently was not one of the twelve wise men. Across the United States, a chain reaction set in, as one newspaper after another repeated this canard. Overnight, the name of Albert Einstein, previously known only to a handful of physicists in the United States, became a household word. The story then bounced back to Europe, but it took a while to gather momentum in Germany.

Reporters from English and American newspapers arrived in Berlin to interview Einstein, and he was invited to write a guest article for *The Times* of London. In this article he expounded in great detail the role of coordinates in physics, which must have come as a disappointment to readers expecting lurid revelations about the warping of spacetime. On that topic, he makes only the cryptic remark, ". . . the laws according to which material bodies are arranged in space do not exactly agree with the laws of space prescribed by the Euclidian geometry of solids. This is what is meant by 'a warp in space.' "[43]

He concluded his article with a joke about his German-Swiss-Jew status: "By an application of the theory of relativity . . . to-day in Germany I am called a German man of science, and in England I am represented as a Swiss Jew. If I come to be regarded as a *bête noire*, the descriptions will be reversed, and I shall become a Swiss Jew for the Germans and a German man of science for the English!"[44] This was a somewhat tactless and ungracious joke. By describing Einstein as Swiss, the English were merely trying to bury the painful memories associated with the more than 2 million casualties inflicted on them by the Germans in the recent war.[45] But Einstein was particularly fond of this joke, and he used in on other occasions, and it was widely reprinted in the international press.

In December, the *Berliner Illustrirte Zeitung* displayed a picture of Einstein across its front cover, with the caption "A new luminary in the history of the world: Albert Einstein, whose researches imply a complete revolution of our conception of nature, equivalent to the discoveries of Copernicus, Kepler, and Newton."[46] In this picture, neatly retouched, Einstein's unruly hair is carefully brushed back, and he looks very distinguished and thoroughly Prussian. All that is missing to make this into a perfect portrait of a Prussian *Junker* is a monocle and a medal around the neck. The missing medal arrived two years later, when Einstein was elected a knight of the Prussian order of *Pour le mérite* for science and arts and received the medal of the order, a pretty bauble, about the size of a hefty pocket watch, with a large blue-and-gold circle, adorned with four gold crowns and a Prussian eagle. But he never wore it, and when Nernst chided him for the absence of this decoration on some formal occasion, Einstein agreed, yes, "Toilettenfehler," error in toilette.

INSTEAD OF ATTEMPTING any serious presentation of Einstein's ideas, the newspaper articles emphasized the mysterious and counterintuitive character of the notions of curved space and four-dimensional geometry. This provided a whiff of esoteric mysteries and gave the reading public some pleasant frisson without demanding any real mental effort. Einstein accurately described this reaction of the public in an interview in which he wondered why so many people showed such excitement over his theories "about which they cannot understand a word." He said, "I am sure that it is the mystery of non-understanding that appeals to them."[47] Charlie Chaplin said much the same when, during a movie premiere that he and Einstein attended in Los Angeles, he commented on the wildly cheering crowds: "They cheer me because they all understand me, and they cheer you because nobody understands you."[48] Puzzled by the screams of the crowd, Einstein asked Chaplin, "What does it mean?" and Chaplin answered, "Nothing."[49]

Celebrities are famous for being celebrities, and the public cheers them for the sake of cheering them. As the German humor magazine *Kladderadatsch* remarked, the public was so befuddled that it commonly confused the physicist Einstein with the physiologist Steinach, a professor at Vienna who had acquired some notoriety by promoting a theory of rejuvenation and enhancement of sexual potency by means of vasectomies. Tongue-in-cheek, the magazine proposed that Einstein-Steinach should fuse their theories into a unified theory of "Relativity-rejuvenation," to be exploited commercially by paid lectures, worldwide tours, and by the sale of postcards, tie clips, and lapel pins.[50] The commercial proposals in *Kladderadatsch* were prophetic—today, Einstein is among the top-earning dead celebrities, and the Hebrew University of Jerusalem rakes in some $18 million a year by the sale of copyright licenses for postcards, posters, Einstein puppets, and other Einstein paraphernalia. It also licenses the Einstein name and image for advertisements and endorsements and collects hefty royalties on the gimmicks the Walt Disney Corporation peddles under the Baby Einstein™ label, by which gullible mothers hope to transmogrify their babies into geniuses.

But the public's ignorance did not prevent them from engaging in enthusiastic and sometimes passionate debates about relativity. Einstein found this amusing: "The world is a madhouse. At the present, every coachman and every waiter debates whether the theory of relativity is right."[51] A French cartoon from the early twenties shows a group of elegantly dressed ladies involved in an animated discussion; the caption reads: "Such songs of praise! These ladies must still be talking about their couturiers.—Not at all, it's about Einstein."[52]

His fame reached even into the unscientific—and often antiscientific— world of poets and songsmiths. William Carlos Williams composed a modernistic poem in honor of Einstein's first visit to the United States in the spring of 1921. Entitled "St. Francis Einstein of the Daffodils," this poem is mostly incomprehensible (at least to physicists, if not to literati). But a few lines seem to hint at length contraction:

> *Einstein, tall as a violet*
> *in the latticearbor corner*
> *is tall as a blossomy*
> *peartree*[53]

In music, Einstein made an appearance in the well-known song "As Time Goes By" from the 1930s, later featured in the film *Casablanca*. In the popular version, the best-known refrain in this song proclaims, "A kiss is just a kiss . . ." But in the full version, as originally written, singers also crooned about "Mr. Einstein's theory" of "speed and new invention . . . things like fourth dimension," which gave "cause for apprehension."[54]

SENSATIONALISM CREATED by the gutter press usually lasts for no more than fifteen minutes. The persistence of Einstein's celebrity status throughout his lifetime and to this day involves factors that go beyond the public mood of the postwar era and the commercial exploitation of the news by newspaper editors and book publishers looking for a quick profit.

A major contributing factor was his personality. He enjoyed the attention he received from the press and the public, and he was always cooperative with reporters, displaying great patience and tolerance for their invariably inane questions—and the reporters loved him for it. He could always be counted on for a catchy soundbite: "Pompous phrases and words give me goose bumps whether they deal with the theory of relativity or with anything else" (*Berliner Tageblatt*, 1920); "I found Princeton lovely: an as yet unsmoked pipe, so fresh, so young" (*The New York Times*, 1921); "Japan is now a kettle without a safety valve" (*The New York Times*, 1925); "National-ism is an infantile disease. It is the measles of mankind" (*Saturday Evening Post*, 1929); "I don't drink, so I couldn't care less" (newsreel, on arrival in the US, then under prohibition, 1930); "Why is it that nobody understands me, yet everybody likes me?" (*The New York Times*, 1944).[55]

In public, Einstein was always good-humored, friendly, and approach-able. When his off-the-cuff remarks gave offense, he would quickly find a way to soothe his audience by a creative reinterpretation or amplification of his remarks, what we today call "spin." Einstein's talents as a spin doc-tor are well illustrated by his handling of a remark he made after a lecture at Princeton, in a discussion of some recent results from a Michelson-Morley experiment that seemed to indicate an ether wind, contradicting the theory of relativity. He dismissed this result, saying, "Raffiniert ist der Herrgott, aber boshaft ist Er nicht," which translates into English as "Cun-ning is the Lord, but malicious He is not."[56] With Einstein's permission, this phrase was carved into the mantel of a fireplace in the professors' lounge in Fine Hall at Princeton University (this was then the mathemat-ics department; recently, the interior of the building was remodeled into a Near Eastern studies department, renamed Jones Hall).

When this rather irreverent comment about the Lord attracted some unfavorable attention, Einstein backpedaled immediately and explained that what he really meant was that "Nature hides her secrets by the gran-deur of her character, not by cunning."[57] This was pretty much the oppo-site of what he had said at first, but it calmed the critics.

There is a sequel to this story. Years later, in Princeton, while frustrated with the lack of progress in his unified field theory, Einstein commented to his assistants Peter Bergmann and Valentin Bargmann, "I have second thoughts. Maybe God *is* malicious."[58] Fortunately, this comment did not become known until long after his death. If Rabbi Goldstein of New York, who sent a telegram to Einstein demanding that he declare whether he believed in God, had known about that comment—oy vey, oy vey!

EINSTEIN SOMETIMES COMPLAINED about the storm of publicity that engulfed him, declaring it "not fair and not even in good taste to select a few individuals for boundless admiration and to attribute superhuman powers of mind and of character to them."[59] But he really loved the attention, and he was a willing participant in the publicity cir-

Fireplace in the professors' lounge in Fine Hall (now called Jones Hall) at Princeton University. The inscription carved into the mantel after Einstein's 1921 visit to Princeton translates as "The Lord is cunning, but not malicious."

cus. In photographs he looks happy and relaxed when surrounded by hordes of journalists. He played to the gallery, and he enjoyed posing for pictures. He loved the camera and the camera loved him. In his posed photographic portraits he looks either intensely soulful or charmingly friendly, with an engaging smile playing around his mouth and his eyes. The public saw this as a welcome change from the haughty and serious posing of most savants (little do they know that the haughtiness seen in savants is mostly the result of nervousness in facing the camera and the public). His long unkempt hair, which made him look like an aged Struwwelpeter, was deliberately conspicuous. He definitely understood the advantages of a signature look.

Some of Einstein's friends and colleagues were scandalized by his tolerance for public attention, and they warned him that his cooperation with journalists would be perceived as despicable self-aggrandizement. To the modern ear, such warnings sound quaint—today every scientist, aided by the public relations office of his or her university, is more than ready for any effort at self-aggrandizement. But in those days, even the inclusion of biographical information about Einstein and—horror of horrors—a photograph of him for the frontispiece in a book on relativity written by his friend Max Born was frowned upon by his other friends and colleagues. Von Laue, though a friend and admirer of Einstein's, complained that he "and many colleagues would take umbrage at the photograph and the biography." Accordingly, the photograph and biography were deleted in later editions.[60]

When Einstein agreed to cooperate with the journalist Alexander Moszkowski in the writing of a full-length biography, Hedi Born, wife of Max Born, demanded that he immediately withdraw permission for the publication of this book, because "This book will constitute your moral death sentence for all but four or five of your friends. It could subsequently be the *best confirmation of the accusation of self-advertisement.*"[61] Hedi's letter gave Einstein second thoughts, and he withdrew permission for publication. But the book was published anyhow, the first of many biographies of Einstein.[62] After its publication, Einstein wrote Born (with some disappointment?) ". . . without any earth tremors so far."[63]

A FINAL FACTOR that contributed to Einstein's fame is found in the premeditated propaganda and advertising campaigns, first by publishers of his books and other entrepreneurs, then by the German postwar republican government, and then by Zionist organizations. Eddington had immediately recognized the political implications of the excitement over relativity; he wrote to Einstein, ". . . all England has been talking about your theory . . . It is the best possible thing that could have happened for scientific relations between England and Germany."[64]

In Berlin, Walther Rathenau, the foreign minister of the Weimar republic reached the same conclusion. Einstein was on friendly terms with Rathenau, a rich Jewish industrialist who, during the war, had worked for the imperial government in the organization of raw materials and industrial production, and who, on the fall of the Kaiser, had ascended to the position of foreign minister. Desperate to restore normal international relations with former enemies of Germany, Rathenau thought of Einstein as a kind of goodwill emissary, who could serve as an icebreaker to initiate a dialogue that might help to overcome the hatred engendered by the war, what we today call "ping-pong" diplomacy. Einstein had many contacts in international scientific circles who were eager to invite him for visits and lectures. His opposition to the war and his pacifist activities were widely known, and this made him into a "good German," acceptable to audiences in the countries of former enemies (besides, in a pinch, he could claim he was really a neutral Swiss).

Rathenau was pleased that Einstein had received an enthusiastic welcome during his 1921 tour of the United States. The English were somewhat less willing to embrace a German, but when Einstein stopped there on his return from the States, he skillfully won them over by his respectful attitude toward Newton, at whose tomb in Westminster Abbey he laid a wreath.

German relations with France were even worse than with England, because of a vindictive French insistence on the payment of harsh war reparations. In the hope of achieving some improvement in scientific and other relations, Rathenau therefore strongly encouraged Einstein to accept an invitation from Langevin to lecture in Paris. Einstein's visit to Paris was a success, though not an unqualified success. Einstein understood that to win the hearts and minds of the French, you have to speak French, even if only badly. At the Collège de France, he lectured in his schoolboy Swiss French to a standing-room only audience, with Langevin hovering nearby, acting as *souffleur*. Newspaper reports of Einstein's visit were favorable, but he had to cancel a further lecture at the French Academy of Sciences when some thirty members refused to remain in the same room with a German and threatened to rise and walk out *en masse* if he were to come to the academy.

Rathenau also favored Einstein's plans for a tour of Japan, planned for the end of 1922. But Rathenau did not live to see the results. He underestimated the violent resentment that his conciliatory and submisssive policy toward France, his ostentatious wealth, and his Jewish origins provoked among the deprived and humiliated masses of ex-officers. He was assassinated by a cowardly gang of nationalist and anti-Semitic ex-officers in a drive-by shooting in June of 1922. Although Einstein thereby lost an influential friend, the Foreign Office continued to support his international tours, and for his South American tour in 1925 they even granted him a diplomatic passport.

SEVERAL OF EINSTEIN'S TOURS were commercial or fund-raising campaigns, which relied heavily on advertising and on orchestrated public events. His first tour of the United States in 1921 was sponsored by Chaim Weizmann's international Zionist organization, with the goal of exploiting the American penchant for celebrity worship to raise money for the foundation of the Hebrew University of Jerusalem. Weizmann reckoned that a Jewish celebrity could extract a large amount of dollars from the large Jewish population of the United States. Arrangements were made for the mayor of New York to greet Einstein on arrival and for a parade along Broadway. Then followed a long series of public meetings in New York, Boston, Chicago, Cleveland, and Washington at which Einstein lectured, played his violin, and encouraged donations. Einstein said, "I had to let myself be exhibited like a prize ox,"[65] but during this tour he collected $750,000, which in today's dollars is equivalent to about $8.5 million. Weizmann was disappointed—he had hoped for more.

Einstein's fund-raising performances occasionally alternated with serious scientific lectures, such as a series of lectures on relativity he gave at Princeton University. These lectures were published in the form of a book, in German and also in an English translation under the title *The Meaning of Relativity*.[66] These lectures illustrate Einstein's expository writing at its best, and they give a good, concise introduction to special and general relativity. However, the lectures repeat many of the mistakes found in his earlier work, such as the mistake about the meaning of synchronization by light signals, the mistake about the proof of $E = mc^2$, and the mistake about the Principle of Equivalence.

Einstein's grand tour of Japan was brought about by an odd recommendation that Bertrand Russell—philosopher, mathematician, and antiwar activist who would later be awarded a Nobel Prize in Literature—had made to a Japanese publisher during his own visit to Japan the year before. Asked to name the three most distinguished personages that might next be invited to lecture in Japan, he had named only two: Einstein and Lenin. The publisher opted for Einstein. The tour operated on the principle of a traveling circus show. Tickets were sold for the lectures at a steep price, and earnest audiences of several thousand sat enthralled while Einstein lectured about the mysteries of relativity in incomprehensible German, followed by a translation into Japanese, which was probably equally incomprehensible. The Japanese book publisher found the lectures so profitable that he prevailed on Einstein to give a few extra performances. A German newspaper reported that the real reason for the great popularity of Einstein's lectures and books in Japan, especially among women, was that the Chinese characters that the Japanese used for "relativity" have the alternative interpretation of "sexual relations." This was somewhat of an exaggeration, but it held a grain of truth—the Chinese characters

相对论

for "theory of relativity" have the alternative interpretation of "theory of personal relationships," which might well have caused some false, and excessive, expectations.

The tour of South America was another Zionist fund-raising campaign, but the German residents of Argentina, Uruguay, and Brazil exploited Einstein's fame for their own purposes, to generate favorable publicity for Germany. Einstein was everywhere called "el sabio alemán" (the German sage) and, as in Japan, the reports that the German ambassadors sent to Berlin were highly favorable. In his diary, he described the meetings with South American dignitaries as "high comedy."[67]

No doubt Einstein would have attained the heights of fame even without these professionally managed promotional tours, but the fact is that no scientist before or after has ever been given such heavy promotion by commercial and political organizations. A good fraction of Einstein's fame among the general public can be attributed to this deliberate effort to generate publicity. And publicity engendered more publicity, in a chain reaction that escalated to a supercritical, runaway condition. What Einstein said or did became irrelevant for the populace—the mere fact of his presence caused mass hysteria. Thus, an eyewitness to one of Einstein's appearances in Vienna reported that the audience was "in a curious state of excitement in which it no longer matters what one understands but only that one is in the immediate neighborhood of a place where miracles happen."[68]

Like any worthy celebrity, Einstein made it a point to be seen in the company of the rich and famous. He became an ornament at all salons of the Berlin elite, whether intellectual, artistic, political, or financial, Jewish or gentile. Among Einstein's close acquaintances were Rathenau, the foreign minister; Gerhart Hauptmann, a celebrated playwright and winner of the Nobel Prize in Literature; Erich Kleiber, conductor and director of the state opera; and Max Liebermann, an impressionist painter and director or the Academy of Arts, who painted Einstein's portrait (about which Einstein commented, ". . . the picture resembled him more than it did me, which was in his favor").[69] He was also often found in the company of Count Harry Kessler, an elegant bon-vivant and man-about-town, whose diaries give many acerbic descriptions of Berlin society, including this description of a dinner party at Einstein's home: ". . . excessively grand dinner on a large industrial scale, . . . to which the really lovable, almost still childlike couple lent a certain naïveté."[70]

On the international scene, he became a fellow traveler of ministers (prime, not prime, and ex-), presidents, kings and queens, an empress of Japan, high and low church dignitaries, writers, playwrights, actors, millionaires, billionaires, the elite of the society of the cities he visited, and, of course, all the known leaders of the Zionist movement.

In Berlin, a new observation tower erected at the Potsdam Observatory was named the Einstein Tower. It was paid for by private donations, and it is symptomatic of the great public interest in Einstein that even in the economically harsh conditions of the postwar years enough funds were quickly raised for this project. A special telescope in this tower was intended to permit the observation of stars near the Sun in daytime, for new measurements of the deflection of light, but this proved unsuccessful.

The architect Erich Mendelsohn designed this tower in a hypermodern style, in an obvious imitation of the curvaceous architecture of Antonio Gaudí. But whereas Gaudí's creations convey a sense of playfulness and fit cheerfully into Barcelona's city landscape, the Einstein Tower looks clumsy and out of place, rather like a gigantic boot planted in the Potsdam forest (rumor has it that Mendelsohn intended to evoke the conning tower of a U-boat). Einstein was the guest of honor at the opening of the observation tower in 1922, and he dryly commented it looked "organic."

IN 1922, EINSTEIN'S long-overdue Nobel Prize was finally awarded to him. He was about to leave on his grand Japanese tour, when a letter from Svante Arrhenius, the chairman of the Nobel Prize committee for physics, gave him a preliminary hint that the award was in the works. Arrhenius had heard about Einstein's travel plans, and he coyly wrote, "It will probably be very desirable for you to come to Stockholm in December, and if you are then in Japan, this will be impossible." Einstein knew what to read between the lines of this message. Unless enticed by the offer of a Nobel Prize, nobody goes to Stockholm in December, a month when the

The modernistic Einstein Tower of the Potsdam Observatory. It was recently restored and is now a museum.

weather turns cold, windy, and gloomy, daylight lasts for only a few misera-ble hours, and the suicide rate reaches a seasonal maximum. He answered equally coyly explaining that he had a binding contract for the Japanese tour, and that he would proceed on his tour "in the hope that this would only delay the prospective invitation, but not cancel it . . ."[71]

Einstein was at sea, somewhere between Hong Kong and Shanghai, when the official announcement was made in December. A telegram from Sweden was delivered to his Berlin apartment: "Nobel Prize for physics awarded to you details by letter. Aurivillius." What the Swedish Academy awarded to Einstein was not the 1922 Prize, but the 1921 Prize, which had been left unawarded and untouched in the preceding year.

The details that Christopher Aurivillius, the secretary of the Swedish Academy, sent in the letter include a curious final sentence:

As I have already informed you by telegram, in its meeting held yesterday the Royal Academy of Sciences decided to award you last year's Nobel prize for physics, in consideration of your work on theoretical physics and in particular for your discovery of the law of the photoelectric effect, but without taking into account the value which will be accorded your relativity and gravitation theories after these are confirmed in the future.[72]

The explicit exclusion clause covering relativity is remarkable—it means that relativity never earned a Nobel Prize, not for Einstein nor for anybody else. This is an amazing gap in the Nobel awards. Relativity is widely recognized as one of the greatest scientific achievements of the twentieth century—but Einstein's work on relativity did not earn an award, nor did any of the original and profound work on relativity by his followers, such as John Wheeler, Stephen Hawking, or Roger Penrose. The Nobel committee brushed against relativity only in a couple of awards involving direct and essential applications of relativity in astrophysics: the award to Chandrasekhar and William Fowler (1983) for their theoretical investigations of the final stages in the evolution of stars, and the award to Russell Hulse and Joseph Taylor (1993) for their confirmation of the theoretical predictions of emission of gravitational radiation by a pulsar. Maybe the good professors on the Nobel committee had some Pavlovian conditioning imprinted on their souls by their authoritarian Swedish-Lutheran schooling that made them shudder at the thought of any kind of *relativitet*?

THE AWARD TO EINSTEIN was the result of a long struggle within the Nobel Committee and the Swedish Academy. Einstein had been nominated for the prize as far back as 1910, by Wilhelm Ostwald, the eminent German chemist, who was a leading opponent of atomic theory. He was nominated several more times in the succeeding years, and, after 1917, he was nominated every year, again and again by Wien, von Laue, Lorentz, Zeeman, Planck, Bohr, Eddington, and other leaders in physics, mostly for his work on special or general relativity. Often, these nominations recommended a joint award to Einstein and to Lorentz, and this probably bewildered the committee, because Lorentz was known to be opposed to Einstein's views on relativity. Maybe the committee feared that if Einstein and Lorentz showed up in Stockholm together for the award ceremony, they would reciprocally accuse each other of not understanding relativity.

Besides, the Nobel committee has always had a prejudice against awarding the prize for purely theoretical work. This prejudice against theory is built into the terms of Alfred Nobel's will, which says quite explicitly that the prize is to be awarded for "the most important discovery or invention" that "during the preceding year, shall have conferred the greatest benefit on mankind." Evidently, it is difficult to argue that relativity or energy quantization have done much to benefit mankind, despite the revolutionary implications of these concepts for the foundations of physics.

And this prejudice was reinforced by the difficulties that the Nobel committee members had in understanding and judging theoretical achievements. They were, of course, uncomfortable with awarding prizes for theories they did not understand, and they found it much easier to make awards for neat, concise experimental tidbits that were well within their range of competence. Examination of the list of Nobel Prize winners in Einstein's lifetime, from 1901 to 1955, reveals there were only a dozen awards for purely theoretical work. The standards for awards in theoretical physics are much higher than for experimental physics. The theoretical Nobelists are all eminently memorable, and any physicist can recite their names; the experimental Nobelists are sometimes eminently forgettable.

In the case of Einstein's theory of relativity, there was also some additional hesitation in the Nobel committee because they felt this theory had not yet received enough observational confirmation—they were afraid to award the prize for a theory that might yet be shown to be wrong. But under the flood of nominations, the committee felt that the pressure in favor of Einstein was becoming irresistible. The way out of this impasse was suggested by the Swedish physicist Carl Wilhelm Oseen, a member of the committee, who proposed that as a compromise the Nobel Prize be given to Einstein not for relativity, but for his discovery of the law for the photoelectric effect. This was, admittedly, a much less grand discovery than relativity, but it was a neat, concise thing, within the range of competence of all the members of the committee—and they gladly approved it.

The committee also approved the award of the 1922 Nobel Prize to

Bohr, for his work on the structure of atoms and their emission of radiation. The awards of the 1919, 1921, and 1922 Nobel Prizes to Planck, Einstein, and Bohr, respectively, for their contributions to quantum theory corresponded neatly with the historical sequence of these contributions. Bohr congratulated Einstein and added that he felt it was appropriate that ". . . your fundamental contribution had to receive public recognition, before I could be considered for such an honor."[73]

EINSTEIN'S NOBEL PRIZE led to an amusing diplomatic pas de deux between the Swiss and German ambassadors in Stockholm. In Einstein's absence, the ambassador of his country was supposed to represent him at the formal award ceremony and receive the Nobel Prize medal and certificate from the hands of the king of Sweden. But was Einstein German or was he Swiss? The German ambassador was instructed by a telegram from the Prussian Academy that Einstein was German. The Swiss ambassador thought this odd, because Einstein was traveling in the Far East under a Swiss passport, but he graciously yielded to the German ambassador, and he let the Germans take possession of the medal.

Meanwhile Einstein had returned to Berlin, and he requested that the medal be delivered to him via the Swiss embassy in Berlin, because he considered himself to be Swiss. The diplomatic imbroglio was resolved by a compromise: the Germans transferred the medal to the Swedish ambassador in Berlin, who presented it to Einstein.

Doubts about Einstein's citizenship remained to be resolved. After lengthy examination of documentation in the German archives, it transpired that, although Einstein had indeed expressed his intention to remain an exclusively Swiss citizen when he came to Germany in 1914, neither he nor the authorities had taken the necessary formal steps to preserve this legal status and prevent his "naturalization" as a German. By accepting membership in the Prussian Academy in 1914, Einstein had automatically become an official of the Prussian State, and therefore a German. In the end, Einstein surrendered, and in a letter to the Prussian Academy he conceded that since the German authorities were firmly of the opinion that "my employment at the Academy was linked with the acquisition of the Prussian citizenship . . . I have no objections against this interpretation."[74]

Although Einstein had a preference for his Swiss citizenship, he had ceased to attach much importance to that. During the war, he had written, "I regard affiliation with a country as a business matter, rather like a relationship with a life insurance company."[75] For him, a passport had become a mere flag of convenience. After 1923, he used Swiss and German passports indifferently; thus for his grand South American tour in 1925, he used a convenient German diplomatic passport provided courtesy of the Foreign Office. The Swiss passport he held in reserve was to prove helpful in 1933, when he renounced all his affiliations with Germany and entered the United States as a Swiss citizen.

11

"Does God play dice?"

*Einstein's question to Niels Bohr during their discussions at
the 1927 Solvay Conference, where Einstein stubbornly argued
against the probabilistic interpretation of quantum theory.*

For several years there had been some simmering criticism of relativity, and in 1920 the hot spotlight of publicity focused on Einstein brought this criticism to a boil, with public denunciations of Einstein's theories. Biographers and historians have tended to see these denunciations as motivated by anti-Semitism, and it is true that some shady characters ("men of darkness," as Planck called them) tried to exploit the scientific dispute for their own ends. However, the physicists who criticized Einstein and his theories in the early 1920s, and who came to be called "antirelativists," thought they had sound scientific arguments against relativity. If these physicists had an ulterior motive, it was academic envy rather than anti-Semitism. The dispute was initially conducted at a proper scholarly level, but it later descended to a personal level, and it finally acquired anti-Semitic overtones. Einstein was as much to blame for this downward spiral as were his critics—he gave as good as he got, and he sometimes threw the first punch.

A public denunciation of Einstein's theories was organized in August 1920 by Paul Weyland, an engineer with political ambitions who hoped to achieve fame by demolishing Einstein, and who appears to have had some clandestine financial backing from persons with a political agenda. For the occasion, he rented the Berlin Philharmonic Hall, and he invited Ernst Gehrcke, an experimental physicist at the Reichanstalt, to give a speech against relativity. Gehrcke was a competent experimenter, but he had no talents as a theorist, and his understanding of the basic concepts of special and general relativity was somewhat shaky. In his confused state of mind he imagined that he could disprove both theories by showing that they contained fatal contradictions, and he had published several papers about this over the preceding ten years.

Gehrcke imagined that one such fatal contradiction was the famous twin paradox, often called the Langevin paradox in those days, because Langevin had first described it. This paradox involves a pair of hypothetical twins, one of whom travels in a fast spaceship from the Earth to a nearby star and returns some years later. According to the time-dilation effect of special relativity, such high-speed travel implies that the body clock of the traveling twin runs slower than that of the twin who stays at home, so when the twins are reunited, the traveling twin will be younger than the twin who stayed at home. The paradox arises when the situation is examined from the point of view of the traveling twin: in the reference frame of that twin, the Earth travels at high speed in the opposite direction, and therefore the time-dilation effect should apply to the twin on the Earth, not the twin in the spaceship.

Most physicists—but not Gehrcke—understood that the time dilation cannot be applied in this simplistic way in the reference frame of the spaceship, because that reference frame is not an inertial reference frame. The spaceship must decelerate and reverse its motion when it turns around upon reaching the star, and during this negative and positive acceleration the relativistic time dilation is modified. In a 1918 paper, written in response to Gehrcke's criticism, Einstein had explained how the acceleration resolves the paradox by canceling the time dilation for the twin on the moving Earth.[1] Although Einstein's explanations in this paper were clear and meticulous, they were presented in a somewhat condescending manner, in imitation of the dialogues of Galileo, and Gehrcke was not amused by the style or satisfied by the content.

EINSTEIN HIMSELF CAME to the public meeting in the Philharmonic Hall, and so did his colleagues Nernst and von Laue. Weyland spoke first, denouncing relativity as a publicity stunt and "scientific Dadaism." Gehrcke then gave a lengthy and earnest analysis about imagined contradictions in relativity and the inadequacies of the theoretical and observational evidence for the perihelion precession and light deflection. He also claimed that Einstein's result for the perihelion precession had already been obtained more than twenty years earlier by Paul Gerber, a physics teacher, who had published a confused calculation with fanciful ideas about the propagation speed of gravitational forces. Einstein did not speak at the meeting, but he published a response in a newspaper a few days later. In his reply to the "Antirelativity Co.," Einstein adopted an insulting tone and made an accusation of anti-Semitism: "I am quite aware of the circumstance that the two speakers are not deserving of a reply from my pen; for I have good reasons to believe that this enterprise is based on motives other than the search for truth. (If I were a German national with or without swastika instead of a Jew with a freedom-loving, international disposition . . .)"[2] He also accused Gehrcke of "personal attack."[3] This was unfair to Gehrcke, who had limited his criticism to purely scientific matters and had punctiliously refrained from any personal comments about Einstein.

What provoked Einstein's irate and ill-considered reply was the presence of some rabble-rousers in the audience who had made crass anti-Semitic remarks within his earshot. Einstein was so incensed that in his reply he not only skewered Gehrcke but also Philipp Lenard, another critic of relativity, who had, however, not been present at the meeting. In earlier years, Einstein had admired Lenard's experimental work on the photoelectric effect, and they had conducted a friendly correspondence. But later Lenard had sharply criticized Einstein's Principle of Equivalence and established himself as a leading opponent of general relativity. In his reply, Einstein launched an inappropriate personal attack against Lenard, saying of him, "I admire Lenard as a master of experimental physics; in theoretical physics he has not yet achieved anything, and his objections to general relativity are of such superficiality that I have until now not considered it necessary to answer about them."[4] Lenard, as a professor at Heidelberg and a Nobel Laureate, felt deeply insulted, and he demanded that Einstein apologize in a public manner. But this Einstein refused to do.

Einstein's friends were appalled by his rash, intemperate outburst. Ehrenfest scolded him: "My wife and I cannot believe that some of the expressions are from your own hand."[5] Einstein soon conceded that he had overreacted and he wrote to Born: "We all must from time to time make a sacrifice at the altar of stupidity, for the entertainment of the deity and mankind. And I did this thoroughly in my article."[6] Despite this admission of his own error, the next year Einstein was to repeat the inflated accusation that all opposition to his theories was rooted in anti-Semitism. In an interview he gave upon arrival in New York, he said, "No man of culture, of knowledge has any animosity toward my theories. Even the physicists opposed to the theory are animated by political motives," and he explained that by this he meant anti-Semitic feelings, adding that the attacks in Berlin were entirely anti-Semitic.[7]

A month after the Berlin affair, at the congress of German natural scientists in Bad Nauheim, near Frankfurt, both Einstein and Lenard lectured about relativity, for and against. It was widely expected that Einstein and Lenard would have a fiery confrontation. To the disappointment of newspaper reporters, both kept their cool and confined their discussion to technical matters. On the advice of his colleagues, Einstein finally allowed a brief notice to be published in the Berlin newspaper expressing "his strong regrets for having directed reproaches against Herr Lenard, whom he values highly."[8]

But for Lenard this notice—signed by two of Einstein's colleagues, but not by Einstein himself—was too little, too late, and he never forgave Einstein's insult. Two years later, in 1922 at the congress in Leipzig, Lenard added an introduction to a reprint of one of his papers against relativity in which he declared that it was typical of Jews to shift objective discussions into the realm of personal quarrels.[9] This was tit for tat, and it was the first step in Lenard's sad downward slide into virulent anti-Semitism. He later became a founder of the movement for "German Physics," which tried to impose a racist, Aryan slant on the physical sciences. When Hitler took power in 1933 and Einstein left Germany permanently for the United States, Lenard viciously applauded the departure of "the relativity Jew, whose mathematically cobbled theory . . . is now gradually falling to pieces."[10]

Einstein did not attend the Leipzig congress. By 1922 anti-Semitism was on the rise in Germany. Acquaintances warned him that he was on a

Nazi "death list," and advised him to stay away from Berlin or appearing in public anywhere in Germany. In view of the recent murder of Foreign Minister Rathenau, he took these warnings seriously. The invitation for the lecture tour of Japan organized by a Japanese publisher provided a welcome excuse for leaving Germany for a while, to avoid the increased dangers he faced there.

GEHRCKE AND LENARD objected not only to technical details and perceived inconsistencies in relativity but also to the style of its physics. They felt that even if relativity were mathematically correct and free of any logical contradictions, it would still be an excessively abstruse theory that failed to explain the physical processes underlying the crucial phenomena, such as length contraction and time dilation. They thought the presence of an ether was needed for a physical explanation of these phenomena. As Gehrcke said in his sober assessment of Einstein's treatment of relativistic electrodynamics: "In Einstein's theory, which is fundamentally no more than an alternative interpretation of Lorentz's theory, the ether is completely eradicated, it appears merely as an unreasonable troublemaker, as something fantastic, of no use to critical reason; instead, from a mosaic of available observations and formulas, a general rule, the Principle of Relativity, is deduced by logical analysis. This method is undoubtedly the most objective and logically unobjectionable, but it does not rest on physical intuition; because, in principle, it examines things only superficially and makes no attempt to gain a feeling for the physical processes and to grasp their inner essence."[11]

Lenard defended the ether at the Nauheim congress, and he was much aggrieved that the German Physical Society ignored his arguments and pronounced against the ether. He ridiculed this pronouncement: "The abolition of the ether was announced in Nauheim at the plenary opening session . . . Nobody laughed. I do not know if it would have been otherwise if the abolition of air had been announced."[12] After the congress, Lenard resigned from the Physical Society, and he posted a warning on the door to his office at the Physics Institute in Heidelberg: "Members of the so-called German Physical Society are not permitted to enter here."[13]

Lenard believed that physics should rest on "healthy common sense," and he strongly objected to the abstract mathematical approach adopted by Einstein in his theory of relativity. Einstein responded to this demand for common sense by pointing out that everyday common sense can be misleading. For instance, according to common sense, the engineer of a locomotive might say that relativity is not valid for the uniform translational motion of his locomotive relative to the countryside, because "he does not have to keep on fueling and lubricating the *countryside*, but the locomotive, and that therefore it must be the latter in whose motion is revealed the effect of his labors."[14]

Gehrcke and Lenard's belief in the ether was shared by a fair number of other physicists, among them several highly regarded professors and Nobel laureates. The most distinguished of these other antirelativists were Johannes Stark and Emil Wiechert in Germany, Sir John Larmor and Sir Oliver Lodge in England, Charles Guillaume in Switzerland, Paul Painlevé and Emile Picard in France, and Abraham Michelson in the United States. In addition to these outspoken opponents of relativity, we must count Lorentz and Poincaré as tacit opponents—they never spoke out against relativity, but they withheld their approval, and that was a clear enough message. In fact, Lorentz never abandoned his belief in the ether, and, to accommodate him, Einstein proposed in a lecture at Leiden that his theory of relativity could perhaps be regarded as a theory with an unobservable ether. Comparing this relativistic ether with that of Lorentz, he said, "As concerns the mechanical nature of Lorentz's ether, one can somewhat jokingly say that immobility is the only mechanical property with which Lorentz has endowed it. One can add to this that the only change in the conception of the ether contributed by the theory of special relativity consisted in taking away from the ether its last mechanical quality, that is, the immobility."[15]

BESIDES DISAPPROVING of the scientific basis of Einstein's relativity, Gehrcke also disapproved of the slanted and sensationalistic publicity that he thought relativity received in newspapers, popular books, and public lectures. He thought that relativity was a fraud and that its acceptance by the public was a case of mass suggestion. To prove his case, he assiduously collected newspaper clippings from German and international sources, which showed that the public was totally befuddled and had the wildest misconceptions about relativity. In this he was undoubtedly right—the public had absolutely no understanding of the theory, and Gehrke's collection of newspaper clippings, which he published in a book in 1924,[16] is good raw material for a study in mass psychology and mass hysteria. But Gehrcke should have recognized that the psychopathology of the public has no bearing on the validity of the theory of relativity—although the public did not understand relativity, the experts in relativity certainly did.

Stark held much the same views as Gehrcke about the publicity surrounding relativity. Stark had initially voiced approval when Einstein published his first paper on relativity in 1905, but he objected to what he considered an excess of abstraction and formalism in the recent developments in general relativity, and he was irritated by the publicity accorded to the theory. He sternly objected to the "outrageous propaganda," and he blamed Einstein for "dragging the theory into the marketplace." Later, after Einstein had left permanently for the United States, Stark's attacks on Einstein became bitterly anti-Semitic. In an article published in the English journal *Nature*, he declared, "The relativistic theories of Einstein constitute an . . . obvious example of the dogmatic spirit," and he announced, "I have . . . directed my efforts against the damaging influence of Jews in German science, because I regard them as the chief exponents and propagandists of the dogmatic spirit."[17]

Gehrcke maintained contacts with a group of scientists and engineers in the United States who had founded an obscure society called the Academy of Nations, with the ostensible objective of fighting the specialization of the different branches of science so as to achieve an overall unity and uniformity of all of science. But the leading participants in this endeavor had an ulterior motive: they believed in daft pseudoscientific theories of their own, which they wanted to include in the mix. The leader of the Academy of Nations was the engineer/achitect Arvid Reuterdahl, of the University of St. Thomas in St. Paul, Minnesota. He presented his objections to relativity in a series of articles in the St. Paul newspapers, with headlines such as "Einstein Branded Barnum of Science; Minnesota Man Calls Relativity 'Bunk.' "[18]

Among Reuterdahl's close associates was Captain Thomas J. J. See, USN, an astronomer at the Mare Island Naval Observatory in California. As an expert on celestial navigation, Capt'n See must have been thoroughly familiar with the curved two-dimensional spherical geometry of the surface of the Earth, but apparently the idea of a curved three- or four-dimensional geometry was beyond him. In articles and long letters he argued heatedly that the curvature of space is a deception invented by Riemann, and that Carl Gauss, Riemann's teacher, never approved of this idea and "never once even spoke of the *curvature of space*." Therefore, concluded Capt'n See, general relativity is a monstrous error: "In fact relativity is definitely and finally disproved—as dead as the Dodo—and no amount of effort will galvanize it into life again."[19]

In 1921, Gehrcke founded a German branch of the Academy of Nations, which promoted the publication of articles disputing relativity. Gehrcke apparently hoped to obtain some financial support from his American friends, because Reuterdahl had hinted that Henry Ford, a known anti-Semite, was eager to promote attacks on Einstein (Ford-owned newspapers indeed attacked Einstein, but it is unclear to what extent Ford was personally involved in this). The American financial support did not materialize, and the German branch of the academy died off within a few years. The American branch died off a bit later, around 1930.

MOST OF THE ANTIRELATIVISTS were bewildered by the counterintuitive behavior of the speed of light and by the (apparent) paradoxes arising from the time dilation and the length contraction, and instead of trying to sort out their confusion, they laid the blame on relativity and pronounced it illogical and mistaken.

But the criticism of some of the more competent physicists was more subtle. They conceded that relativity was logically consistent, but they judged it unsatisfactory because it provided no physical explanation of the time dilation and the length contraction—it treated these effects as mathematical consequences of the constant speed of light, and not as mechanical effects arising from the internal dynamics of clocks and measuring rods. This was the main reason why Lorentz and Poincaré disapproved of Einstein's formulation of relativity.

Poincaré had reservations about Einstein. He perceived his genius and his creativity, but he thought him somewhat rash and injudicious: "Monsieur Einstein is one of the most original minds I have known; in spite of his youth he already occupies a very honorable position among the leading scholars of his time . . . He does not remain attached to the classical principles and, faced with a physics problem, promptly envisages all possibilities . . . I would not say that all his expectations will resist experimental check when such checks will become possible. Since he is probing in all directions, one should anticipate, on the contrary, that most of the roads he is following will lead to dead ends . . ."[20]

In a memorable confrontation at the first Solvay Conference in 1911, Poincaré asked Einstein, "What mechanics are you using in your reasoning?" and Einstein replied, "No mechanics,"[21] which left Poincaré speechless. Einstein was not pleased with Poincaré's reaction; he said of him, "Poincaré was simply generally opposed and showed little understanding of the situation, despite all his acuteness."[22]

But in asking, What mechanics? Poincaré was displaying more understanding than Einstein, and he put the finger on a weak link in Einstein's relativity. All that Einstein's formulation of relativity says by way of an explanation of length contraction and time dilation is that these phenomena are required to keep the speed of light constant. This comes dangerously close to the teleological explanations of Aristotelian philosophy, where stones were said to fall downward because they *want* to move to the center of the Earth. Are we to believe that a moving rod contracts because it *wants* to keep the speed of light constant?

This failure of Einstein's theory to provide physical explanations for several of its basic assertions was what had led Sommerfeld to complain, with some justification, about "unvisualizable dogmatics" and "the conceptually abstract style of Semites."[23] Today, physics students are brainwashed in their relativity courses at an early age, so they take the strange consequences of relativity in stride and ask no questions. This substitution of familiarity for understanding brings to mind an anecdote about the famous mathematician John von Neumann, who answered the complaint of a student about not being able to understand the arcana of group theory with the magisterial pronouncement, "In mathematics you don't understand things, you just get used to them."[24]

Poincaré was quite right in asking, What mechanics? and Einstein was wrong in dismissing this question as obtuse. On other occasions Einstein declared that he wanted to understand how nature works: "I want to know how God created this world . . . I want to know his thoughts."[25] But he never seems to have shown any interest in God's thoughts on length contraction and time dilation—these he was quite happy to take on faith.

LORENTZ AND POINCARÉ preferred a different formulation of relativity, first proposed by Lorentz in 1904 and then expanded by Poincaré in 1905. Lorentz and Poincaré insisted that there must be an ether, and they believed that the failure of the Michelson-Morley experiment and other experiments to detect the motion of the Earth through the ether must be attributed to the length contraction and time dilation, which they proposed to explain as mechanistic physical effects arising from the internal dynamics of measuring rods and clocks.

Lorentz showed that if a solid body, such as a measuring rod, is regarded as an array of positive and negative electric point charges held in equilibrium by their mutual electric forces, then the FitzGerald-Lorentz contraction of this body has a simple explanation. Maxwell's equations imply that when the body is in motion, the electric fields associated with the array of charges suffer a kind of contraction,[26] and this alters the forces that the charges exert on one another. The alteration of the forces causes a shift in the equilibrium positions of the charges, so they adopt new equilibrium positions, closer to each other, which makes the body shorter.[27]

But Lorentz's simple and clear explanation had an unfortunate defect: it assumed that the electric charges in a solid body are in *static* equilibrium, that is, they are at rest relative to one another. By 1911 it became clear that the electric charges in atoms are *not static*—the experimental investigations of Ernest Rutherford and the theoretical investigations of the Danish physicist Niels Bohr established that the electrons in atoms are not at rest, but instead are orbiting around the atomic nuclei at high speed.

The motion of electrons is controlled by the laws of quantum mechanics instead of the laws of classical mechanics, and the problem of the size of atoms and the separation between atoms has to be reexamined by quantum theory. This made the analysis of the length contraction much more difficult, because the behavior of atoms in a solid body moving at high speed is controlled by the laws of *relativistic* quantum mechanics, and these were not discovered until 1928, by the young British physicist Paul Dirac.[28] Thus, the theoretical tools needed for a realistic physical explanation of the length contraction were not available to Lorentz and Poincaré. Einstein's abstract axiomatic mathematical treatment of length contraction was the only game in town, and it became widely accepted—but not by the antirelativists who, not altogether unreasonably, clamored for a more physical, intuitive treatment.

When the equations for relativistic quantum mechanics were finally formulated by Dirac, Lorentz's attempt to calculate the length contraction from atomic physics had nearly been forgotten, and neither Dirac nor any of his contemporaries thought to apply his new equations of relativistic quantum mechanics to the problem of length contraction. It was not until 1941 that the American physicist W.F.G. Swann revisited Lorentz's arguments in the context of relativistic quantum mechanics and showed that, indeed, the length contraction emerges from a quantum-theoretical calculation of the length of a solid body when the length of a moving solid body is compared with the length of a similar body at rest.[29] Swann's arguments also apply to the time dilation, and they can be used to show how the internal dynamics of an atom or of any kind of clock lead to a time dilation, that is, a reduction of the frequency of vibration or ticking of a moving atom or a moving clock. But Swann's arguments were as quickly forgotten as Lorentz's.

In 1976, J. S. Bell, a brilliant theoretical physicist at CERN, tried once more to draw attention to the advantages of Lorentz's treatment of length contraction, and he calculated in detail how the dimension of the orbit of an electron around a nucleus is distorted and acquires a length contraction when the atom is accelerated to some new, high speed.[30] Bell recommended that this physical explanation of the length contraction be included in the teaching of relativity. His recommendation fell on deaf ears—no current textbook on relativity uses this approach, and only very few even mention that length contraction and time dilation have a physical explanation.[31]

EINSTEIN CALLED HIS FORMULATION of relativity a *principle*, and he adhered to this terminology for several years, using *Relativitätsprinzip* in the title of several papers.[32] In contrast, Lorentz called his formulation a *theory*,[33] and this distinction is important. Although in everyday usage, the word *theory* often indicates an opinion or a conjecture (e.g., "My theory is that Casanova was a castrato"), in physics this word indicates an explanation. When physicists speak of the *theory* of the tides, the *theory* of superconductivity, or the *theory* of nuclear fission, they really mean the *explanation* of tides, superconductivity, or fission. To prevent confusion of the public, it would be best to replace "theory" with "explanation" in all physics textbooks (and also in textbooks in other sciences, especially in evolutionary biology).

So when Lorentz described his treatment of relativity as a *theory*, he meant that he was offering an explanation of relativity; that is, he was offering an explanation, on the basis of mechanics and electrodynamics, of why we cannot detect the motion of any reference frame by any experiments within that reference frame. In contrast, when Einstein described his treatment of relativity as a *principle*, he meant that he was not offering an explanation, but a prescription; that is, he was telling us how to synchronize clocks and how to impose a time dilation on clocks and a length contraction on measuring rods (by unspecified means), so that we would not be able to detect the motion of any reference frame.

This fundamental difference in methodology between Einstein's treatment of relativity and that of Lorentz-Poincaré's needs to be emphasized: *Einstein's treatment is prescriptive, whereas that of Lorentz-Poincaré's is explanatory.* In medical language, we might say that the Michelson-Morley experiment had revealed a disease in the ether model, and that Einstein cured this disease by prescribing potent medicines and surgical excision of the ether; whereas Lorentz-Poincaré simply explained this disease away by showing that it was a perfectly healthy, though surprising, consequence of Maxwell's dynamics of the electric and magnetic fields, so there was no need to eradicate the ether, although the laws of physics conspire to prevent us from identifying the ether and its rest frame observationally.

Einstein's contemporaries failed to grasp this fundamental difference. Over Einstein objections, they changed the name of his version of relativity from *principle* of relativity to *theory* of relativity, and they imagined Einstein's version was a refinement or a generalization of the Lorentz-Poincaré version. Predictably, the name *theory* of relativity became a source of confusion. Physicists and nonphysicists alike expected that Einstein's "theory" would provide an explanation of the relativity of motion, and when they found a prescription instead of an explanation, the result was puzzlement and frustration. And often this frustration led to a strong reaction against the theory, as seen in the publications of Gehrcke and other antirelativists of Einstein's days. Even today, books and articles are occasionally published that challenge relativity or the conclusions drawn from it.[34]

In Germany, relativity went by the name of "Lorentz-Einstein theory" for some twelve years, and when the name of Lorentz was finally dropped after the end of World War I, this was not because of some recognition or agreement that Einstein deserved the full credit for the invention of special relativity, but because Lorentz was pushed into the background by the stellar fame that Einstein acquired from his theory of general relativity. As Mehra said, Einstein became the great cat of relativity, and everything in relativity was attributed to him, rightly or wrongly.

Lorentz was lucky that his name was not deleted from the Lorentz trans-formations and replaced by that of Einstein. Poincaré was not so lucky. He first stated the principle of relativity that Einstein repeated in his 1905 paper, without giving credit. Poincaré's name vanished from relativity, except in technical treatises, where the combination of a Lorentz transfor-mation with a translational transformation (that is, a change of the loca-tion of the origin of coordinates) is called a Poincaré transformation.

THE FAILURE TO UNDERSTAND the difference in method-ology between Einstein's and Lorentz-Poincaré's treatment of relativity led to a dispute in 1953, when Whittaker published his history of twentieth-century physics in which he awarded most of the credit for the invention of special relativity to Lorentz and Poincaré. Whittaker demoted Einstein to the role of a minor player, saying, "Einstein published a paper which set forth the relativity theory of Poincaré and Lorentz with some amplifica-tions, and which attracted much attention."[35]

Einstein and his defenders reacted with annoyance. In a letter to Born, Einstein pouted, "If he manages to convince others, that is their own affair. I myself have certainly found satisfaction in my efforts, but I would not consider it sensible to defend the results of my work as being my own 'property,' as some old miser might defend the few coppers he had labori-ously scraped together. I do not hold anything against him . . . After all, I do not need to read the thing."[36] And Einstein's friend and biographer Abraham Pais peevishly said of Whittaker, "His treatment of the special theory of relativity . . . shows how well the author's lack of physical insight matches his ignorance of the literature."[37]

But Whittaker's views cannot be dismissed that easily. In his history of twentieth-century physics he gave full credit to Einstein for the discovery of general relativity, which tends to show that he had no prejudice against Einstein. And Born, who knew Whittaker personally and respected his scholarship, and who had also attended some of Poincaré's early lectures on relativity, admitted, "The reasoning used by Poincaré was just the same as that which Einstein introduced in his first paper of 1905 . . . Does this mean that Poincaré knew all this before Einstein? It is possible."[38] Pais was just as dismissive of Born as he was of Whittaker; he said of Born, "He did not acquit himself very well."[39]

The real defect in Whittaker's account of the invention of relativity is his lack of awareness of the profound differences between Einstein's treatment and Lorentz-Poincaré's. To ask, Who invented relativity? is like asking, Who invented the sail? The question is badly posed and meaningless—the answer depends on what kind of sail. The square sail was invented by the Egyptians, but the lateen sail (the first fore-and-aft sail) was invented by the Arabs or maybe by the Polynesians.

Likewise, one way of treating relativity was invented by Einstein, but another way was invented by Lorentz-Poincaré. Both ways of doing relativity lead to the same physical results, but they reach these results by different pathways. Today, physicists prefer to do relativity à la Einstein, because it is more concise and direct, without the extra metaphysical baggage of an undetectable ether and its preferential reference frame absolutely at rest. Besides, as Pauli pointed out long ago, the Lorentz-Poincaré approach is too narrowly focused on electromagnetism: "Lorentz and Poincaré had taken Maxwell's equations as the basis of their considerations. On the other hand, it is absolutely essential to insist that such a fundamental theorem . . . should be derivable from the simplest basic assumptions. The credit for having succeeded in doing this goes to Einstein."[40]

Thus, today, "relativity" is usually taken to mean relativity à la Einstein. However, modern cosmology identifies a preferential rest frame, that is, the rest frame of the cosmic microwave radiation, or the thermal radiation left over from the Big Bang. And calculations about the cosmic microwave radiation are most conveniently done à la Lorentz, in the preferential reference frame suggested by cosmology. Lorentz would undoubtedly have regarded this as confirmation of his ideas about the ether.

So any dispute over which way of doing relativity is better is ultimately futile. Einstein's is better most of the time, but Lorentz-Poincaré's is better sometimes. And both are right—there is more than one way to skin the cat.

THEORETICAL PHYSICISTS SUFFER from the curse of

Knabenphysik ("boyish physics").[41] According to this curse, physicists are condemned to make their great discoveries before age thirty or not at all. There are some exceptions to this curse—for instance, Planck discovered quantization at the ripe old age of forty-two—but it applies to most of the great discoveries in theoretical physics. The curse does not apply to experimental physics, probably because experimenters usually need expensive equipment to make their discoveries, and this is rarely placed in the hands of youngsters. Nor does it apply to artists, many of whom produce their most remarkable work late in life. From this, one might conclude that some mental deterioration is artistically advantageous (an extreme example of this is seen in van Gogh, who was a rather unremarkable painter until he went bonkers at age thirty-five).

A list of post-Planck contributions to quantum physics, all of them crowned with Nobel Prizes, shows the prevalence of *Knabenphysik*: quanta of light, by Einstein at age twenty-six; quantum theory of atomic orbits, by Niels Bohr at age twenty-seven; wave properties of particles, by Louis deBroglie at age thirty-one; matrix formulation of quantum mechanics, by Werner Heisenberg at age twenty-four; wave equation of quantum mechanics, by Erwin Schrödinger at age thirty-nine; exclusion principle, by Wolfgang Pauli at age twenty-five; relativistic wave equation for electrons, by Paul Dirac at age twenty-two.

Dirac was acutely aware of his lack of remarkable contributions after age thirty—even his splendid textbook on quantum mechanics was written before he reached that age. Late in life he said, "My own contributions since the early days have been of minor importance."[42] Dirac's lament was included in a parody of Goethe's *Faust* that a group of brilliant young physicists wrote and performed at a conference at Niels Bohr's Institute in Copenhagen during in the spring of 1932. The actor portraying Dirac complains:

> *Certainly, old age is a cold fever*
> *That every physicist suffers with!*
> *When one is past 30,*
> *He is as good as dead!*[43]

Although some of the most striking examples of *Knabenphysik* belong to the twentieth century (as in the case of the three youngest *Knaben*: Heisenberg, Pauli, Dirac), there are plenty of examples from earlier centuries. Galileo, Newton, and Lorentz all made their greatest discoveries before age thirty, and Maxwell followed not far behind (he formulated his celebrated equations at age thirty-three). Newton was quite aware of the advantage of youth; he said that at twenty-four years he was "in the prime of my age for invention & minded Mathematicks and Philosophy more than any time since."[44]

Einstein attained the cursed age of thirty years in 1909, but the creative phase of his life extended beyond that—he was thirty-five when he produced general relativity, his greatest achievement. The curse finally caught up with him when he reached his forties and was at the height of his fame. His last great discovery was made in 1924, at age forty-five, when he predicted a strange condensation in gases at extremely low temperature (the "Bose-Einstein condensation").

Perhaps the delayed onset of the curse is explained by Einstein's youthful demeanor—he did not look or act his age. After meeting the forty-three-year-old Einstein during a visit to Paris, the French astronomer Charles Nordmann said of him, ". . . the strongest impression is that of a stunning youthfulness, very romantic and at certain moments irresistibly reminiscent of the young Beethoven . . . And then suddenly laughter breaks out and you have a student before you."[45]

After 1924, there was a perceptible decline in Einstein's work, a decline that became more pronounced with time and that gradually isolated him from the new developments in physics. It finally led him into an obsessive pursuit of a delusional unified theory of electromagnetism and gravitation with which he hoped to surpass the triumphs of his theory of general relativity.

IN PART THIS DECLINE in the quantity and quality of Einstein's scientific work can be attributed to the loss of enthusiasm, malleability, and inventiveness expected with advancing age. Einstein recognized this himself: "Discovery in the grand manner is for young people, and hence for me a thing of the past."[46] But the changing circumstances of Einstein's life also played a significant role. His celebrity status put heavy demands on his time, and he became deeply involved in pacifist and in Zionist causes and in the "Jewish question" in Germany. He wrote and lectured on these issues, and he often was the main attraction at fund-raisers for the assistance of Jewish refugees from Eastern Europe, for Isreali settlers, and especially for the Hebrew University of Jerusalem. He traveled extensively, on several lecture tours and fund-raising tours to South America, Southeast Asia, and, repeatedly, to the United States.

Besides, he had administrative duties as secretary of the German Physical Society, to which had had been elected, and as director of the Kaiser-Wilhelm Institute for Physics. This institute existed only on paper, as an

Einstein's summer house in Caputh near Potsdam, now a museum. He kept his sailboat on the nearby Havel, a river that at Caputh broadens into a lake.

organization for dispensing research funds. Only later, after Einstein left Germany, did the institute acquire a building of its own, when John D. Rockefeller, Jr., then the richest man of the world and a Nazi sympathizer, made generous donations to the Kaiser-Wilhelm Society in the 1930s.

Einstein also devoted more time to his private amusements: sailboats and women. First he rented a summer cottage and a sailboat on the river Havel, near Berlin, and later he sailed on the Baltic Sea, at Kiel, where Hermann Anschütz-Kaempfe, a rich industrialist and inventor of a gyrocompass, placed a boat at his disposal. In 1929 he acquired a summer home of his own in the small village of Caputh, near Potsdam, outside Berlin, and his friends gave him a pretty little sailboat as a birthday gift. There he would while away long hours on his sailboat on the Havel or in the company of various women not quite as bad as *demimondaines* but not quite as good as what Balzac would have called *honnête femmes*. The names of several of these paramours have surfaced—Betty, Toni, Margarete, Estella, Ethel—sometimes in letters that Einstein wrote to his wife and to his stepdaughter Margot, openly bragging about his conquests. As the serving woman at Caputh later said in an interview, the professor "had an eye for the ladies. He had a weakness for the ladies."[47] At Caputh, Einstein serviced a small harem.

BESIDES, THERE WERE also some other distractions, somewhat less frivolous, related to Einstein's inclination to tinker with bits of machinery. He had the instincts of an inventor, which is why he had found his employment at the patent office so congenial. During his days in Berne he had invented a *Maschinchen* (little machine) for the measurement of electricity, which he had hoped to produce commercially for sale to physics laboratories. Later, in cooperation with the Hungarian physicist Leo Szilard, he designed and patented a refrigerator that was cooled by the evaporation of methyl alcohol; but, like his *Maschinchen*, this device had no commercial success.

And, during World War I, his desire to tinker led to his attempt to design a new airplane wing, which, if successful, would have had important military applications. This excursion into the world of aeronautical engineering ended rather suddenly, when Einstein's new wing was installed on a small aircraft, for a test flight. The aircraft took off well enough, but then proved unstable in flight. As the test pilot described his experience: ". . . after take-off, I was hanging in the air like a 'pregnant duck,' and was mighty glad when after a painful straight-ahead flight I had the wheels back down on firm ground, just short of the fence at the end of the runway."[48]

Einstein's involvement with the military-industrial complex became more serious in the 1920s, when he participated in the development of a gyrocompass by the German company of Anschütz-Kaempfe. Because of Einstein's background in patents, he had been appointed as an expert witness in a lawsuit over patent infringement that Anschütz had brought against the rival US company of Sperry-Rand, which had copied some features of the Anschütz gyrocompass. After winning the lawsuit, Anschütz hired Einstein as a consultant, offering him not only a generous fee but also the use of a villa and a sailboat at Kiel, where the company was located (today, the Anschütz company is still to be found at Kiel, but it is now owned by Raytheon).

Einstein contributed some crucial improvements to the design, which not only earned him his fee, but also a 1 percent royalty on sales of the gyrocompass. The fee of 20,000 Mark was delivered to him in cash, because payment by check "would merely result in questions regarding taxation."[49] Einstein continued to work on the development of this gyrocompass for several years, until it passed its test on a German torpedo boat and went into production.

The helmsman's station in the control room of the German submarine U 505. This U boat was captured by the US Navy in 1944, and it is on display at the Museum of Science and Industry in Chicago. The vertical compass card on the bulkhead is electrically controlled by the Anschütz gyrocompass installed in the rounded cylindrical container to the right of the helmsman's station. (*Photo J. B. Spector / Museum of Science and Industry, copyright © 2008 J. B. Spector / Museum of Science and Industry*)

The Anschütz gyrocompass was intended primarily for ships of the navy and for submarines, in which the large mass of steel armor plays havoc with a magnetic compass. It was widely adopted by almost all the navies of the world—in World War II, it was standard equipment on German, French, Italian, and Japanese battleships and submarines. Only the American and the British navies refused to adopt it, for purely chauvinistic reasons. Because the Versailles Treaty forbade the manufacture and sale of such military items in Germany, Anschütz set up a front company in Holland that handled the sales. Einstein's royalties were paid into a secret bank account in Holland, an account that he also used for secret deposits of royalties from his books published outside Germany and of fees he earned from guest lectures.

Ehrenfest supervised this account for him, and in letters to Einstein he would report the income received into the account in a secret code designed to fool the German tax authorities. The receipt of fees from Holland and England would be described as "Results that you and I here in Leiden have achieved with the concentration of Au-ions." To a report of the receipt of 6700 guilders (encoded as a concentration of 6.7×10^{-3}), Einstein replied with delight, "Your notice about the high concentration of Au-ions is most gratifying, especially since it appears from the status of our local investigations that such a high concentration is quite desirable."[50] Einstein had probably become acquainted with such tax-evasion maneuvers in Switzerland, where tax evasion is a national sport, and secret accounts (those infamous numbered accounts at Swiss banks) are widely used to hide income from the black market, what the Swiss call *Schwarzes Geld*, or black money.

Royalties were paid into his account at regular intervals, even after 1933, when Einstein moved to the United States. In 1938, his royalties suddenly stopped, and when he sent a letter of complaint to the front company in Holland, he was curtly informed that Anschütz was now operating openly in Germany. The front company had been liquidated—and no more royalties were paid to him.[51]

EINSTEIN'S PARTICIPATION in military projects continued during World War II, when he was hired as a consultant for the US Navy's Bureau of Ordnance on a project for improving the detonators of torpedoes, which had proven notoriously unreliable. But the navy was not much impressed by Einstein's performance on this project and soon canceled his contract.

The intermittent flirtation with the military-industrial complex was manifestly inconsistent with the pacifist stance that Einstein had ostentatiously adopted since the beginning of Word War I. Ever since those days he had been a member and activist in various pacifist and humanitarian organizations, and for this he was much admired, especially by people unqualified to appreciate his scientific contributions. Thus, the celebrated Spanish cellist Pablo Casals said of him: "Although I never had the good fortune to get to know Albert Einstein, I developed the highest esteem for him. Certainly he was a great scholar, but beyond that he was also a pillar

Detail from the west portal of the Riverside Church on Riverside Drive in New York City. Einstein is portrayed as a mustachioed saint, at the center.

of the human conscience in a time when so many civilized values seemed to be tottering. I was perpetually grateful to him for his protest against the injustice to which my homeland was sacrificed . . ."[52]

The public perception of Einstein as an icon for world peace and justice led to some kind of beatification in 1930, when a statue of Einstein was included among the other statues of philosophers and saints in the decorative carving above the west portal of the Riverside Church in New York City. This gigantic, pseudo-Gothic Baptist church had been built with donations from John D. Rockefeller, perhaps with a nod toward the biblical verse, "It is easier for a camel to go through the eye of a needle than for a [very] rich man to enter the kingdom of God." Einstein inspected the church during one of his visits to New York, and he quipped, "I might have imagined that they could make a Jewish saint of me, but I never thought I'd become a Protestant one!"[53]

If Einstein's pacifist sympathizers had inquired a bit deeper, they would have been surprised to learn that their saintly hero was also somewhat of an opportunist, not above making some profits from military projects. Einstein never explained this inconsistency in his behavior.[54] He kept his participation in his military projects very quiet, and the facts emerged only years after his death, from documentation preserved by the Anschütz company and from letters in the Einstein Archives. With one hand Einstein worked on weapons of mass destruction, and with the other hand he beat the pacifist drum. And speaking with a forked tongue, he declared in an interview in 1947, "Non-cooperation in military matters should be an essential moral principle for all true scientists . . . who are engaged in basic research."[55]

IN SPITE OF ALL THE DISTRACTIONS, Einstein continued to produce a fair number of papers throughout the 1920s and early 1930s. Only when he left Germany in 1933 did his output finally decline. Although Einstein always published a sufficient quantity of papers to meet the criteria of "publish or perish," his work lacked the quality of his early period. The deep creative originality that characterized his earlier work is conspicuously absent—it is as though Einstein had run out of ideas. Whereas his earlier work was entirely his own, as sole author, his later work was often a cooperative effort written with various colleagues or assistants.

In contrast to his early period, which yielded a long list of memorable contributions to physics—photoelectric effect, special relativity, theory of Brownian motion, energy-mass relation, theory of latent heats, light deflection, general relativity, cosmology—his late period yielded only two memorable contributions: the quantum statistics for a gas (including the Bose-Einstein condensation) and a critique of the quantum theory of measurement (the Einstein-Podolsky-Rosen, or EPR, paradox).

And neither of these was an original, exclusive Einstein contribution: the first was a gloss on original work by the Indian physicist Bose, and the second was produced in cooperation with Podolsky and Rosen. Neither of these contributions attracted much attention at the time, and their importance emerged only many years after Einstein's death. The contribution that did attract much attention was his unified theory. This was indeed an original, exclusive Einstein contribution—and it proved an unmitigated disaster.

In the paper that Bose sent to Einstein in 1924, asking for help with its publication, he treated the photons in blackbody radiation as a gas, following Einstein's first such attempt of nineteen years earlier. At that time, Einstein had only been partially successful—his gas treatment of photons worked at high frequencies, but not at low. Bose had hit upon the felicitous and very original idea of treating photons as absolutely indistinguishable. This alters the statistical arguments that are used to discover the most likely state of the photon gas when in thermal equilibrium, and Bose was able to obtain the Planck formula not only for high frequencies but also for all frequencies.

Einstein recognized that Bose's statistical argument might also be applicable to a gas of identical, indistinguishable atoms. Soon after the publication of Bose's paper, he published a paper of his own in which he used Bose's statistical methods to examine the implications of the indistinguishability of atoms for the behavior of a gas, such as hydrogen or helium. He found that at low temperatures, the behavior of the gas is quite different from that expected on the basis of Boltzmann's classical statistical theory. In a second paper he showed that when the temperature is extremely low, a disproportionate fraction of the atoms settles into the quantum state of lowest energy, so this state becomes much more heavily populated than what is expected on the basis of Boltzmann's theory. As he wrote to Ehrenfest, "From a certain temperature on, the molecules 'condense' without attractive forces, that is, they accumulate at zero velocity. The theory is pretty, but is there also some truth to it?"[56]

Einstein thought that hydrogen gas or helium gas might be the best candidates for an experimental investigation of this effect, which became known as the Bose-Einstein condensation. But until 1928, there was no experimental evidence for this condensation, and it acquired "the reputation of having only a purely imaginary character."[57] Then it was found that at a temperature of 2.18 degrees above absolute zero, liquid helium exhibits a strange alteration, which suggests a Bose-Einstein condensation. However, because this involved helium liquid, rather than helium gas, the situation was, and remained, rather murky.

The first clear, convincing example of an authentic Bose-Einstein condensation was not found until 1995, in preparations of very cold rubidium-87 and sodium-23 gases. For this achievement, the 2001 Nobel Prize was awarded to the American physicists E. A. Cornell, C. E. Wieman, and W. Ketterle. For Einstein, this experimental confirmation of his prediction came too late. If it had happened fifty years earlier, while Einstein was still alive, he would have deserved a share of this Nobel Prize.

AT THE 1909 MEETING of the German society of natural scientists and physicians in Salzburg, Planck gave a remarkably prescient vision for the future of quantum theory:

Only for long time intervals can we say what the laws [of motion] are. But for short time intervals and large accelerations we face a gap, the filling of which requires new hypotheses. Perhaps we have to assume that a swinging oscillator does not have a continuously variable energy, and that instead its energy is a simple multiple of an elementary quantum. I believe that by means of this proposition, one can arrive at a satisfactory theory of [blackbody] radiation. Then there is the question: how can one imagine such a thing? That is to say, one seeks a mechanical or electrodynamical model of such an oscillator. However, in the mechanics and electrodynamics of today, we do not have discrete quanta, and we therefore cannot assemble a mechanical or electrodynamical model. Mechanically this thus appears impossible, and one will have to get used to it . . . We also wanted to imagine the electric current mechanically by analogy with a flow of water, but we had to give up on that, and just as we got used to that, we will also have to get used to such oscillators. Of course, this theory will have to be treated in much more detail than has been done so far; perhaps somebody else will be more lucky with this than I.[58]

Planck's perception that new laws of motion would have to be found to describe nature on a small scale and his insistence that physicists would have to get used to a breakdown of the familiar, classical laws of Newton and Maxwell are an accurate preview of the later development of quantum theory. (And, sadly, he was also right in guessing that his luck had run out—the new theoretical developments would be contributed not by him, but by other physicists.)

Planck has often been described as a very conservative physicist, but, as the above passage shows, he was willing to accept change. In contrast, Einstein was to prove an archconservative, fighting an endless, and ultimately futile, battle against the new laws of quantum mechanics. Einstein accepted two of the early developments in quantum mechanics. He admired the quantization of atomic orbits proposed by the Danish physicist Niels Bohr in 1913, and he said that Bohr's work was the "highest musicality in the realm of ideas."[59] He also accepted that every particle has an associated quantum wave, as proposed by the French physicist Louis deBroglie in 1924, although neither deBroglie nor Einstein understood the significance of this wave—deBroglie called it a "fictitious associated wave,"[60] and Einstein called it a "ghostly field."[61]

But Einstein could not accept that the new quantum theory implied a lack of definite, deterministic predictions about the motion of quantum-mechanical particles, such as the motion of electrons in atoms. He felt this violated the basic conception of causality at the root of all physics: the initial information about a physical system should completely determine its future development. In a letter he complained to Max Born, "The quantum theory provokes in me quite similar sensations as in you. One ought really to be ashamed of the successes, as they are obtained on the basis of the Jesuitic rule: 'One hand must not know what the other does . . .' "[62]

Einstein rejected all the developments in quantum mechanics after Bohr and deBroglie. When Werner Heisenberg, a young student of Born's, discovered his matrix formulation of quantum mechanics in 1925, Einstein dismissed it immediately: "Heisenberg has laid a large quantum egg. In Göttingen they believe in it (I do not)."[63]

A year later, when the Austrian physicist Erwin Schrödinger discovered his wave equation for the evolution of the quantum-mechanical wavefunction, Einstein was at first enthusiastic. He described it to Ehrenfest as "Not some kind of infernal artifact, but a clear idea—and compulsory in its application,"[64] and he wrote to Schrödinger himself, "Your idea attests to true genius."[65] But then his enthusiasm quickly evaporated, and when Born introduced the probabilistic interpretation of the wavefunction, Einstein preferred to listen to his inner voice, his intuition: "Quantum mechanics is certainly imposing. But an inner voice tells me that it is not yet the real thing. The theory says a lot, but does not really bring us any closer to the secret of the 'old one.' I, at any rate, am convinced that *He* is not playing dice."[66]

Einstein could not accept that the wavefunction merely determines the probabilities for different outcomes of measurements, and that it is impossible to calculate with certainty the future development of a physical system from the initial information. He declared, "In that case, I would rather be a cobbler, or even an employee of a gaming-house, than a physicist."[67]

His rejection of the basic probabilistic features of quantum mechanics prevented his participation in the development of the core of the theory, and his only contributions dealt with problems that were tangential, although sometimes important. For instance, in 1916 Einstein made a new attempt at a derivation of Planck's law from general considerations of equilibrium of the emission and absorption process of photons, and he invented the so-called A-B coefficients to describe the emissions and absorption rates.[68] Although his new derivation was only partially successful and did not yield the complete form of Planck's law, the A-B coefficients were to prove useful many years later in the study of light emission by lasers.

AT THE FIFTH SOLVAY CONFERENCE in 1927, Einstein and Bohr engaged in an intensive discussion of the fundamental ideas of quantum mechanics. Although there were sharp disagreements, the tenor of the discussions always remained friendly, imbued with humor rather than animosity. Einstein argued that the probabilistic interpretation was inadequate and he mockingly asked Bohr whether he really believed that God plays dice (". . . ob der liebe Gott würfelt"). To which Bohr replied earnestly that since ancient times philosophers have advised caution in ascribing everyday attributes to God.[69]

Einstein thought that he could undermine quantum mechanics by attacking the Uncertainty Principle, and he invented several *Gedankenexperimente* with which he sought to demonstrate that he could overcome the limitations on the precision of measurement dictated by this principle. Every day Einstein would come down to breakfast in the splendid dining room of the Hotel Metropole and announce he had invented a new clever *Gedankenexperiment* that contradicted the Uncertainty Principle.

Pauli and Heisenberg, who where there, did not pay much attention, dismissing Einstein's arguments with a superficial, "Never mind, it's all right, it's all right."[70] But Bohr took the arguments seriously. He was a deep and ponderous thinker, slow to reach a conclusion. He would think about it all day, and in the evening, at dinner, he would invariably point out a mistake in Einstein's argument. Ehrenfest, who participated in these discussions, reported to his students in hurried, incoherent sentences, "It was wonderful for me to be present at the discussions between Bohr and Einstein. Like a chess game. Einstein always new examples. So to speak *perpetuum mobile* of the second kind, to break down the uncertainty relations . . . Einstein like jack-in-the-box. Every morning jumping out anew. Oh, that was precious . . . He conducts himself toward Bohr exactly like the defenders of absolute simultaneity have conducted themselves toward him."[71]

The discussion took a dramatic turn three years later, at the 1930 Solvay Conference, when Einstein contrived a rather complicated *Gedankenexperi-*

ment that seemed to defeat the energy-time uncertainty relation. Einstein considered a closed container filled with light or other radiation, to which is attached an alarm clock, which momentarily opens a shutter in the wall of the container and releases some amount of light at a precise, preset time. Einstein proposed to use a balance to measure the mass of the container before and after the release of the light and to calculate the precise energy released by means of the mass-energy relation, $E = mc^2$.

This example troubled Bohr. He was at a loss on how to deal with it. "During the whole evening he was extremely unhappy, going from one to the other and trying to persuade them that it couldn't be true, that it would be the end of physics if Einstein were right; but he couldn't produce any refutation. I shall never forget the vision of the antagonists leaving the club: Einstein, a tall majestic figure, walking quietly, with a somewhat ironical smile, and Bohr trotting near him, very excited."[72]

But by the next morning, after much thought and little sleep, Bohr had the answer. He pointed out that Einstein had failed to take into account the time-dilation effect of general relativity. When the container is placed on a balance, the balance pan must be free to move up and down, and thus there will be a (small) uncertainty in the vertical position of the pan, the container, and the attached alarm clock. Correspondingly, according to the time-dilation effect of general relativity, the clock will acquire an uncertainty in its rate and an uncertainty in the time that it registers. Bohr showed that this time uncertainty is related to the energy uncertainty in the way required by the energy-time uncertainty relation.

We do not know what Einstein's reaction was to this refutation of his argument. He must have appreciated the irony of the situation—Bohr defeated him with his own theory of general relativity. We can guess that Einstein conceded with his customary good sense of humor and perhaps a *bon mot*. But this defeat left a permanent impression on him, and he never again made any attempt to challenge the Uncertainty Principle. However, he continued to oppose quantum mechanics, and "God does not play dice" became his battle cry.

"The graveyard of disappointed hopes"

*Einstein, in a 1938 letter to a friend, describing the failure
of the latest of his several futile attempts at constructing a
unified theory of electricity and gravitation.*

The most grandiose mistakes of Einstein's career were his several
unified theories of electricity and gravitation. For nearly thirty
years, from 1926 until his death in 1955, these were the central
focus of his research. They were a grand delusion—they led to papers and
more papers of abstruse mathematics, but they never yielded anything
of lasting interest in physics. Born described the weak point in Einstein's
work in those final years: ". . . now he tried to do without any empirical
facts, by pure thinking. He believed in the power of reason to guess the
laws according to which God has built the world."[1]

Guesswork inspired by God and unsupported by facts is perhaps suitable
for theology and theocracy, but it is not suitable for physics. Not surprisingly, all of Einstein's several attempts at unified theories were trash, and
it is the crowning tragedy of Einstein's scientific career that this was obvious to all his close colleagues, but out of compassion and respect for the
great old man only a few could bring themselves to tell him so. When the
physicist Lee Smolin arrived in 1979 at the Institute for Advanced Study in

Princeton, where Einstein had spent his final years, he was eager to make contact with Einstein's "living legacy," and he asked Freeman Dyson, one of Einstein's surviving colleagues at the institute, to tell him what Einstein was really like. He got a revealing answer:

> Dyson explained that he too had come to the institute [in 1947] hoping to get to know Einstein. So he went to Einstein's secretary, Helen Dukas, to make an appointment. The day before the appointment, he began to worry about not having anything specific to discuss with the great man, so he got from Dukas copies of Einstein's recent papers. They were all about Einstein's efforts to construct a unified-field theory. Reading them that evening, Dyson decided they were junk.
>
> The next morning he realized that although he couldn't face Einstein and tell him that his work was junk, he couldn't *not* tell him either. So he skipped the appointment and, he told me, spent the ensuing eight years before Einstein's death avoiding him.[2]

Einstein's publication of this junk was pathetic. He was publishing his failures and he was admitting that his theories were incomplete, but again and again he deluded himself into believing that he was within striking distance of success. Any physicist reading his papers could see from a cursory inspection that Maxwell's equations were conspicuously absent from these papers and that there was no way that they could emerge from Einstein's mathematical framework. But Einstein was blind to the obvious defects of his lofty mathematical creations, and he would usually take years to come to the recognition that the latest version of his unified theory had to be dumped into that large wastebasket next to his desk.[3]

Robert Oppenheimer, the head of the research team that produced the first atomic bomb, visited the institute before the war and later became its director. He said, "Einstein is completely cuckoo,"[4] and in an essay published in the UNESCO volume *Science and Synthesis* he emphasized the pathos inherent in Einstein's futile struggles with his unified theories: "In the last years of Einstein's life, the last twenty-five years, his tradition in a certain sense failed him. They were the years he spent at Princeton and this, though a source of sorrow, should not be concealed. He had a right to that failure."[5]

Oppenheimer may have been somewhat of an expert in matters

"cuckoo." In the 1920s, while a postgraduate student at Cambridge, he tried to poison his physics tutor with an apple laced with a toxic chemical. The Cambridge dons treated this incident with remarkable leniency, merely placing Oppenheimer on probation, under the supervision of a psychiatrist.[6] Perhaps Edward Teller was thinking of this incident when he testified against granting Oppenheimer a security clearance at the notorious 1953 Atomic Energy Commission Hearing, saying, "I feel that I would like to see the vital interests of this country in hands which I understand better, and therefore trust more."[7]

DREAMS OF A UNIFICATION of electricity and gravitation date back to the middle of the nineteenth century, when Bernhard Riemann speculated that the mechanical properties of the ether might be the cause of both electric and gravitational effects.[8] The first serious, and unsuccessful, attempt at such a unification was made by Hilbert in 1915, in the course of which he discovered, or rediscovered, Einstein's equations (see Chapter 9).

The next attempt was made in 1918 by the Swiss mathematician Hermann Weyl, who proposed a modification of Einstein's theory of general relativity involving weird changes of the length of measuring rods and changes of the rates of clocks when they are transported from one place to another. According to Weyl, when a meterstick or a clock is transported through an electric or magnetic field the length or the rate suffers a permanent change. This means that electric and magnetic fields affect the geometrical measurements . in spacetime performed with metersticks and with clocks, and that therefore the measured curvature of spacetime depends not only on the gravitational fields, but also on the electric and magnetic fields. The participation of all these fields in the measured curvature of spacetime implies a geometric unification of electromagnetism and gravitation.

At first Einstein was enthusiastic about Weyl's theory, and he wrote to him, "And now you have even given birth to the child that I absolutely was not able to foster, the construction of the Maxwell equations from the $g_{\mu\nu}$ [the metric tensor]."[9] But he soon realized that there was a fatal flaw in Weyl's proposal: atoms can be regarded as little clocks and, according to Weyl's theory, if two otherwise identical atoms have traveled through different electric and magnetic fields sometime in their past, they should vibrate at different rates, that is, the vibration rate should depend on the individual electric and magnetic history of each atom. Einstein pointed out that this is contrary to experience—when we examine the atoms of any given chemical element in our laboratories, we find that they all vibrate at exactly the same rate, even though the individual magnetic histories of these atoms might be quite different.[10] Weyl's theory became a text-book example of a pretty theory slain by ugly facts. (However, some of the ideas that Weyl developed during his investigations of his theory were later exploited in a different context, and they play an important role in the relativistic quantum theories of today.)

A NOVEL FORM of a unified theory was contrived in 1921 by the German mathematician Theodor Kaluza and shortly thereafter improved on by the Swedish physicist Oskar Klein (no relation to the mathematician Felix Klein). Kaluza achieved his unification by adding an extra dimension to spacetime—he used a five-dimensional spacetime with four space dimensions and one time dimension, and a corresponding metric tensor for these five dimensions. The gravitational field and the electric field (more precisely, the electromagnetic potentials) are included as different components of the five-dimensional metric tensor. The Einstein equations (p. 216) then have extra components for the fifth dimension, and these extra components are the Maxwell equations.

Thus, this unified theory reproduces all the features of Einstein's theory of gravitation and Maxwell's theory of electromagnetism. There is only one catch: the physical space we observe has *three* spatial dimensions and one time dimension, not *four* spatial dimensions and one time dimension, as postulated by Kaluza and Klein. Somehow, the fifth dimension, if it exists, remains hidden from our view. To explain this apparent absence of the fifth dimension, Kaluza proposed that the fifth dimension is curled up in a small circle, so a particle that moves in the fifth direction merely runs round and round this circle, like a dog on a short leash, without ever managing to get far away. If the circle is very small, maybe a billionth of a billionth of a billionth of a centimeter, we would never notice this motion at all, and the particle would seem to us to remain at rest in three-dimensional space.

But, although the Kaluza-Klein unification was clever and mathematically elegant, it was a purely formal unification, rather like the pro forma marriages of prepubescent children sometimes performed in the Middle

Ages for dynastic reasons. Those marriages remained unconsummated and produced no offspring, and likewise the Kaluza-Klein unification produced no offspring, that is, it produced no new physics, no new results, no new predictions. The Kaluza-Klein unification was merely a mathematical trick for placing Maxwell's equations and Einstein's equations in a five-dimensional mathematical framework, but it did not truly join these equations in a real, physically meaningful way. The Kaluza-Klein formalism was elegant, but it was too superficial. In contrast to the unification of space and time in special relativity, which required a drastic modification of Newton's space and time and led to new, interesting physics, the Kaluza-Klein unification of Maxwell's equations and Einstein's equations was accomplished painlessly without any modification of these equations whatsoever. Thus, it was no more than a sterile mathematical exercise—no pain, no gain.

Pauli declared, ". . . [it] is in no way a 'unification' of the electromagnetic and gravitational field."[11] Einstein at first seemed to like the idea, writing to Kaluza, "The idea of achieving [a unified theory] by means of a five-dimensional cylinder world never occurred to me and would seem completely new. As of now I like your idea very much."[12] But then his enthusiasm became variable; sometimes he spoke in favor of the theory, sometimes against.

There is only one interesting physical consequence of the Kaluza-Klein unification. In 1926, Klein[13] showed that the quantization of electric charge can be regarded as a direct consequence of the quantization of the circular motion in the fifth dimension. To obtain the observed value of the electric charge, the circumference of the fifth dimension must be about a millionth of a trillionth of a trillionth of a centimeter (10^{-30} centimeter). For many years, this remained the only explanation of the quantization of electric charge. Later, it was found that the existence of magnetic monopoles requires quantization of the electric charge, and today we know how to derive this quantization from the so-called unified gauge theory of strong, weak, and electromagnetic interactions (gravitation excluded!).

IN 1922, EDDINGTON made a different attempt at unification by trying to exploit a dichotomy in the definition of the curvature of spacetime. There are various ways to detect the curvature of a space or of spacetime. For instance, we can measure the sum of the interior angles of a triangle or the area of a circle, and if the results of these measurements differ from what is expected for flat space [180 degrees, $\pi \times$ radius squared], then we know that the space is curved. All these ways hinge on laying out a straight line, such as the side of the triangle or the radius of the circle. And there are two distinct methods for doing that: we can lay out the straight line by constructing the path that has the shortest distance between two points or by constructing the path that is the "straightest." A mason or a gardener uses the first method when she stretches a tight string between two points, and a draftsman uses the second method when he slides two short rulers alternately one along the other to prolongate a line step by step. Whereas the first method relies on a distance measurement, the second method relies on parallel transport, which is, in essence, a purely directional procedure.

In a flat space, both of these methods obviously give the same result. But in a curved space this concordance of the two methods is not so obvious. It turns out that in a Riemannian geometry, such as Einstein used in his general relativity, the two methods do indeed give the same result. And that is where Eddington recognized an opportunity: he decided to construct a new kind of geometry, more general than Riemannian geometry, in which the two methods do *not* agree, that is, a geometry in which the shortest path does not coincide with the straightest path. To unify electromagnetism with gravitation, Eddington inserted both gravitational and electromagnetic terms into the equations that govern parallel transport, the so-called connection equations.

So far, so good. But then Eddington had to contrive some way to bring the metric tensor and distance measurements into his scheme, because physicists want to know not only what the "straightest" lines look like, but also what their lengths are. Unfortunately, Eddington had not the foggiest idea about how to do this. After some fumbling he finally adopted an equation for the metric tensor more or less at random, but this lacked any physical motivation, and was mathematically inconsistent.[14] Weyl said of Eddington's theory that it was "not fit for discussion."[15]

EINSTEIN'S OWN OBSESSIVE QUEST for a unified theory

began in 1926. His aims in constructing such a theory were ambitious. He not only wanted to unify electromagnetism and gravitation by joining electromagnetic and gravitational fields into a new gravitoelectromagnetic hyperfield, but he also wanted the new field equations to provide an explanation for the structure of charged and uncharged particles, which were supposed to emerge as solutions of the field equations. Besides, in the back of his mind, he held the vague hope that maybe the new field equations might provide some explanation for the quantization conditions of quantum mechanics.

In a critical assessment of the program of unified theory, Oppenheimer questioned Einstein's judgment, but admired his tenacity:

> He also worked with a very ambitious programme, to combine the understanding of electricity and gravitation in such a way as to explain what he regarded as the semblance—the illusion—of discreteness, of particles in nature. I think it was clear then, and believe it to be obviously clear today, that the things that this theory worked with were too meagre, left out too much that was known to physicists, but had not been known much in Einstein's student days. Thus it looked like a hopelessly limited and historically, rather accidentally conditioned approach. Although Einstein commanded the affection or more rightly the love of everyone for his determination to see through his programme, he lost more contact with the profession of physics, because there were things that had been learned which came too late in life for him to concern himself with them.[16]

Altogether, Einstein's program was both too ambitious and not ambitious enough. It was too ambitious in that the unification of electromagnetism and gravitation proved too difficult—every one of Einstein's attempts ended in failure. And it was not ambitious enough in that even if Einstein had succeeded, it would have been an empty gesture—by 1926 physicists knew that besides electromagnetism and gravitation there was at least one more force, the nuclear force, acting within the nucleus of the atom, and any worthwhile unification would have to include this new force. Furthermore, physicists knew that any attempt at an explanation of the structure of particles had to be based on quantum theory, not on the classical physics that Einstein insisted on using. Thus, Einstein's entire program was an exercise in futility. The program might have made some sense ten years earlier, but by 1926 it was obsolete from the start.

ALL OF EINSTEIN'S ATTEMPTS at unified theories took as their basis Eddington's idea of contriving some geometry in which the shortest path does not coincide with the straightest path, and the difference between shortest and straightest was to be associated with electromagnetic fields.[17] In his new theories, Einstein replaced the ten-component metric tensor of general relativity with a new, sixteen-component tensor. Ten components of this enlarged tensor described the measurements of lengths and distances in the curved geometry of spacetime, and the other six components supposedly described the electric and magnetic fields.

In 1925, Einstein announced his first unified theory with fanfare: "After incessant search during the last two years, I now believe I have found the true solution."[18] But the paper announced with such a flourish stops at a dead end—Einstein was not able to extract the Maxwell equations (or a modification thereof) from his calculations. Soon thereafter, in a letter to Ehrenfest, he voiced some doubts, "This summer I wrote a very beguiling paper about gravitation-electricity . . . but now I doubt again very much whether it is true,"[19] followed almost immediately by the confession, "My work of last summer is no good."[20]

In 1928, while laid up in bed because of a cardiovascular collapse, he came up with another version of a unified theory. Again, he was ecstatic: "In the calm of my illness I have laid a wonderful egg . . . Whether the bird emerging from this will be viable and long-lived lies in the lap of the Gods. But for now I bless the illness that has favored me so."[21] His innovative idea was that parallel transport can be used not only to construct a "straightest" line (by transport along the line, in the longitudinal direction) but can also be used to generate one "straightest" line from another (by sliding in the transverse direction, away from the first line). This is obviously true for straight lines in flat space, where we can slide a straight line to one side or another, to generate a parallel straight line; but it is not true in curved space, unless this curved space has some rather exceptional properties.[22]

Einstein called this innovation distant parallelism, or teleparallelism. In regard to the straightest lines, a spacetime with distant parallelism has no curvature at all—parallel lines are related to other parallel lines by transverse sliding, as in flat spacetime. Einstein assumed that the curvature of the space is revealed only in the construction of the shortest lines. Again, the electromagnetic fields are associated with the difference between the "straightest" lines and the shortest lines.

This theory created a sensation, at least in the eyes of the public and the popular press. A thousand copies of the dry-as-dust journal of the Prussian Academy containing Einstein's paper were sold out instantly, and several thousand extra copies had to be printed. Eddington wrote to Einstein, "You may be amused to hear that one of our great department stores in London (Selfridges) has posted on its windows your paper (the six pages pasted up side by side) so that passers-by can read it all through. Large crowds gather around to read it."[23]

In the United States, *The New York Times* had anticipated Einstein's publication with the headlines "EINSTEIN ON VERGE OF GREAT DISCOVERY; RESENTS INTRUSION," and "EINSTEIN RETICENT ON NEW WORK; WILL NOT 'COUNT UNLAID EGGS.' "[24] And when Einstein's paper appeared, the newspaper gushed, "The length of this work—written at the rate of half a page a year—is considered prodigious when it is considered that the original theory of relativity filled only three pages."[25] The *New York Herald Tribune* outdid the *Times* by printing in its pages a translation of Einstein's paper, including all those incomprehensible mathematical formulas. The *Tribune* had prearranged for the transmission of Einstein's paper from Berlin to New York via Telex, using a special code for the transmission of the mathematical formulas (and, amazingly, they got the formulas right).

Einstein contributed to the newspaper furor by offering a lengthy explanation of his new theory in the Sunday edition of *The New York Times* in which he called it the third stage in the development of relativity and claimed that his "new paper yields unitary field laws for gravitation and electromagnetism," and that ". . . now do we know that the force that moves electrons in their elliptical paths around atomic nuclei is the same as that which lets the planets circle in their paths around the Sun."[26]

BUT THE NEW THEORY was another dismal failure. Einstein had written down a set of equations that made no sense—they made no sense to the Londoners breathlessly reading the six pages posted at Selfridges, and they made no sense to Einstein's colleagues. Pauli pointed out that the new equations did not yield the bending of light by the Sun or the precession of the perihelion of Mercury, and he predicted that within a year Einstein would abandon this new theory. Pauli was right on all counts, except for his time estimate—it took Einstein three years to recognize that his theory of distant parallelism was another dead end. Einstein could have said about his own endeavors what he had said about Eddington's: "Above stands the marble smile of implacable nature, which has endowed us with more longing than intellect."[27] In a letter to Pauli, he finally conceded, "You were right after all, you rascal."[28]

Pauli ridiculed Einstein's serial production of unified theories, saying about Einstein, "His never-failing inventiveness as well as his tenacious energy in the pursuit of his target blesses us in recent times with about one such theory per year, on the average—in which it is psychologically interesting that the current theory is usually regarded by its author for a while to be the 'definitive solution," and Pauli proposed that, in a paraphrase of the traditional words spoken upon the death of a French king and the accession of the next, each new unfied theory might be greeted with the exclamation, "Einstein's new field theory is dead. Long live Einstein's new field theory!"[29]

Pauli's sarcastic remarks about Einstein's half-baked unified theories were well-known in the physics community, and Pauli's scoffing was cleverly captured in some verses included in the parody of Goethe's *Faust* performed at Niels Bohr's institute in Copenhagen during the physics meeting in the spring of 1932. In this skit, Einstein is represented as the king of Berlin (a parody of the king of Thule in Goethe's original) and his half-baked unified theories are represented as an infestation of "half-naked" fleas, which are pets of the king and enjoy his protection. Pauli, whom Ehrenfest had sometimes called the scourge of God because of his acerbic wit, is represented as Mephisto, and in the drinking song he sings:

Half-naked, fleas came pouring
From Berlin's joy and pride,
Named by the unadoring:
"Field Theories—Unified."

Now, Physicists, take warning,
Observe this sober test . . .
When new fleas are a-borning
Make sure they're fully dressed![30]

The author and the actors of this skit elected to remain anonymous; they called themselves the Task Force of the Institute for Theoretical Physics. But it is known that most of the skit was written by the physicist Max Delbrück.

Ehrenfest is represented as Faust, the titular character of the play. Perhaps Delbrück made this choice because he sensed that Ehrenfest, like Faust, was afflicted by self-doubts and by feelings of inadequacy about the value of his scholarly contributions. Einstein said that he had "an almost morbid lack of self-confidence."[31] At the end of the skit, Faust-Ehrenfest dies while enjoying the attentions of a press photographer, and Mephisto-Pauli pronounces a funeral oration:

> *No pleasure was enough; no luck appeased him*
> *The changing forms he wooed have never pleased him.*
> *The poor man clung to those who would evade him.*
> *All's over now. How did his knowledge aid him?*[32]

There is much poignancy in this, because just a year later Ehrenfest did indeed die, by his own hand. Ehrenfest's younger son, Vassilji, had Down's syndrome, and years earlier Einstein had advised Ehrenfest to abandon the boy in an institution, because "valuable persons should not be sacrificed to hopeless causes, not even in this case."[33] But, in contrast to Einstein—who had followed this advice in abandoning his son Tete—Ehrenfest could not let go, and he remained much pained by his son's condition. In September 1933, he acquired a revolver, picked up the boy at the institution in Amsterdam where he was confined, took him to a nearby park, and shot him through the head. Then he shot himself.

EINSTEIN'S POSITION AS KING OF PHYSICS in Berlin came to a sudden end in 1932. During a stay at Pasadena earlier that year, he had been approached by Abraham Flexner, an influential American educator who had initiated far-reaching reforms of the nation's medical schools with funding from the Rockefeller Foundation. Flexner was then in the process of founding a new research institute, the Institute for Advanced Study, to be built in Princeton, with Flexner as its first director. He had collected a hefty endowment for the institute from private donors, chief among them Louis Bamberger and his sister Caroline Bamberger Fuld, owners of the

Bamberger department stores, who had had the great good luck of selling their stores just before the Great Crash of the stock market in 1929.

Flexner wanted Einstein as a trophy for the institute. After some of those negotiations over salary at which Einstein was so adept, he agreed to come to Princeton for five months each year, starting in 1933, for a salary of $10,000 tax-free (about $160,000 in current dollars) plus travel expenses for him and his wife.

Einstein kept quiet about this arrangement when he returned to Berlin, and his colleagues at the Prussian Academy were quite surprised when newspaper reports of these arrangements began to emanate from the newly founded institute. Einstein told them not to worry; he declared that he would not abandon Germany, and that he regarded Berlin as his permanent place of residence. He also offered to accept a reduction of his academy salary by one-half, in compensation for the planned lengthy absence from Berlin, an offer he had made on some other occasions, when absent from Berlin for travel or lectures in foreign countries.

In December 1932 Einstein, his wife, and his secretary left Germany for another trip to the United States, first to Pasadena, then to Chicago and New York. He intended to return a few months later, but the rise to power of the Nazi Party made him change his mind, and he never set foot in Germany again. Elsa later claimed that during their last stay at their summer home in Caputh, Einstein felt a premonition and said to her, "Take a good look. You will never see it again."[34] But this story is inconsistent with the fact that just some weeks earlier Einstein had bought a vacant lot of land adjacent to the summer home, which clearly indicated he expected to return.

On January 30, 1933, Adolf Hitler became Chancellor of Germany, and a few weeks later the Reichstag building in Berlin was burned down, probably by secret operatives of the Nazi Party. This led to a suspension of the constitution and of civil rights, "For the protection of the people and the nation." The Nazis exploited this state of emergency to transmogrify the German republic into an authoritarian, fascist state, with the suppression of all other political parties, dissolution of unions, elimination of freedom of the press, and assassination or confinement to concentration camps of "enemies" of the regime, that is, members of opposition parties, journalists, and intellectuals. Upon the death of the old and doddering President Hindenburg a few months later, Hitler became Führer and Chancellor and commander-in-chief by a plebiscite in which 90 percent of the voters gave their enthusiastic approval.

Einstein was in Pasadena when Hitler took power, and he wrote to one of his Caputh paramours: "In view of Hitler, I do not dare to return to German soil."[35] In a widely publicized statement he gave to the press, he announced, "I will reside only in a country dominated by political freedom, tolerance and equality of all citizens before the law . . . These conditions are not at present satisfied in Germany."[36] Instead of returning from New York to Germany, Einstein disembarked in Belgium, where he rented a small villa on the coast for a few months. He surrendered his German passport and renounced his German citizenship. And he also resigned from the Prussian Academy of Sciences, the German Physical Society, and the numerous other German scientific societies to which he belonged.

The Nazis confiscated his accounts at German banks, his summer house in Caputh, and his beloved sailboat. Einstein tried to get the Swiss Foreign Office to intervene, on the grounds that he was the holder of a Swiss passport and that thereby he and his property were entitled to Swiss protection. He admitted that in the press he had been widely described as a German scientist, but claimed that he himself had never used this terminology, and that he "was not to be equated with the kind of person designated in Switzerland with the descriptive name of 'paper-Swiss.'" In support of his contention that he was a true Switzer, he oddly listed the fact that he had a divorced wife living in Switzerland.[37] Perhaps the divorced wife impressed the Swiss Foreign Office—they agreed to make some discreet inquiries in Berlin, although, as a dual Swiss-German citizen, Einstein technically was entitled to Swiss protection only *outside* of Germany. Elsa also tried to get the Belgian Foreign Office to support the Swiss efforts, by mentioning to Belgian officials that Einstein was well acquainted with their queen. But none of this made the slightest impression on the Nazis, and Einstein's confiscated property remained confiscated.

Fortunately, the German bank accounts represented only a small fraction of Einstein's assets. For years, he had been depositing substantial amounts of royalties from his books in secret accounts in Holland and the United States, so the loss of his German bank accounts was no great hardship. He wrote to Planck that he had no need of financial assistance "because I was careful and took precautions." And he wrote to Born, who had meanwhile emigrated to England, "In Germany, I have been promoted to an evil beast, and they have taken all my money. But I console myself in that the latter would soon have been spent."[38]

THE NAZIS HAD INTENDED to subject Einstein to a humiliating investigation for supposed participation in anti-German propaganda over war atrocities, what they called *Greuelhetze* (atrocity-mongering), and they had planned to dismiss him from the academy as part of their program of elimination of Jews from German universities. They were furious that Einstein's resignation aborted these plans. They tried to retaliate by releasing an announcement to the press stating that "on these grounds [anti-German propaganda], the academy has no reason to regret the resignation of Einstein."[39]

In a plenary meeting of the academy, von Laue proposed to disavow the Nazi announcement, but this proposal was greeted with silence by the academicians. When Einstein sent a letter to the academy denying the charge of *Greuelhetze*, the academy replied that, indeed, he had not participated in anti-German propaganda, but that he had remained silent about the anti-German propaganda and therefore given his consent to it. Only Planck later entered a declaration in praise of Einstein into the official records of the academy: "Prof. Einstein is not only one among many distinguished physicists, but Prof. Einstein is that physicist by whose published works our physical knowledge has attained a greater depth, of a significance that can be measured only against the achievements of Johannes Kepler and Isaac Newton."[40]

Einstein was disappointed by the lack of support from his colleagues at the academy (although he could hardly have expected their favor after *he* had rejected them by his preemptive resignation). With a touch of sour grapes, he wrote in a letter to Haber, "They were not able to disappoint me, because I never had respect or sympathy for them—with the exception of a few pure personalities (Planck 60 percent noble and Laue 100 percent)."[41] This was patently unfair to Planck, who had nobility to spare. As Lise Meitner later said in her sketch of Planck's character, "He had an unusually pure disposition and inner rectitude, which corresponded to his outer simplicity and lack of pretension . . . Again and again I saw with admiration that he never did or avoided doing something that might have been useful or damaging to himself. When he perceived something to be right, he carried it out, without regard for his own person."[42]

Planck's strong sense of duty was to put him on a collision course with Hitler and the Nazis, which ultimately resulted in the tragic judicial murder of his younger son Erwin and only remaining child (his older son had fallen at Verdun, and both his twin daughters had died of complications of childbirth). When the Nazis issued their decree for the "improvement of public officials" in 1933, which resulted in the summary dismissal of hundreds of Jews from academic posts, Planck decided to speak in favor of these Jews during a personal audience with Hitler, at which he had to present himself in his official capacity as president of the Kaiser-Wilhelm Institute. He tried to convince Hitler that these Jews had made important contributions to German science, and he focused on the case of Haber, arguing that without Haber's nitrogen-fixation process Germany would have lost World War I in a matter of months. Hitler threw a screaming fit, which was his usual (and quite deliberate) theatrical technique for aborting arguments that he could not win. Planck described that Hitler ". . . forcefully slapped his knee, talked faster and faster, and roused himself to such a fury, that in the end I could only remain silent and take my leave."[43]

This run-in with the Führer was not forgotten by the Nazis when in 1944 they sentenced Planck's younger son to death for treason, merely because he had been acquainted with some of the conspirators of the July 20 assassination attempt against Hitler. Planck begged for clemency for his son, but his request was ignored and in January 1945, three months before Germany's surrender, his son was hanged on a wire in the infamous Plötzensee prison in north-central Berlin (now a memorial to the ninety conspirators executed there).[44] The execution was carried out in secret, and when Planck was finally informed several days later, he did not say a word. He retired to his piano and spent several hours playing all of his son's favorite melodies. "My pain cannot be expressed in words," he later wrote to a friend.[45]

IN OCTOBER 1933, Einstein, accompanied by his wife Elsa, his secretary Helen Dukas, and his mathematical assistant Walther Mayer left for the United States. Einstein never returned to Europe. His half-year appointment at the Institute for Advanced Study became a full-time, permanent appointment. When Einstein arrived in Princeton in 1933, the institute was still under construction, and Einstein was temporarily assigned an office at Fine Hall, then the mathematics department of Princeton University. When he moved into his office, he asked that it be equipped with a desk, a chair, and a "large wastebasket . . . so I can throw away all my mistakes."[46]

Princeton in those days was a small college town, known for its university and its theological seminary. Princeton University had only recently started to acquire a reputation in physics and mathematics. In the early 1900s, Princeton had been more like a sleepy college than a university, and it had not then enjoyed the high renown for academic excellence it enjoys today. It was regarded as a school for Southern playboys, who deemed it bad manners to earn more than the gentlemanly C in their courses. The writer F. Scott Fitzgerald, a member of the class of '17, described Princeton as "the pleasantest country club in America." He had nothing favorable to report about the intellectual milieu at Princeton, but he had much praise for the sartorial achievements of the students: "The men—the undergraduates of Yale and Princeton—are cleaner, healthier, better-looking, better dressed, wealthier and more attractive than any undergraduate body in the country."[47]

Yale and Princeton physicists and mathematicians were expected to teach long hours of high school physics and mathematics to their nattily dressed students, but there were very few resources available for research. At Yale, the physicist Willard Gibbs, who had achieved worldwide recognition for his work on thermodynamics, was refused a salary because his studies were deemed irrelevant to the teaching of students. At Princeton, the British physicist Owen Richardson, who later gained a Nobel Prize, was shown into a basement when he arrived to take up an appointment in 1906: "I remember getting quite a shock when I was first introduced to the part where I was expected to set up a research laboratory. This was a kind of dark basement, ventilated by a hole in the wall, apparently accidental in origin, and inhabited by an impressive colony of hoptoads which enjoyed the use of a swimming pool in one corner."[48]

Princeton's ascent to a leading institution of higher learning had its roots in reforms initiated by Woodrow Wilson during his tenure as president of the university from 1902 to 1910, before he became president of the country. Wilson drastically increased the size of the faculty in the liberal arts, raised academic standards, and gave his friend Henry Fine, professor of mathematics and later dean of science, a free hand to recruit first-rate mathematicians and scientists. After Wilson's departure, Fine gradually assembled a cluster of young, bright stars—viewed with some trepidation

The south side of Fuld Hall at the Institute for Advanced Study on the outskirts of Princeton. Einstein's office was on the ground floor, at the right, near the tree.
(Courtesy of the Institute for Advanced Study, Princeton, New Jersey, USA)

Einstein's house at 112 Mercer Street in Princeton, a short walk from the Institute for Advanced Study. He added a study on the second floor at the rear of the house.

by the undergraduates and dubbed "Fine's research men"—whose achievements soon raised Princeton's academic standing in the world of science.

Fine died in an unfortunate bicycle accident on Princeton's main street in 1928, but his program for promoting research in mathematics and the sciences survived and expanded. In the 1930s Princeton benefited from the generosity of the Rockefellers, who wanted to improve higher education in America and used the Rockefeller Foundation to channel a large amount of cash into programs for the support of students and faculty in the sciences. At first, the foundation provided grants for American students to continue their graduate or postgraduate education in Europe—Compton, Oppenheimer, and Wheeler all went to various European universities with Rockefeller purses. But then the foundation decided that it would be even better to import the best and brightest European mathematicians and scientists, by enticing them to selected American universities with offers of research professorships at extravagant salaries. Among these European imports were the Hungarians John von Neumann and Eugene Wigner, who arrived in Princeton in 1930.

Von Neumann's name brings to mind an irresistible anecdote, which is an amusing illustration of the relativity of motion. Von Neumann was a notoriously bad, accident-prone driver, and one day, while driving along a road near Princeton, he let his car drift to the side and smashed into a tree. In his accident report, he described the event in the reference frame of the car: "I was proceeding down the road. The trees on the right were passing me in orderly fashion at 60 MPH. Suddenly, one of them stepped out in my path. BOOM!"[49]

The Institute for Advanced Study imitated the policy of the Rockefeller Foundation, enticing European scientists and mathematicians with even more extravagant salaries and the promise of complete independence and freedom from all teaching duties. Furthermore, by 1933, the institute was able to take advantage of the flow of émigrés from Europe to America triggered by the gathering storm of Nazism. In addition to Einstein, the institute imported Herman Weyl from Germany and Kurt Gödel from Austria. All those European newcomers spoke German, and students wandering into Fine Hall must have thought they had come upon the German department of the university.

Einstein acquired a comfortable house on Mercer Street, at the south end of town, within walking distance of both the university and the institute. There he lived with his wife, his secretary, and his stepdaughter Margot, who had left Germany shortly after Einstein. With the help of the French Embassy in Berlin, Margot had arranged for the packing and shipment of Einstein's papers and most of his furniture from the Berlin apartment, except for some rugs and a few trinkets stolen by Nazi thugs. The furniture—most of it heavy, Biedermeier pieces that came from Elsa's family—stayed at the Mercer Street house until recently and then was donated to the Princeton Historical Society, where some of it is now on display. Theories come and go, but Biedermeier is forever.

IN PRINCETON, EINSTEIN soon settled into a routine. Every day he would walk to the institute and spend the morning in his office, over calculations or in discussions with his assistants. Then he would wander back for lunch and spend the rest of the day at home, where he had a private study. Occasionally, he would sail on the small Lake Carnegie south of the campus.

Elsa died in 1936, just three years after her arrival in Princeton. Soon after Elsa's death, Einstein's sister Maja came to live with him (she died in Princeton in 1951). The death of Elsa resulted in a rapprochement and reconciliation between Einstein and his older son, Hans Albert, who came to visit for several months in 1937, liked the United States, later immigrated, and ended up as a professor of engineering at the University of California, Berkeley.

In 1940, Einstein became an American citizen, and so did his secretary Helen Dukas and his stepdaughter Margot. With the eager cooperation of the federal judge in charge of the proceedings, the swearing-in ceremony turned into a media circus, including a radio interview. However, not all branches of government were equally delighted to have Einstein as a U.S. citizen. The FBI, under the directorship of the sinister and bizarre J. Edgar Hoover—who, according to some sources, was wont to cruise the Washington bars in drag for extraofficial purposes of his own—had accumulated a thick file on Professor Einstein. He had first come under suspicion for his pacifist activities, which the FBI tended to confuse with un-American and possibly communist activities: If you don't like war, what kind of American are you anyhow? A Commie?

Einstein's "confidential" file ultimately grew to more that 1400 pages filled with irrelevant and often wrong information, wild rumors, bits and pieces that FBI agents had sifted out of his household trash, and a great many accusations from anti-Semitic sources. But with all their diligent inquiries and rumor-mongering, the Keystone Cop FBI agents missed the one incident in Einstein's life that might have been real cause for concern. In 1941, Einstein became intimately involved with Margarita Konenkova, wife of the Russian sculptor Sergei Konenkov, who had carved the impressive bust of Einstein now on display at the Institute for Advanced Study. Konenkova was a genuine spy and "honey pot" working for the Soviet intelligence services, and she had been given the assignment of seducing Einstein in the hope of extracting some information about American military projects. She easily succeeded in seducing Einstein but found that he had no information to give her. Letters from Einstein to her were published half a century after the fact, and they show that their affair lasted until the end of the war, when she returned to Moscow.

THE FINAL MEMORABLE CONTRIBUTION to physics in Einstein's declining years was the paper on the Einstein-Podolsky-Rosen (EPR) paradox published in 1935, in which he tried to throw doubts on the conventional, probabilistic interpretation of the quantum-mechanical wavefunction. This paper was the result of a collaboration with Boris Podolsky and Nathan Rosen, both of them colleagues of Einstein's at the Institute for Advanced Study. The EPR paradox deals with a puzzle concerning measurements performed on two widely separated particles that are, however, in correlated quantum states. Today, such a correlation is called quantum entanglement or, as the Oxford mathematician-physicist Roger Penrose prefers to call it, quanglement.

Einstein-Podolsky-Rosen (EPR) imagine that two particles of, say, equal masses, are initially in a quantum state of precisely known interparticle separation and precisely known total momentum. For instance, the separation between the particles might be exactly 100 meters, and the total momentum, that is, the sum of the momenta of the two particles, might be exactly four units of momentum. This does not conflict with the Uncertainty Principle, because a precise value (with zero uncertainty) for the sum of the particle momenta requires a large uncertainty for the *sum* of the particle positions, but it does not impose any conditions on the uncertainty of the particle separation, that is, the *difference* between the positions.

Given such a pair of particles, EPR imagine that they measure the position of the first particle precisely by means of some detector of small size. Since they already know that the second particle is 100 meters away, the measurement of the precise position of the first particle permits them to deduce the precise position of the second particle. EPR then argue that since the detector is small and operates only in the vicinity of the first particle, it in no way disturbs the second particle, and that therefore this second particle must have had this deduced position even *before* the measurement was performed. Accordingly, EPR assert that a precise (although unknown) value of the position of the second must have existed even before any measurement was performed.

Next, EPR imagine that instead of measuring the position, they measure the momentum of the first particle. Since they already know the total momentum, the measurement of the momentum of the first particle permits them to deduce the momentum of second particle; and, again, they argue that since this measurement in no way disturbs the second particle, a precise value of the momentum of the second particle second must have existed even before the measurement was performed.

Thus, both the position and the momentum of the second particle must have had precise values before the measurement, in contradiction to quantum mechanics, which demands that the position and momentum of any particle can never be precisely defined at the same time. From this paradoxical result, EPR conclude that quantum mechanics is incomplete—it gives a description of particles that is not as precise as can be. In the view of EPR, the quantum-mechanical description of particles must be supplemented by some extra "hidden variables," which determine the precise values of the position and the momentum whose existence are indicated by their argument.

THE EPR ARGUMENT hinges on the reality of the attributes of particles and on the local character of measurements performed on a particle. The position and momentum of the second particle are supposed to exist, in themselves, even if we do not measure them; and the measurement performed on the first particle is supposed to have no disturbing effect on the second particle. Bohr and other defenders of quantum mechanics disputed the EPR paradox by denying both of these suppositions. They asserted that particles do not have attributes in themselves, but only in relation to a measurement procedure; and they asserted that when a measurement is performed on one of a pair of entangled particles, this affects the other particle, even if it is very distant from the site of the measurement.

When the EPR paper was published, it attracted some attention from the press, as did most of Einstein's publications of those days. Thus, *The New York Times* announced "Einstein attacks quantum theory."[50] On the title page of the paper, Einstein's name takes precedence over those of Podolsky and Rosen; but this merely reflects the alphabetical order of names that is customary in physics publications in the United States—it does not indicate who did most of the work. The historical record indicates that Einstein's own role in the preparation of this paper was minimal: the main idea for the paper originated from Rosen, and the paper was written by Podolsky (and Einstein complained, "It did not come out as well as I had originally wanted").[51]

The EPR paper was a far-reaching contribution to quantum mechanics, and disputes about the EPR paradox and quantum entanglement continue to this day, among physicists and also among philosophers. This has led to some recent experiments on hidden variables, quantum entanglement, and "teleportation," a form of quantum magic, whereby measurements at one site create changes in a quantum-mechanical state at another, remote site, beyond the reach of signals from the first site.

APART FROM THE EPR PAPER, Einstein's research during his years in Princeton led to no valuable, lasting contributions to physics. He was a trophy for the institute but, as Oppenheimer used to say, he was regarded as more of a memorial than a guiding light. His work at the institute was second-rate, and many of the papers he published were irrelevant or wrong or both, but he paid scant attention to criticism of his work, and he sometimes would take offense when criticism was offered.

Thus, when the editor of *The Physical Review*, the leading physics journal in the United States, sent one of Einstein's manuscripts to an anonymous referee for review, Einstein haughtily refused to deal with the various comments and criticisms made by the referee. He furiously replied to the editor that he had submitted the manuscript for *publication* and that he had not authorized that it be shown to specialists before it was printed. And he added, "I see no reason to address the—in any case erroneous—comments of your anonymous expert. On the basis of this incident I prefer to publish the paper elsewhere."[52] Thereafter, Einstein never again sent any manuscript to *The Physical Review*—he preferred journals that were more likely to kowtow to the Einstein name, without interference from meddlesome referees.

But the sequel to this story revealed that it was not the referee who was in error, but Einstein. The title of the disputed paper was "Do Gravitational Waves Exist?" and Einstein thought his paper proved they do not. He submitted the paper to a different journal, where it was immediately accepted. But then some of his own colleagues at the institute pointed out mistakes, and Einstein had second thoughts. He revised the paper drastically, and he reached the opposite conclusion: gravitational waves *do* exist, which was exactly what the referee had said in the report that Einstein had dismissed so haughtily. He finally published the revised paper with a new title, "On Gravitational Waves." In the meantime, he had promised to give a lecture at Princeton, about the *non*existence of gravitational waves, as claimed in his original proof. He discovered the mistake in this proof the day before this lecture, when he had not yet found the proof of the opposite conclusion. So he gave a lecture about his mistake and ended with these words: "If you ask me whether there are gravitational waves or not, I must answer that I do not know. But it is a highly interesting problem."[53]

Among his irrelevant papers was a series of several tedious calculations, published with Peter Bergmann and other assistants, on the motion of particles in a gravitational field. In the standard formulation of general relativity, particles move along geodesics, that is, along the lines of shortest length in the curved geometry. Einstein wanted to regard particles as knots, or "singularities," in the geometry of spacetime, that is, as concentrations of gravitational fields and, if so, then the motion of a particle merely represents the propagation of a disturbance in the gravitational field. A good analogy is the motion of a smoke ring in air, which is merely the propagation of a disturbance in the air itself, and not the motion of some kind of extraneous object distinct from the air.

Since the Einstein equations for the gravitational field completely determine the evolution of all aspects of the gravitational field, they should also determine the motion of knots (just as, in the smoke-ring analogy, the hydrodynamic equations for air determine the motion of the smoke ring). In laborious calculations, Einstein and Bergmann found that the motion of knots in the geometry indeed agrees with geodesic motion.

Although these calculations were an impressive tour de force, they were of more mathematical interest than physical interest. If particles are treated as small knots in the geometry, then it is actually necessary to apply quantum mechanics to the behavior of these knots, and a purely classical calculation, such as performed by Einstein and Bergmann, is merely a pretty, purely hypothetical, mathematical exercise.

ANOTHER PROBLEM THAT EINSTEIN tackled during these years was a long-standing puzzle about the Schwarzschild solution. As far back as 1917, when Karl Schwarzschild constructed the first exact solution of Einstein's equations for the gravitational field surrounding a spherical mass, it had been noticed that this solution has a singularity, or a mathematical defect, at a small radius. The metric tensor misbehaves at a small radius, called the Schwarzschild radius, where the magnitude of the metric tensor increased toward infinity. For instance, if the mass is equal to the mass of the Sun, then the magnitude of the metric tensor increases toward infinity at a radius of 3 klicks, the Schwarzschild radius for the Sun.

At first, this problem was dismissed as an irrelevant quirk. The radius of the Sun is much larger than 3 klicks, and since the Schwarzschild solution is valid only for the empty space outside the Sun, a singularity in the solution at 3 klicks is merely a mathematical artifact, physically irrelevant. However, it was later recognized that when a star such as the Sun ages, it will ultimately shrink in size, and perhaps it might even collapse to a size of less than 3 klicks. If so, the Schwarzschild radius and the Schwarzschild singularity would take on a real significance; they would be exposed to view, hovering at some distance above the shrunken surface of the Sun.

To get rid of this pesky problem, Einstein tried to prove that a star will never shrink to a size smaller than its Schwarzschild radius. In a paper published in 1939, he treated the star as a swarm of particles (that is, a swarm of atoms), and he assumed that these particles are in circular orbits around a common center. He assumed that the circular orbits have different inclinations (different planes), so the swarm of particles is spherical. With this simple model, he found that the circular orbits cannot be smaller than 1.5 times the Schwarzschild radius, or 4.5 klicks if the total mass of the swarm is equal to the mass of the Sun. Accordingly, Einstein concluded that the star would always be larger than the Schwarzschild radius, and he claimed to have achieved a "clear understanding as to why these 'Schwarzschild singularities' do not exist in physical reality."[54]

Oppenheimer and his student Hartland Snyder quickly proved him wrong. Instead of considering particles that are locked in circular orbits, they relied on a more realistic model consisting of a spherical conglomeration of neutrons, like a gigantic nucleus, whose collapse is resisted by a repulsive nuclear force between the neutrons. With realistic value for the magnitude of the nuclear force, they found that if the mass of the configuration is larger than three-fourths of the mass of the Sun, the nuclear force cannot resist the inward gravitational force. The spherical conglomeration inexorably collapses on itself under its own weight—it does not stop at 1.5 times the Schwarzschild radius, but continues shrinking and shrinking forever.

Oppenheimer and Snyder found that this gravitational collapse has a strange feature, in that the collapse proceeds more and more slowly, but never stops. This odd behavior and the final fate of the collapsing mass were not fully understood until the 1960s, when a careful analysis of the Schwarzschild geometry revealed that the Schwarzschild singularity was no singularity at all—it merely was an artifact resulting from a bad choice of coordinates.[55] The collapsed mass forms a region of strongly curved spacetime surrounded by a "one-way membrane" at the Schwarzschild radius, which prevents the escape of anything—even light—from its interior.

John Wheeler gave such a collapsed mass, hidden within its own one-way membrane, the name "black hole." The theoretical study of black holes, and, later, the observational study of collapsed stars that have formed such black holes in supernova implosions, became the hottest and most exciting topic in general relativity. Einstein did not get a chance to become acquainted with these stunning developments—he died just a few years too soon.

EINSTEIN OFFICIALLY RETIRED from his position at the

institute in 1946, although he was permitted to retain his office there indefinitely. His retirement occasioned one of those financial skirmishes that were one of his trademarks. In accord with the contractual arrangements, the institute granted Einstein a pension of one-half of his salary. However, Einstein demanded that he be paid a pension equal to his full salary. The institute at first refused this cheeky demand, but then caved in when Einstein threatened to move away from Princeton and complain publicly about how badly he was being treated: "I threatened to leave Princeton upon retirement, which they did not wish because of my popularity."[56] Thus, his retirement was in name only—he kept his office and his salary, and even his assistant.

During the last ten years of his life, Einstein returned once more to his failed first attempt at a unified theory, and he tried to modify this attempt in several ways. In a letter to a friend, he had admitted, "Most of my offspring end up very young in the graveyard of disappointed hopes,"[57] but now he tried to resurrect one of these buried offspring.

Einstein's longtime, long-suffering mathematical assistant Walther Mayer had decided some years earlier that enough was enough, and that he no longer wished to be associated with Einstein's stubborn and futile attempts at a unified theory. Several assistants came and went in quick succession, and most seem to have had an aversion to working on Einstein's unified theory. For instance, Peter Bergmann, who later became a well-known expert on general relativity, stayed with Einstein for five years, but avoided involvement with the quagmire of Einstein's unified theory; and John Kemeny lasted for only one year and asked not to be reappointed rather than deal with Einstein's messy equations. Finally, for the last five years of his life, Einstein secured the assistance of Bruria Kaufman, who, as a woman mathematician, was a rarity in those days (she later had a successful career as a professor of mathematics in Israel).

He included his final field equations for the unified theory in an appendix to the 1949 edition of one of his earlier books, *The Meaning of Relativity*, and *The New York Times* promptly reprinted the equations on its front page with the headline "NEW EINSTEIN THEORY GIVES A MASTER KEY TO THE UNIVERSE." This was wishful thinking, but the *Times* was just as stubborn as Einstein, and when the 1952 edition of the same book appeared, the *Times* greeted it again with the headline "EINSTEIN OFFERS NEW THEORY TO UNIFY LAW OF THE COSMOS."[58]

The trouble with Einstein's "master key to the universe" was that it was not actually a key, but only a dream about a key. Einstein again failed to extract the Maxwell equations from his unified theory.[59] Furthermore, he failed to incorporate electric charges into his theory, and he proposed that what we perceive as electric charges are merely some kind of knots in the field lines. But because he never found any exact solution of the complicated field equations he had constructed, it remained unclear just what kind of unification of electromagnetism and gravitation he had accomplished.[60] As he wrote to a friend, "The unified field theory has been put into retirement. It is so difficult to employ mathematically that I have not been able to verify it," and he tried to lay the blame on his colleagues, adding, "This state of affairs will last for many more years, mainly because physicists have no understanding of logical and philosophical arguments."[61]

Einstein himself admitted that nature might or might not obey his equations, but in his final words on this topic he put the best face on it he could: "In my opinion, the theory presented here is the logically simplest relativistic field theory which is at all possible. But this does not mean that nature might not obey a more complex field theory."[62] It was a flat and sad note on which to end a brilliant career.

DURING HIS FINAL YEARS, Einstein became at least vaguely aware of what his colleagues at the institute really thought about him. In 1949 he wrote in a letter to Born, "I am generally regarded as a sort of petrified object, rendered blind and deaf by the years."[63] In his work, he had always been a loner, and the scorn of his colleagues for his junky unified theories and for his stubborn opposition to quantum mechanics encouraged his natural inclination to isolation.

He avoided contact with anybody who might criticize or dispute his ideas, and he especially avoided any discussion of quantum mechanics. As he said himself, "I must seem like an ostrich who forever buries its head in the relativistic sand in order not to face the evil quanta."[64] When Bohr visited the institute in 1948, Einstein refused to meet with him. In a comical incident during this visit, Einstein sneaked into an office in which Bohr was having a discussion with Pais, and found himself suddenly face to face with Bohr—but he merely wanted to borrow some tobacco for his pipe from a tin sitting on a shelf.[65]

Leon Rosenfeld, Bohr's assistant, recorded his impressions of Einstein during this visit:

> Einstein was only a shadow of his former self. He remained locked in his office all day long and talked only to his two assistants, oddly named Bergmann and Bargmann. Only once within these four months did he announce a lecture; it involved one of his innumerable attempts to establish a unified field theory . . . During these four months Bohr and Einstein met only once during an afternoon reception, but the

conversation did no go beyond banalities: Einstein made it clear that he preferred to avoid any discussion with Bohr. With this, Bohr was profoundly unhappy.[66]

Einstein died in April 1955 at age seventy-six of a burst aneurism of the abdominal branch of the aorta. He had had exploratory surgery nine years earlier, during which the surgeon attempted some reinforcement of the aorta. When his condition became acute, Einstein refused any further surgery.

His secretary, Helen Dukas, reported that he faced his end calmly, like a physically necessary event. And as a physicist, Einstein would have had a clear understanding that the progressive, and ultimately fatal, ballooning of his aorta was indeed a physically necessary event—he probably had a clearer understanding of this event than his own doctor. The expansion of an aneurism is a simple physical process, governed by two well-known laws of fluid dynamics: the continuity equation and the Bernoulli equation. When a weak segment of the aorta expands due to the internal blood pressure, the continuity equation dictates a reduction in the speed of blood flow at the site of the expansion. The Bernoulli equation then dictates an increase of the local blood pressure. This causes a further expansion of the aorta, which leads to a further reduction of speed of flow, and a further increase of pressure, etc., etc. In the absence of surgical intervention, this step-by-step process escalates inexorably . . . until the expanding aorta bursts.

On the day before his death in the hospital in Princeton, he felt somewhat better, and he spent some hours on several pages of calculations of this latest version of the unified theory. His son Hans Albert was present, and Einstein joked about his calculations, saying, "If only I had more mathematics."[67] He died at night, in his sleep.

He had left instructions that his body was to be cremated, and the ashes scattered at an undisclosed location, so his final resting place would not become a tourist attraction. And so it was done, except for the macabre circumstance that the pathologist who performed the autopsy on Einstein's body removed his brain and kept it, sliced into pieces and pickled in two jars of formaldehyde. He occasionally gave away slices of the brain to various researchers, and finally returned the remaining pieces to the Princeton hospital in 1998.

AFTER HIS DEATH, access to Einstein's papers was tightly controlled by Helen Dukas. Freeman Dyson of the Institute for Advanced Study knew Dukas well, and he said of her, "She fought like a tiger to keep out people who tried to intrude upon Einstein's privacy while he was alive, and she fought like a tiger to preserve the privacy of his more intimate papers after he died . . . Underneath Helen's serene surface we could occasionally sense the hidden tensions. She would sometimes mutter darkly about unnamed people who were making her life miserable."[68]

Her interest in preserving the privacy of intimate papers may not have been completely selfless. Einstein's older son Hans Albert believed she had been one of his father's mistresses.[69] Some of Einstein's claqueurs have rejected this accusation as preposterous, but Hans Albert had lived for some months with Einstein and Helen in the Mercer Street house, and he was in a position to know.

Besides, Einstein's last will suggests that Helen was much more than a devoted secretary. Einstein's liquid assets amounted to some $65,000 at his death (adjusted for inflation, this corresponds to $490,000 in current dollars). He left $10,000 to his son Hans Albert, $15,000 to his son Eduard (for his keep at the Zurich mental institution), $20,000 to his stepdaughter Margot (who also received the house, which had already been put under her name), and $20,000 plus "all of my personal clothing and personal effects, except my violin" to Helen. In addition, the will placed Helen and Margot in control of the trust that received the income from his royalties. Although this trust would ultimately benefit the Hebrew University, the will specifically stated that the trust arrangements had as "primary object" not the Hebrew University, but ". . . to make further provision for the care, comfort, and welfare for my said secretary, Helena Dukas."

Of all the bequests in the will, the income from royalties was by far the most valuable. Einstein was entitled to very substantial royalties from his books and from the use of his name and his image. When the control and ownership of these royalties was inherited by the Hebrew University of Jerusalem after Dukas's death, the profits from royalties rose into the millions—today the yearly profit to the Hebrew University is somewhere around $18 million (and on every postcard, doll, or other Einstein gadget, you will find the copyright label of the Hebrew University or its agent in Beverly Hills).

Helen Dukas, jointly with Otto Nathan, was also given control over his papers, which were ultimately destined to go to the Hebrew University. Helen spent the next twenty-five years sorting through Einstein's papers, and she is under strong suspicion of having removed and destroyed items detrimental to Einstein's (and her?) reputation.

In 1981, a few months before her death, she finally arranged for the transfer of the papers to Israel under rather mysterious circumstances. By accident, Freeman Dyson witnessed the removal operation:

> Then, one night around Christmas, when most of the Institute members were on holiday, there was a sudden move. It was a dark and rainy night. A large truck stood in front of the Institute with a squad of well-armed Israeli soldiers standing guard. I happened to be passing by and waited to see what would happen. I was the only visible spectator, but I have little doubt that Helen was also present, probably supervising the operation from her window on the top floor of the Institute. In quick succession, a number of big wooden crates were brought down in the elevator from the top floor, carried out of the building through the open front door, and loaded onto the truck. The soldiers jumped on board and the truck drove away into the night. The next day, the archive was in its final resting place in Jerusalem.[70]

Postmortem

instein made so many mistakes in his scientific work it is hard to keep track of them. There were mistakes in each of the papers he produced in his miracle year 1905, except for the paper on Brownian motion. And there were mistakes in dozens of the papers he produced in later years. The list of Einstein's scientific papers assembled by Helen Dukas after his death comprises about 180 original items.[1] Of these, about 40 contain mistakes (counting each of the several papers on the unified theory as a mistake). It's a bad scorecard.

Despite all of these mistakes, Einstein was unquestionably the greatest physicist of the twentieth century, and he was the second-greatest physicist ever, outranked only by Newton. (In a poll conducted in 2007, scientists of the Royal Society were asked, Who made the more important contributions to science, Newton or Einstein? Their vote was 86 percent for Newton, 14 percent for Einstein.)

Einstein's mistakes did not affect his rank because they did not prevent him from making his groundbreaking discoveries. He might have been confused and mistaken about some details, but he had a clear grasp of the larger picture. We could say he was a poor tactician, but a superb strategist, so he could afford to commit some mistakes in his skirmishes and nevertheless win every war—except the wars of quantum mechanics and of the unified theory, which he lost resoundingly.

In fact, many of Einstein's mistakes were amazingly fruitful—they played a seminal role in leading Einstein to his revolutionary theories. As James Joyce said, "The man of genius makes no mistakes. His errors are volitional and are the portals of discovery."[2] Although Einstein's errors were volitional, or willful, they were willed at a subconscious level, not at a rational, conscious level. He committed his mistakes unwittingly in moments of inspiration, while relying on intuition rather than on logical thought.

Arthur Koestler, in his bold and incisive book *The Sleepwalkers*, characterized the great discoverers in science—such as Copernicus, Kepler, Galileo, Newton, and Einstein—as mental somnambulists. They wander through the realms of the mind toward their goal without being consciously aware of what they are doing. They proceed intuitively and by inspirations, guided by an inner compass, which keeps them on a true path. They make mistakes, but instead of tripping over these mistakes, they dance around them, and they often use these very mistakes as stepping stones and shortcuts to reach their goal. The unpredictable, erratic path that these sleepwalkers follow in their wanderings defies logic and often seems totally incomprehensible, and yet in the end it leads to a perfectly sensible, logical result.

Koestler regarded Kepler as a sleepwalker par excellence. In the lengthy calculations that led him toward the laws of planetary motion, Kepler made mistakes in copying data, mistakes in arithmetic, and mistakes in the assumptions he adopted for variations of the speeds of motion along the orbits. Yet in the end, all these mistakes canceled out—"as if by miracle," as he himself said—to give him the correct final result. With his penchant for introspection, Kepler later analyzed his own mistakes, and he described in detail how he "unconsciously" repaired his mistakes. And he commented: "The roads that lead man to knowledge are as wondrous as that knowledge itself."[3]

AMONG EINSTEIN'S MISTAKES, the most productive were the two that launched his theories of special and of general relativity. His misconceptions about clock synchronization led him to introduce the postulate about the constant speed of light, which gave him a shortcut to the Lorentz transformations, the length contraction, the time dilation, and a wealth of other new results. His mistaken, or at least defective, Principle of Equivalence led him to the discovery of general relativity. As Synge said, he needed the Principle of Equivalence as a "midwife" to give birth to the insight that, in the presence of gravitation, spacetime is curved. Without such a midwife, he might never have thought of such an outrageous thing.

Likewise, Einstein's mistakes in his derivations of $E = mc^2$ made it possible for him to claim credit for this equation. His first derivations of this equation, although flawed, convinced physicists that the energy-mass relation ought to be of general validity. In 1905, Einstein did not have available the mathematical tensor techniques needed for a correct derivation of this formula, but he instigated the theoretical investigations that later led von Laue and Klein to the correct derivation.

Einstein's remarkable talent for making fruitful mistakes was rooted in his mental habits and his personality—his Swabian habit of *grübeln*, his deeply intuitive, inspirational, and mystical approach to problem solving, and his independent, rebellious, and stubborn disposition.

Einstein himself gave us a clue to the psychopathology of his mistakes. In an interview with the psychologist Max Wertheimer, he said: "I rarely think in words at all. A thought comes, and I may try to express it in words afterwards."[4] And in his *Autobiographical Notes* he said, "For me it is unquestionable that our thinking proceeds to a large extent without the use of symbols (words) and also to a considerable extent unconsciously."[5] When thinking in words, we are compelled to think logically and mathematically —we argue with ourselves, and often even speak with ourselves, "thinking aloud." But when we think in ideas and images, and when we think "unconsciously," logic and its syllogisms are left behind, and the mind soars into an irrational realm of inspiration and intuition.

Inspiration opens a much wider range of possibilities to the imagination, but it also increases the risk of mistakes. Ideas harvested in the realm of inspiration need to be placed in quarantine, so they can be rigorously inspected and cleared of any infestation of errors. But geniuses are liable to become infatuated with an "incomparably beautiful" idea. They are liable to embrace it without subjecting it to adequate quarantine and inspection— and they then fail to notice that this lovely idea is infested with errors.

Koestler concluded that what geniuses have in common is "on the one hand scepticism, often carried to the point of iconoclasm, in their attitude towards traditional ideas, axioms and dogmas, towards anything that is taken for granted; on the other hand, an open-mindedness that verges on naïve credulity toward new concepts which seem to hold out some promise to their instinctive gropings."[6] In geniuses (as well as in political leaders and revolutionaries), this naïve credulity often leads to the blind acceptance of astonishing—and even outright silly—mistakes. Thus, Koestler explains that in the case of Kepler it was as though "his conscious, critical faculties were anaesthetized by the creative impulse."[7]

ONCE A GENIUS FALLS into a seductive mistake, it is liable to become an idée fixe. He is then blind to the mistake, and if this blindness is compounded by stubbornness, he will be deaf to criticism, and he will cling to his mistake forever, *Dickkopf, Dummkopf* (stubborn head, stupid head), as the German proverb says.

We find this infatuation and blindness in Galileo's and Newton's mistakes about the tides, and, above all, we find it in several of Einstein's mistakes in special and general relativity, such as his misconceptions about the synchronization of clocks, his misbegotten attempts at derivations of the energy-mass relation, and his defective Principle of Equivalence. He perceived these ideas as "the key to the problem," "jolly and beguiling," "the most fortunate thought of my entire life," and fell head over heels in love with them. He became deaf to criticism, as is evident from his lack of attention to Eddington's criticism of clock synchronization, Eddington's and Ehrenfest's criticism of the Principle of Equivalence, and Planck's, Laue's, and Klein's criticisms of his work on the energy-mass relation.[8]

Einstein's stubborn persistence in mistakes is displayed in the book *Relativity, The Special and the General Theory* (which triggered Crowhurst's madness), where he still clings to his misconceptions about synchronization, almost fifty years after the publication of his 1905 paper.[9] And it is also displayed in his repetitious mistakes in the derivation of the energy-mass relation, where thirty and forty years after his first mistaken proof, he repeated essentially the same mistake in his last two "elementary" proofs, again assuming, without justification, that the energy of a system is particle-like.

Einstein was aware that irrational processes play a role in scientific discoveries. He wrote in one of his autobiographical notes, "Invention is not the product of logical thought, even though the final product is tied to a logical structure."[10] However, Einstein did not have Kepler's inclination to self-criticism, and he lacked Kepler's compulsion to dissect his own mistakes. Einstein thought that by tying his final product to a logical structure, he was purging his invention of all mistakes, and he overlooked the mistakes that served as midwifes in the birth of his theories. As Synge later recommended, after completing his theories, Einstein should have given these midwives honorable burial. But Einstein could not let go. He doted on his midwives—he was more faithful to them than to his wives.

Although Einstein's stubbornness made him persist in his mistakes, it sometimes proved beneficial. It gave him the perseverance he needed for his relentless pursuits of special relativity and general relativity, each pursuit stretching over eight long years, undeterred by disappointments and setbacks. He credited his success in these pursuits to his stubbornness, "All I have is the stubbornness of a mule; no, that's not quite all, I also have a nose." But the same obstinate perseverance later served him badly in his futile, obsessive pursuit of the unified theory, stretching over the last thirty years of his life.

HOW MUCH OF AN ADVANTAGE did Einstein gain over his colleagues by his mistakes? Typically, about ten or twenty years. For instance, if Einstein had not introduced the mistaken Principle of Equivalence and approached the theory of general relativity via this twisted path, other physicists would have discovered the theory of general relativity some twenty years later, via a path originating in relativistic quantum mechanics.

In the 1928, Dirac formulated the relativistic quantum theory of electrons, and in the 1930s he and other quantum theorists proceeded to formulate relativistic quantum theories for all other particles. Electrons have spin, that is, they have an intrinsic rotation, vaguely analogous to the spin of a Frisbee about its axis, but quantized. Most other elementary particles also have spin, either of the same magnitude as the electron spin or larger. For instance, the magnitude of the electron spin is ½ (in suitable units), that of the proton is also ½, and that of the photon is 1. Theorists concluded that drastically different quantum-mechanical equations were needed to describe particles of different spins and of different masses.

When they formulated the equations for particles of spin 2 and mass zero, they found, to their surprise, that these equations coincided approximately with Einstein's equations for the gravitational field. And when they modified these equations so as to ensure that all forms of energy gravitate, they found that the result coincided *exactly* with Einstein's equations.[11] Thus, if Einstein had not discovered his equations for the gravitational field in 1915, then quantum theorists would certainly have discovered them in the mid-1930s, some twenty years later.

Much the same can be said of all of Einstein's other great discoveries. In the absence of Einstein, all these discoveries would have been made somewhat later—mostly by an entirely different path—and physics today would have been pretty much the same as it is.

The lead of ten or twenty years that Einstein held over his contemporaries may not seem much, but we have to remember that he achieved this lead again and again, in more than half a dozen of his seminal publications. This is what made him the preeminent physicist of the twentieth century.

WHAT LESSONS CAN WE EXTRACT from Einstein's mistakes? Not many. The art of sleepwalking and of distilling great (or even small) discoveries from fruitful mistakes is a gift that cannot be learned—you either have it or you don't, and most of us don't. And nothing remarkable can be learned from Einstein's *unfruitful* mistakes. They were perfectly mundane, careless, and sometimes stupid lapses in logic and in mathematics, such as we all commit, and wish we didn't. Knowing that Einstein did the same can offer all of us some consolation—instead of kicking yourself for such a mistake, just remind yourself that you are in good company—but nothing more.

But perhaps we can extract a lesson from Einstein's stubborn and futile search for his unified theory, on which he spent thirty years, whereas the successful searches for special and for general relativity took only eight years. Maybe this can serve as a cautionary tale for today's string theorists.

String theory was first introduced in the 1970s, when the physicists Yoichiro Nambu, Holger Nielsen, and Leonard Susskind (at the University of Chicago, the Niels Bohr Institute, and Stanford University, respectively) made the cute proposal that the ultimate, subatomic entities—at the bottom of the downward progression from atoms to electrons and nuclei, to protons and neutrons, to quarks and gluons—are not pointlike particles, but tiny bits of string vibrating in a spacetime of 10 dimensions. For a great many years, a great many theorists labored over this string model with obsessive zeal, hoping that from the equations governing the dynamics of the strings they would be able to deduce and explain the properties of

quarks and gluons and all the elementary particles and forces. They called
this insanely grandiose program the Theory of Everything, or TOE.

But in the 1990s—after twenty years of hard work and innumerable
papers, dissertations, seminars, and conferences—string theorists had
to concede that their Theory of Everything was a theory of too much of
everything. The study of strings gave not just one theory, but five different kinds of theories, in each of which there were millions of possible
solutions of the string equations, each representing a different possible
world of physics. String theorists were at loss on how to choose amongst
all these possibilities—they were faced with the proverbial problem of
finding a needle in a (very large) haystack, and they were in hay well
over their heads. To get rid of this hay, Edward Witten, the leading string
theorist, proposed a unification of string theories, so that all the five theories could be regarded as different aspects of a single overarching theory,
called M-theory (where, Witten explained, "M stands for magic, mystery,
or membrane, according to taste").[12] In this new theory, the elementary
entities are not strings, but membranes, like tiny little handkerchiefs. The
dynamics of membranes are much more complicated than the dynamics
of strings, and string theorists fervently hope that this complication will
get rid of the hay in their haystack (as Big Brother might say, "Complication is Simplicity"). This radical proposal has breathed fresh life into the
failed Theory of Everything. Today, string theorists (now transmogrified
into membrane theorists) are again hard at work, laboring over membrane models with obsessive zeal.

The search for string/membrane theory has an uncanny resemblance
to Einstein's stubborn and futile search for his unified theory. After thirty
years of labor, unified theory and string/membrane theory have remained
just as elusive as Bigfoot. Born faulted Einstein, because " . . . he tried to do
without any empirical facts, by pure thinking. He believed in the power of
reason to guess the laws according to which God has built the world." The
fault of string/membrane theorists is much the same. With supreme arrogance, they imagine that they can guess a Theory of Everything by relying
on vague criteria of beauty and simplicity. So far, they have failed. Maybe
string theorists need to learn from Einstein's experience and bow to what
we might call the eight-year rule: It is possible to give birth to a beautiful
theory after eight years of gestation, but there is little hope for a live birth
after thirty years of unsuccessful labor.

IN A LETTER to Lorentz written in 1915, Einstein summarized his views on the kinds of mistakes committed by theorists:[13]

> There are two ways that a theorist goes astray
> 1) The devil leads him by the nose with a false hypothesis.
>   (For this he deserves our pity)
> 2) His arguments are erroneous and sloppy.
>   (For this he deserves a beating)

Einstein often made the first kind of mistake and sometimes the second (his views on clock synchronization and the Principle of Equivalence belong in the first category; his misbegotten proofs of the energy-mass relation and his series of missteps in the search for general relativity belong in the second; and his unified theories belong to both). So sometimes he deserves our pity, sometimes he deserves a beating, and sometimes a bit of both.[14]

But above all he deserves our admiration for his incredible skill in unwittingly exploiting his own mistakes. With his sure intuition and his deep mystical insights into the world of physics, he performed wonders of sleepwalking and he snatched victory from the jaws of defeat, again and again.

In a speech at a formal banquet honoring Einstein, Bernard Shaw compared him to the great conquerors in history and called him a maker of universes: "Napoleon and other great men of his type, they were makers of empire, but there is an order of men who get beyond that. They are not makers of empires, but they are makers of universes. And when they have made those universes, their hands are unstained by the blood of any human being on earth . . . Ptolemy made a universe, which lasted 1400 years. Newton, also, made a universe, which lasted 300 years. Einstein has made a universe, and I can't tell you how long that will last."[15] So far, Einstein's universe has lasted 100 years, and counting.

IN THE GARDEN of the US National Academy of Sciences, at the National Mall in Washington, DC, you can find the Albert Einstein Memorial. Dedicated in 1979, it displays a monumental impressionistic bronze sculpture of Einstein by the artist Robert Berks. At first sight, this sculpture seems all wrong. The head is much too large, out of proportion to

The Einstein Memorial in the garden of the National Academy of Sciences at the National Mall in Washington, DC. The sheet of paper in Einstein's hands displays $E = mc^2$ and his equations for the photoelectric effect and for the gravitational field. *(Courtesy of Dan Smith, copyright © Dan Smith)*

the body. It creates the impression that the sculpture represents a gigantic child, an impression that is reinforced by the clumsy posture of the body and the sloppy clothing. And yet, in this apparently clumsy and misproportioned sculpture, Berks has captured the essence of Einstein. Because a genius has much in common with a child. The genius—like a child—has a rebellious, questioning attitude and also an uncanny ability to look at things from a fresh perspective to achieve unexpected insights. As Baudelaire expressed it:

> Genius is no more than childhood recaptured at will, childhood equipped now with man's physical means to express itself, and with the analytical mind that enables it to bring order into the sum of experience . . .[16]

And, of course, children do make mistakes.

Notes

CPE stands for volumes in the series *The Collected Papers of Albert Einstein* published by Princeton University Press (Princeton, NJ, 1987–2006; ten volumes under various editors, with further volumes to come). Brackets [ ] indicate information intended for readers with mathematical expertise.

Prelude: "I will resign the game"

1. Einstein, *Relativity, The Special and the General Theory*, p. 27.
2. Tomalin and Hall, *The Strange Last Voyage of Donald Crowhurst*, p. 206.
3. Tomalin and Hall, op. cit., pp. 239 et seq. Tomalin and Hall, reporters for the *Sunday Times*, investigated the possible causes of Crowhurst's madness, and they eliminated chemical causes, such as drugs, alcohol, vitamin deficiency, or moldy tea. They concluded the cause was probably stress, "the loneliness, the hostile environment, the strain of lying" (op. cit., p. 234). My contention that Einstein's book was the trigger rests on the time coincidence—Crowhurst began to write irrational nonsense when he began his attempt to analyze and imitate Einstein.
4. Nichols, *A Voyage for Madmen*, p. 237.
5. Koestler, *The Sleepwalkers*, p. 335.

Chapter I: "A lovely time in Berne"

1. Einstein, in Schilpp, *Albert Einstein, Philosopher-Scientist*, vol. I, p. 14.
2. Neffe, *Einstein*, p. 411.
3. Calaprice, *The New Quotable Einstein*, p. 199.
4. Ibid., p. 202.
5. Ibid., p. 194.
6. Fölsing, *Albert Einstein, Eine Biographie*, p. 791.
7. Calaprice, op. cit., p. 226.
8. Litz and MacGowan, *The Collected Poems of William Carlos Williams*, p. 130.
9. French, *Einstein, A Centenary Volume*, p. 31.
10. French, op. cit., p. 31.
11. Fölsing, op. cit., p. 698.
12. Ibid.
13. Einstein, op. cit., p. 8.
14. Ibid., p. 14.
15. Ibid., pp. 2, 4.
16. Ibid., p. 16.
17. Schwarzenbach, *Das Verschmähte Genie*, p. 17.
18. Fölsing, op. cit., p. 87.
19. Ibid., p. 65.
20. Overbye, *Einstein in Love*, p. 43.
21. It was customary for German-speaking professors to address each other as "Herr," but students were expected to use "Herr Professor."
22. Schwarzenbach, op. cit., p. 49.
23. CPE1, p. 332.
24. CPE1, p. 285.
25. Rosenkranz, *Albert Einstein, privat und ganz persönlich*, p. 49.
26. Schwarzenbach, op. cit., p. 77.
27. Rosenkranz, op. cit, p. 55.
28. Michelmore, *Einstein: Profile of the Man*, p. 139.
29. By an odd twist, this paper was to become the most referenced of all of Einstein's writings, perhaps because it was the least controversial and was immediately accepted in the community of physicists and chemists.
30. Overbye, op. cit., p. 158.
31. Fölsing, op. cit., p. 282.
32. Pais, *'Subtle is the Lord . . .'*, p. 115.
33. Recent, more precise versions of the Michelson-Morley experiment do not use interferometry; instead, they use a comparison of the frequencies of standing waves in the two arms.
34. Pais, op. cit., pp. 116, 117.
35. Everitt, F., *Physics World* 19, 32 (2006).
36. The amount of the time dilation depends on the direction of travel relative to the rotation of the Earth. The number given assumes eastward travel.
37. Audoin and Guinot, *The Measurement of Time*, p. 94.

38. Pais, op. cit., p. 123.
39. Fölsing, op. cit., p.232.
40. Ibid., p. 233.
41. Ibid., p. 310.

Chapter 2: "And yet it moves"

1. Rowland, *Galileo's Mistake*, p. 253.
2. Santillana, *The Crime of Galileo*, p. 306 et seq.
3. Ibid., p. 312.
4. Galilei, *Dialogue Concerning the Two Chief World Systems*, p. xxvii.
5. Castiglione, *The Book of the Courtier*, p. 105.
6. Viviani, quoted by Drake, *Galileo at Work*, p. 19.
7. Galilei, *Two New Sciences*, p. 66. Galileo neglected to give the unit of weight for his musket ball. Some translators insert "a pound" instead of "an ounce." The former is perhaps more plausible.
8. Koestler, *The Sleepwalkers*, p. 368.
9. Ibid., p. 467.
10. You can easily make your own telescope by borrowing two sets of reading glasses (with convex lenses) from farsighted family members, preferably one strong set and one weak. Place the weaker lens at the front of your telescope, and the stronger at the back. Eyeglasses have been in use since the thirteenth century. It is puzzling that the invention of telescopes took 300 years.
11. Santillana, op. cit., p. 99.
12. A. M. Clerke, *Encyclopaedia Britannica*, 11th ed., vol. 11, p. 407.
13. Santillana, op. cit., p. 119.
14. Ibid., p. 121.
15. Koestler, op. cit., p. 464.
16. Ibid., p. 471.
17. Ranke, *History of the Popes*, vol. II, p. 371.
18. Galilei, *Dialogue Concerning the Two World Systems*, p. 538.
19. Ibid., p. 260.
20. Galilei, *Dialogue on the Great World Systems*, p. 199. This is the 1661 Salusbury translation; the Drake translation of this passage is faulty.
21. Galilei, *Dialogue Concerning the Two Chief World Systems*, p. 220.
22. Ibid., p. 132.
23. The inability of the uniform translational motion of the Earth to affect the behavior of the oceans can also be understood in terms of the relevant forces on a parcel of water. These forces are the gravitational attraction of the Earth, the Moon, and pressure forces. These forces and the response of the water to them do not depend in any way on the translational motion of the Earth.
24. The tides in Venice are small compared with the tides in the English Channel, but they are among the largest in the Mediterranean. The city of Venice maintains an excellent Web site with complete information about the tides (www2.comune.venezia.it/maree).
25. Galilei, op. cit., p. 536.

26. Ibid., p. 538.
27. Santillana, op. cit., p. 190. Italics added. The claim of "falsity" and the reference to "several arguments" suggest that Urban had perhaps not merely quibbled on points of theology.
28. Koestler, op. cit., p. 483.
29. Santillana, op. cit., pp. 192, 193.
30. Ibid., p. 195.
31. Ibid., pp. 255, 256.
32. Urban issued this decree as head of the congregation of cardinals of the Holy Office. Technically, the decree represented the joint views of Urban and the cardinals. But most of Urban's cardinals were his toadies, always ready to do his bidding; as the Venetian ambassador said of them ". . . the cardinals make it their only thought to speak in such a manner as might satisfy the pope" (Ranke, op. cit., vol. III, p. 341). There can be no doubt that the decree represented the wishes of the pope.
33. In fact, the commission had made very little progress, in part because of lethargy of its leaders, and in part because—despite repeated requests—it had not been granted access to the archives of the Inquisition. See the review by G. V. Coyne, S. J., in McMullin, *The Church and Galileo.*
34. Ratzinger, *A Turning Point for Europe? The Church in the Modern World—Assessment and Forecast,* p. 98.
35. Santillana, op. cit., p. 305.
36. Weissmann, *Galileo's Gout,* p. 20.
37. The right index finger and thumb as well as one of his vertebrae were also removed. The finger and thumb bones were lost, but the vertebra is preserved at the University of Padua.

Chapter 3: "If I have seen farther . . ."

1. W. J. Broad, *Science* **213**, 1341 (1981).
2. More, *Isaac Newton: A Biography,* p. 387.
3. Klawans, *Newton's Madness,* p. 38.
4. H. M. Taylor, *Encyclopaedia Brittanica,* 11th ed., vol. XIX, p. 589.
5. Ibid.
6. Ackroyd, *Isaac Newton,* p. 62.
7. Westfall, *Never at Rest,* p. 534.
8. Ibid.
9. Gribbin, *The Scientists,* p. 164.
10. Westfall, op. cit., p. 596.
11. White, *Isaac Newton,* p. 351.
12. Ibid., p. 268.
13. Westfall, op. cit., p. 143.
14. Gleick, *Isaac Newton,* p. 2.
15. Westfall, op. cit., p. 468.
16. White, op. cit., p. 217.
17. Newton, *Principia,* p. xvii.

18. R. Westfall, *Science* **179**, 751 (1973); see also Westfall, op. cit., p. 732 et seq.
19. Derham's value for the speed of sound is remarkably close to the modern value, 1130 ft/s. The actual source of the discrepancy between Newton's theoretical value of the speed of sound and experiment lies in a subtle effect that Newton failed to consider: In a sound wave, not only do the density and the pressure of air oscillate, but also the temperature [the oscillations are adiabatic, not isothermal, as Newton assumed]. This enhances the effective elasticity of air and increases the speed of sound, as was recognized by Laplace, who corrected Newton's calculations, but declared (perhaps tongue-in-cheek?) that the manner in which Newton had proceeded "was one of the most remarkable strokes of his genius" (Newton, *Mathematical Principles of Natural Philosophy*, vol. 2, Appendix by F. Cajori, p. 662).
20. Westfall, op. cit., p. 735.
21. The first two of these, and the calculation of the speed of sound, are discussed in detail in R. Westfall, *Science* **179**, 751 (1973), and also in Westfall, op. cit., p. 735 et seq.
22. Actually, Euler wrote Newton's Second Law as $2ma = F$, because he thought it would be more accurate to measure time (especially short time intervals) by means of the distance that a mass falls under the influence of standard gravity. In changing over to this new method for time measurement, he rescaled the time coordinate, introducing a factor of 2 in Newton's law.
23. Newton, *Mathematical Principles of Natural Philosophy*, vol. 1, pp. 20, 21.
24. De Haas-Lorentz, *H. A. Lorentz, Impressions of His Life and Work*, p. 102.
25. Ibid., p. 8.
26. Ibid., p. 89.
27. Ibid., p. 124.
28. Fölsing, *Albert Einstein, Eine Biographie*, p. 327.
29. CPE5, p. 301.
30. Lorentz, *The Theory of Electrons*, p. 230 and Note 72*.
31. Pais, '*Subtle is the Lord . . .*', p. 134.
32. Born, *Physics in My Generation*, p. 104.
33. For a discussion of this controversy, see R. Cerf, *Am. J. Phys.* **74**, 818 (2006).
34. Calaprice, *The New Quotable Einstein*, p. 228.

Chapter 4: "A storm broke loose in my mind"

1. CPE1, p. 282.
2. Levenson, *Einstein in Berlin*, p. 23.
3. Einstein, in Schilpp, *Albert Einstein, Philosopher-Scientist*, vol. I, p. 52.
4. Brian, *Einstein, A Life*, p. 61.
5. Fölsing, *Albert Einstein, Eine Biographie*, p. 179.
6. Ibid.
7. Einstein, *Relativity, The Special and the General Theory*, p. v.
8. Galison, *Einstein's Clocks, Poincaré's Maps*, p. 291.
9. This statement is included in many collections of Einstein quotations, but there seems to be no paper trail for it.
10. Dick, *Sky and Ocean Joined, The U.S. Naval Observatory*, p. 473.

11. Einstein, op. cit., p. 27.
12. Fölsing, op. cit., p. 228.
13. Hermann, Max Planck, p. 92.
14. Born, *The Born-Einstein Letters*, p. 156. See also Born, *Experiment and Theory in Physics*, p. 1.
15. Born, *The Born-Einstein Letters*, pp. 149, 156. The label "Jewish physics" was coined by the Nazis, with defamatory intent, and one might expect that Born and Einstein would react against it; but they make no objection to this label and readily adopt it in their letters.
16. www.lorentz.leidenuniv.nl/history/Einsteins_poem/Spinozas_Ethik.jpg, accessed December 2, 2007.
17. Sommerfeld, *Electrodynamics*, p. 212.
18. Sommerfeld assumed that Maxwell's equations are valid in all inertial reference frames and from this deduced the constant speed of light (Sommerfeld, *Electrodynamics*, Section 27). But this is a questionable approach, because Maxwell's equations might not be exactly valid.
19. Eddington, *The Mathematical Theory of Relativity*, p. 29.
20. See Aharoni, *The Special Theory of Relativity*, p. 2; and Bohm, *The Special Theory of Relativity*, p. 32.
21. By a curious twist of history, the new Doppler-shift experiments were direct descendants of the very first determination of the speed of light about three centuries earlier. In 1676, the Danish astronomer Olaf Römer, a contemporary of Newton's, had unwittingly exploited the Doppler shift to measure the speed of light. Astronomical observations of the moons of Jupiter had shown that the observed periods of revolution of the moons seemed slightly shorter when the Earth is moving toward Jupiter and slightly longer 6 months later, when the Earth is moving away from Jupiter. Römer correctly interpreted this as a change in the travel time of the light from the moon to the Earth: when the Earth is moving toward Jupiter, the light given off by the moon as it emerges from the shadow of Jupiter needs somewhat less time to intercept the Earth than the light given off during the previous revolution of the moon, and when the Earth is moving away from Jupiter, it needs somewhat more time. In essence, this is a Doppler shift, although Römer, two centuries before Doppler, did not call it that. From the measured apparent lengthening and shortening of the periods of revolution, Römer deduced that light takes 22 minutes to traverse the diameter of the Earth's orbit; expressed in metric units, this gives 230,000 klicks per second for the speed of light (too small by about 23 percent, but nevertheless an impressive achievement). This was the first quantitative measurement of the speed of light, and it gave the earliest evidence that light does not propagate instantaneously.

   Römer's method requires a measurement of the speed of the Earth, and some scientists contend that this merely shifts the burden of speed measurement and clock synchronization elsewhere, without coming to grips with it. But the Earth is in a circular orbit (or, more precisely, an elliptic orbit) of known size, and for such, the speed can be measured by merely timing the orbital *period* with a single (hypothetical) stationary clock, provided we assume (as Einstein did) that Newtonian mechanics is applicable for low-speed motion. [For more on the relationship between Newtonian mechanics and synchronization, see H. C. Ohanian, *Am. J. Phys.* **73**, 4 (2005).]

All the new Doppler-shift experiments of the 1960s and later incorporated Römer's basic ideas, although the new experiments were laboratory experiments, with sources and receivers of light installed in a laboratory on Earth, instead of Römer's arrangement, with a source on Jupiter's moon and a receiver on Earth. And the experimenters were not interested in the absolute value of the speed of light, but merely in a comparison of the one-way speeds of light in different directions.

22. More precisely, the required delay is half a *sidereal* day, that is, half of 23 hours 56 minutes.

23. Fölsing, op. cit., p. 688.

24. Ibid., p. 783.

25. See Jackson, *Classical Electrodynamics*, p. 507, and Rindler, *Essential Relativity*, p. 7.

26. Sommerfeld, *Electrodynamics*, Section 27.

27. [If photons have small, but nonzero, mass, then the speed of light depends on the photon energy, and light can have a speed much smaller than c if its frequency is very low. The second postulate then applies only to light of very high frequency.]

28. The engineer would have had to include a length contraction and a time dilation in his game.

29. Pais, *'Subtle is the Lord . . . '*, p. 173.

30. The modified Michelson-Morley experiment with arms of unequal lengths is known as the Kennedy-Thorndike experiment. Robertson combined the results of the Michelson-Morley experiment, the Kennedy-Thorndike experiment, and a time-dilation experiment to deduce the Lorentz transformation equations; he thereby placed relativity on an entirely empirical basis [H. P. Robertson, *Rev. Mod. Phys.* **21**, 378 (1949) and later publications listed in D. W. MacArthur, *Phys. Rev.* A **33**, 1 (1986)].

31. Recent versions of the Michelson-Morley experiment compared the frequencies of standing waves in the two arms of the interferometer, rather than the travel times of light. These recent versions of the experiment were operated at liquid-helium temperatures, to eliminate thermal disturbances, and they achieved precisions about 1000 times better than those achieved by Michelson and Morley. See Ohanian, *Classical Electrodynamics*, p. 176.

32. Einstein, in Lorentz et al., *The Principle of Relativity*, p. 63.

Chapter 5: "Motions of inanimate, small, suspended bodies"

1. F. H. Neville, *Encyclopaedia Britannica*, 11th ed., vol. 2, p. 874.

2. Weaver, *The World of Physics*, vol. I, p. 634.

3. [The product of Boltzmann's constant and Avogadro's number is the gas constant, $kN = R$.]

4. Lenard, *Über Relativitätsprinzip, Äther, und Gravitation*, p. 29.

5. Fölsing, *Albert Einstein, Eine Biographie*, p. 148.

6. Mason, *A History of the Sciences*, p. 501.

7. J. Bernstein, *Am. J. Phys.* **74**, 863 (2006).

8. Ibid.

9. Cercignani, *Ludwig Boltzmann, The Man Who Trusted Atoms*, p. 36.

10. CPE2, p. 171.

11. Fölsing, op. cit., p. 93.
12. CPE1, p. 278.
13. Ibid., p. 284.
14. Ibid., p. 289.
15. Fölsing, op. cit., p. 97.
16. Ibid.
17. CPE1, p. 212..
18. Fölsing, op. cit., p. 98.
19. Ibid.
20. CPE2, p. 44.
21. Ibid., p. 96n.
22. Ibid.
23. Ibid., p. 175.
24. Unbeknownst to Einstein, this method for the determination of molecular size had been considered before, but it was rejected as afflicted with too many uncertainties; see CPE2, p. 178n
25. [Toward the end of his calculation, he recognizes that the presence of a large number of sugar molecules "renormalizes" the flow velocity. Such renormalizations are familiar to theoretical physicists today, but they were a novelty then. Einstein may have been led to this insight by one of the mistakes in his calculation, which gave the (wrong) result that the viscosity decreases when there is only *one* sugar molecule; the renormalization ensures an increase of the viscosity when there are many.]
26. Seelig, *Albert Einstein, A Documentary Biography*, p. 107.
27. A list of Einstein's assistants and collaborators is given in the Appendix of Pais, *'Subtle is the Lord . . .'*
28. CPE2, p. 180.
29. Ibid., p. 181.
30. Ibid., p. 348n.
31. Ibid., p. 176.
32. [In a later paper on Brownian motion (CPE2, p. 498) Einstein concedes that for Stokes's law, the moving body must be large compared with water molecules, but he does not comment on the implication of this for his treatment of sugar molecules.]
33. CPE2, p.176n.
34. Pais, op. cit., p. 90.
35. CPE2, p. 235.
36. Ibid., p. 218.
37. Cercignani, p. 209.
38. Pais, op. cit., p. 95.

Chapter 6: "What is the light quantum?"

1. Hermann, *Max Planck*, p. 9.
2. Einstein, in Schilpp, *Albert Einstein, Philosopher-Scientist*, vol. I, p. 19.
3. The name was invented by Ehrenfest. The ultraviolet catastrophe is often thought to imply that the total amount of radiation is infinite, because the increasing intensity

extends to higher and higher frequencies. But, in practice, the possible frequencies for a blackbody cavity do not extend to infinity, because, at X-ray frequencies, the walls of the cavity become transparent, and thermal equilibrium cannot be achieved. Thus, the ultraviolet catastrophe does not extend toward infinite frequencies, and it does not imply that the total amount of radiation is infinite (which would be a nonsensical result).

4. In German: "Sein unvergleichliches thermodynamisches Feingefühl."
5. Planck, *Physikalische Abhandlungen und Vorträge*, vol. III, p. 125.
6. Ibid.
7. M. J. Klein, *Physics Today*, November 1966.
8. Hermann, *Max Planck*, p. 29.
9. CPE5, p. 602.
10. Hermann, op. cit., p. 35.
11. Einstein, op. cit., p. 44.
12. Jammer, *The Conceptual Foundations of Quantum Mechanics*, p. 54.
13. CPE5, p. 31.
14. CPE2, p. 151.
15. Newton, *Opticks*, Book III, Part I, Query 29.
16. Fölsing, *Albert Einstein, Eine Biographie*, p. 165.
17. CPE1, p. 59.
18. CPE2, p. 165.
19. CPE5, p. 37.
20. Schönbeck, *Albert Einstein und Philipp Lenard*, p. 9.
21. Pais, *'Subtle is the Lord . . .'*, p. 383.
22. Pais, op. cit., p. 384.
23. CPE5, p. 527.
24. Dürrenmatt, *Albert Einstein*, p. 35.
25. [In his paper, Einstein dismisses this issue in a single sentence. Immediately after adopting Wien's law for his calculation, he adds, ". . . keep in mind that our results are valid only within certain limits" (CPE2, p. 157). If we interpret this literally, Einstein is claiming that high-frequency radiation consists of quanta, but low-frequency radiation (where Rayleigh's radiation law applies) consists of waves. He never addresses this inconsistency.]
26. The mistake was finally spotted by M. Nauenberg; see letter in *Physics Today*, October 2005, p. 17.
27. K. Wali, *Physics Today*, October 2006, p. 46.
28. Wali, op. cit.
29. Pais, op. cit., p. 399.
30. Jammer, op. cit., p. 59.
31. Bethe, in Cook, *Faces of Science*, p. 32.

Chapter 7: "The argument is jolly and beguiling"

1. Lorentz, in Lorentz et al., *The Principle of Relativity*, pp. 23, 24, and Whittaker, *A History of the Theories of Aether and Electricity*, vol. II, p. 51.

2. Inspection of several dozen textbooks on relativity reveals that only a few give von Laue's proof of $E = mc^2$, and of these only very few mention von Laue by name.

3. CPE5, p. 31.

4. Ibid., p. 33.

5. In the early years of relativity, it was not known whether the formulas for the kinetic energy and momentum of a single *particle* would also be applicable to an *extended body*, or a system of particles. This is evidently true at low speeds, where Newtonian physics is valid, but it was unclear whether it would also be true at high, relativistic speeds. Thus, Ehrenfest and others speculated that the momentum and kinetic energy of a high-speed nonspherical body might depend on is orientation relative to its direction of motion. To deduce the result $E = mc^2$, Einstein first needed to prove that the mathematical form of the energy-momentum for an extended body is exactly the same as for a particle, at all speeds for the overall translational motion and also the individual motions of the particles in the body. But in his 1905 paper Einstein did not address this issue—he simply took this for granted. In several later papers, he made the same unwarranted assumption. [For details, see H. C. Ohanian, arXiv:0805.1400[physics.hist-ph].]

6. M. Planck, *Ann. d. Physik* **26**, 1 (1908); see footnote on p. 29.

7. Michelmore, *Einstein, Profile of the Man*, p. 46. Based on interviews with Einstein's son, Hans Albert.

8. Einstein, *Out of My Later Years*, p. 52.

9. Pais, *'Subtle is the Lord . . .'*, p. 149.

10. Miller, *Albert Einstein's Special Theory of Relativity*, p. 254.

11. Pais, op. cit., p. 150.

12. CPE2, p. 379.

13. CPE5, p. 50.

14. Hermann, *Max Planck*, p. 128.

15. In his series of papers, Einstein does not explain his motives, except to say that the last two derivations [*Bulletin Am. Math. Soc.* **41**, 223 (1935) and *Technion Journal* **5**, 16 (1946), also published in Einstein, *Out of My Later Years*, p. 112] have the advantage of avoiding the "formal machinery" of relativity (however, relativity enters by the back door, because these papers rely on the momentum carried by light pulses, which is a relativistic effect).

16. CPE2, p. 417; later, he proves that pressure contributes to the mass, CPE2, p. 469.

17. *The New York Times*, December 29, 1934.

18. CPE2, p. 472 and correction on p. 494.

19. M. Planck, *Ann. d. Physik* **26**, 1 (1908).

20. CPE5, p. 99.

21. Hermann, *Einstein*, p. 147.

22. Ibid., p. 152.

23. Fölsing, *Albert Einstein, Eine Biographie*, p. 340.

24. CPE8, p. 17. The editors of CPE naively assume that *Onkels* in Einstein's letter literally means "uncles," and they pedantically list the names of his two uncles in Berlin. But the word is often used in German in the sense of older, avuncular friends, which can be translated as "older fellows," and it probably refers to Haber, Nernst, and Planck.

25. Overbye, *Einstein in Love*, p. 146.

26. Ibid.

27. Fölsing. op. cit., p. 353.

28. Whittaker, *A History of the Theories of Aether and Electricity,* vol. II, p. 64 et seq.

29. Fölsing, op. cit., p. 283.

30. CPE5, p. 445.

31. Einstein, E*instein's 1912 Manuscript on the Special Theory of Relativity,* pp. 112–66 (also published in CPE4, pp. 65–101); in the earlier pages of this manuscript, he stubbornly repeats his first error on $E = mc^2$.

32. Ibid., p. 158.

33. CPE4, p. 569 and CPE4, p. 575. These two papers deal with energy and momentum in general relativity, but they treat the energy-mass relation in passing.

34. [That is, the energy-momentum is a four-vector.]

35. Einstein, op. cit., p. 158.

36. Lombroso, *The Man of Genius,* p. 25.

37. For an extended system with a changing energy, for instance, a molecule being pushed by an external force, $E = mc^2$ is not valid (except in an approximate sense, if circumstances permit us to treat the system as a point particle). In fact, there is no well-defined mass for such a system. [Technically: when the energy and the momentum are time-dependent, we can define an energy and a momentum in each inertial reference frame at any given instant of time, but these energy-momenta do not form a four-vector under Lorentz transformations, and they have no invariant "length" that can be regarded as the mass.]

38. CPE8B, p. 917.

39. H. Ives, *Journal Opt. Soc. Am.* **42,** 540 (1952). As a replacement for Einstein's supposedly erroneous proof, Ives offered a variant of Einstein's argument. This variant contains the same mistake as Einstein's original argument: it tacitly assumes that kinetic energy of a system is the same as that of a particle, for which no proof is offered.

40. Whittaker, op. cit., vol. II, pp. 51, 52.

41. Pais, op. cit., p. 168.

42. Mehra, *Einstein, Hilbert, and the Theory of Gravitation,* p. 86, letter to Eugene Wigner. Mehra actually used this explanation in connection with the discovery of the equations of general relativity, but it also applies here.

43. Cohen and Cohen, *A Dictionary of Modern Quotations,* p. 68.

44. French, *Einstein, A Centenary Volume,* p. 48.

45. Wheeler and Ford, *Geons, Black Holes & Quantum Foam,* p. 42. Only late in the war did the Germans try to build a graphite reactor; see Richelson, *Spying on the Bomb,* p. 58.

46. Segrè, *From X-rays to Quarks,* p. 58.

47. Joliet-Curie, in Weaver, *The World of Physics,* vol. III, p. 71.

48. Segrè, op. cit., p. 205.

49. There are several versions of this, with slightly different wordings. The version quoted is given at the Argonne National Lab Web site, www.anl.gov.

50. Segrè, op. cit., p. 58.

51. Hoffman, *Otto Hahn: Achievement and Responsibility,* p. 17.

52. Sime, *Lise Meitner,* pp. 234, 235. My take on this exchange of letters is the opposite of Sime's—she interprets the correspondence as indicating that Hahn did not recog-

nize the importance of his barium finding until *after* he received Meitner's response. This is clearly contradicted by the timeline, because two days *before* receiving Meitner's response Hahn wrote again, "We cannot suppress our results, even if perhaps they are physically absurd," and by that time he had already mailed a paper on the barium finding to *Naturwissenschaften*. In fact, the correspondence makes it clear that Meitner did not contemplate fission until Hahn asked her to think about it. Maybe if Meitner had stayed in Berlin, she would have preceded Hahn in thinking about fission—but that is pure speculation. In a letter written after Hahn received the Nobel Prize, Meitner herself conceded that she was not the discoverer of fission: "Surely Hahn fully deserved the Nobel Prize in chemistry. There is really no doubt about it. But I believe that Frisch and I contributed something not insignificant to the *clarification* of the process of uranium fission—how it originates and that it produces so much energy, and that was something very remote from Hahn . . ." (italics added; Sime, op. cit., p. 327)

53. Ibid., p. 236.
54. Ibid.
55. Fermi, in the discussion of fission in his *Nuclear Physics* shows a clear grasp of this distinction. He begins his discussion of fission by stating, "The most useful model for explaining fission is the liquid drop model . . . This model permits calculation of the change of potential energy when the nuclear drop suffers an ellipsoidal deformation from spherical shape . . ." (Fermi, *Nuclear Physics*, p. 164). [More generally, the energy is what appears in the Lagrangian and the Hamiltonian; all else follows from the dynamical equations.]
56. Sime, op. cit., p. 244.
57. Segré, op. cit., p. 208.
58. *The New York Times*, January 31, 1939.
59. There is a tangential mention of energy contributing to the inertial mass in his discussion of the theory of gravitation; see Einstein, in Schilpp, *Albert Einstein, Philosopher-Scientist*, vol. I, p. 65.

Chapter 8: "Suddenly, I had an idea"

1. CPE5, p. 152.
2. CPE 5, p. 84.
3. Hermann, *Einstein*, p. 149.
4. Fölsing, *Einstein, Eine Biographie*, p. 274.
5. Ibid., p. 285.
6. Ibid., p. 289.
7. Ibid., p. 453.
8. Ibid., p. 299.
9. CPE3, p. 316 et seq.
10. Fölsing, op. cit., p. 298.
11. A. Harris, *Physics Today*, November 2005, p. 12.
12. Fölsing, op. cit., p. 286.
13. Ibid., p. 285.
14. Ibid., p. 286.

15. H. Medicus, *Isis* **85**, 456 (1994).

16. Fölsing, op. cit., p. 301.

17. CPE5, p. 227.

18. Fölsing, op. cit., p. 304.

19. Ibid., p. 309.

20. Ibid., p. 316.

21. CPE5, p. 274n.

22. Fölsing, op. cit., p. 319.

23. Ibid., p. 324.

24. CPE5, p. 349.

25. J. Bernstein, *Am. J. Phys.* **74**, 870 (2006). My interpretation of *chopin*, which differs somewhat from Bernstein's, is taken from *N'ayons pas peur des mots: Dictionaire du Français Argotique et Populaire* (Larousse, Paris, 1988). Transcribed into regular French, the pun becomes *La bonne affaire de la polonaise*, which is nowhere near as clever.

26. CPE5, p. 345.

27. CPE5, p. 544. See the Preface for the explanation of my translation of *Häringsseele*.

28. Fölsing, op. cit., p. 343.

29. Fölsing, op. cit., p. 344.

30. Verne, *From the Earth to the Moon*, p. 285.

31. CPE4, pp. 492, 493.

32. [For what is probably the best solution of this synchronization problem, see Rindler, *Essential Relativity*, Section 2.16.]

33. In an accelerated reference frame, say, an accelerated train, the round-trip slow transport of a clock alters its synchronization, as it does in the case of round-trip slow transport in a gravitational field. This means it is meaningless to speak of a distant synchronization of clocks in these cases. In practice, in the accelerated rotating reference frame of Earth, metrologists make use of an auxiliary nonrotating reference frame to synchronize clocks; see Audoin and Guinot, *The Measurement of Time*.

34. Pais, *'Subtle is the Lord . . .'*, p. 180. Pais was as blithely unaware of the mistakes in this paper as he was of the mistakes in the 1905 papers.

35. CPE3, p. 486.

36. Ibid., pp. 491–93.

37. Ibid., pp. 493, 494.

38. [For a quantitative calculation of the total light deflection by these means, we have to imagine a sequence of parallel elevators, each with an acceleration equal to the component of the gravitational acceleration at right angles to the direction of the ray.]

39. Fölsing, op. cit., p. 320.

40. Wheeler, *A Journey into Gravity and Spacetime*, pp. 11, 12.

41. CPE4, p. 162. [Cognoscenti will recognize this statement as false. The correct condition (stated in Einstein's paper) is that the spacetime path has extremum proper time, not minimum length. But it is hopeless to explain this to the mathematically uninitiated, and "shortest length" is only readily comprehensible facsimile for extremum proper time.]

42. Pais, op. cit., pp. 211, 212.

43. Ibid., p. 210.

44. Ibid., p. 211.

45. Ibid., p. 212.

46. Einstein, *Relativity, The Special and the General Theory*, p. 63.

47. Synge, *Relativity, The General Theory*, p. ix.

48. CPE4, pp. 304, 324.

49. The argument was later refined and improved by Levi-Civita in an appendix to his book *The Absolute Differential Calculus*.

50. Grossmann seems to have been ignorant of the crucial identities of Bianchi, published some ten years before.

51. [He obtained the equation of motion (or the "geodesic" equation, although Einstein did not call it that) by imposing the extremum condition on the integral of *ds*. Pais, op. cit., p. 220, claims that Einstein did not find the geodesic equation. In fact, he did, but he did not write it in the conventional form, with Christoffel symbols.]

52. CPE4, p. 252n. [Why Einstein thought so is a mystery. It is true that if the only space-dependent component of the metric tensor is g_{00} (and the gravitational field is static), most of the velocity dependence in the second term of the geodesic equation $du^\mu / d\tau + \Gamma^\mu_{\alpha\beta} u^\alpha u^\beta = 0$ disappears, but there is still a residual velocity dependence in the proper time τ, so particles with different speeds in the radial direction still have (apparently) different accelerations, and Einstein's attempt to fix this by insisting that only g_{00} is space-dependent is nonsense. Actually, the velocity dependence contributed by the second term in the geodesic equation is an illusion that results from the parallel displacement of the velocity vector, not from a genuine difference in acceleration between two particles of different velocity at a given point. The equality of accelerations becomes immediately obvious in local geodesic coordinates, for which $\Gamma^\mu_{\alpha\beta} = 0$, regardless of which components of the metric tensor are space-dependent.]

53. CPE4, p. 337. [Grossman's proposed field equation $R^{\mu\nu} = $ constant $\times\ T^{\mu\nu}$ is inconsistent with the conservation condition $T^{\mu\nu}_{,\nu} = 0$ derived by Einstein in the first part of the paper. With the proposed field equation, the conservation condition implies $R^{\mu\nu}_{,\nu} = 0$, so the Bianchi identity implies $R_{,\nu} = 0$, and the trace of the field equation then requires $T_{,\nu} = 0$; integration of this differential equation yields $T = 0$, which is evidently *not* valid for an arbitrarily given energy-momentum tensor. The only way to avoid this snag is by imposing the condition $T = 0$ (which is valid for electric and magnetic fields and suggests the speculation that all matter consists of such fields) or by adopting the field equation $R^{\mu\nu} - \frac{1}{2} g^{\mu\nu} R = $ constant $\times\ T^{\mu\nu}$, as Einstein finally recognized three years later.]

54. Hermann, op. cit., p. 216. [Einstein's mistakes in his early attempts to find the field equations of general relativity have been analyzed in detail by J. Norton in Kox and Eisenstaedt, eds., *The Universe of General Relativity*, pp. 67–102.]

55. Levenson, *Einstein in Berlin*, says it was a red rose, not a white handkerchief; apparently he confuses Einstein with the *Rosenkavalier*.

56. CPE5, p. 569.

57. De Botton, in Cahill, *The Best American Travel Writing 2006*, p. 76.

58. CPE5, p. 574.

59. Hermann, *Max Planck*, p. 128.

60. CPE8, p. 17.

61. Fölsing, op. cit., p. 374.

62. CPE5, p. 457.
63. CPE5, pp. 570, 571.
64. CPE10, p. 23.

Chapter 9: "The theory is of incomparable beauty"

1. Fölsing, *Albert Einstein, Eine Biographie*, p. 380.
2. CPE5, p. 505.
3. CPE6, p. 20.
4. Ibid., p. 23.
5. Pais, '*Subtle is the Lord . . .*', p. 242.
6. Ibid., p. 239
7. CPE5, p. 418.
8. Fölsing, op. cit., p. 355.
9. CPE5, p. 547.
10. Ibid., p. 563
11. Thorne, *Black Holes & Time Warps*, p. 117.
12. Hermann, *Einstein*, p. 207
13. CPE5, p. 558.
14. Ibid., p. 560.
15. Ibid., p. 573.
16. Ibid., p. 574.
17. Ibid., p. 534.
18. Ibid., pp. 545, 546.
19. CPE10, p. 23.
20. CPE8A, p. 45n.
21. CPE8B, p. 974.
22. CPE8A, p. 44.
23. Ibid., p. 45.
24. Ibid., p. 47.
25. Ibid., p. 50.
26. Ibid., p. 51.
27. *Kladderadatsch*, November 12, 1911.
28. Fölsing, op. cit., p. 364.
29. CPE10, p. 25.
30. CPE8A, p. 116.
31. CPE8A, p. 91.
32. Ibid., p. 233.
33. Ibid., p. 206.
34. CPE6, p. 226.
35. Ibid., p. 215.
36. CPE8A, p. 207.
37. CPE6, p. 216.
38. More precisely, all the gravitating matter has an energy-momentum tensor of zero trace.
39. [In proper mathematical notation, the equation is $R_\mu{}^\nu - \frac{1}{2} \delta_\mu{}^\nu R = T_\mu{}^\nu$. Since it would be

futile to try to explain to the uninitiated why the unit tensor needs to be included as a factor in the second term on the left-hand side, I have omitted it, with apologies to Riemann and Ricci.]

40. CPE8A, p. 217.

41. Ibid., p. 244.

42. D. Hilbert, *Nachrichten von der Königlichen Gesellschaft der Wissenschaften zu Göttingen*, November 20, 1915, p. 405.

43. CPE8A, p. 195.

44. Ibid., p. 205.

45. Pais, op. cit., p. 260.

46. Bjerknes, *Anticipations of Einstein*, p. 22, includes several choice epithets, such as "foulmouthed syphilitic whoremonger" and "incestuous adulterer."

47. CPE8A, p. 145.

48. Ibid., p. 201.

49. L. Corry, J. Renn, and J. Stachel, *Science* **278**, 1270 (1997).

50. T. Sauer, *Archive for History of Exact Sciences* **53**, 529 (1999).

51. [Hilbert's theory assumes that matter is entirely in the form electromagnetic fields; for this kind of matter, the trace of the energy-momentum tensor is zero, and so is the Riemann scalar. This means that in Hilbert's equation $R_{\mu\nu} - \frac{1}{2} g_{\mu\nu} R = \text{constant} \times T_{\mu\nu}$, the R terms drops out, and the equation for the gravitational field then reduces to $R_{\mu\nu} = \text{constant} \times T_{\mu\nu}$, which is exactly what Einstein had proposed in early November (and later rejected). To obtain something new, Hilbert would have had to introduce matter with an energy momentum tensor of nonzero trace. But he was obsessed with the unification of electricity and gravitation, and his goal was to interpret all other forms of matter as electromagnetic and gravitational fields. Thus, his treatment of gravitation was merely an elegant rewrite, in the language of variational calculus, of Einstein's earlier paper. Hilbert could easily have obtained the full Einstein equations by applying his variational calculation to forms of matter other than electromagnetic fields, *but he did not.*]

52. Seelig, *Albert Einstein, A Documentary Biography*, p. 165.

53. CPE8A, p. 222.

54. H. A. Medicus, *Am. J. Phys.* **52**, 206 (1984), describes the spat between Einstein and Hilbert and mentions this apology.

55. Mehra, *Einstein, Hilbert, and the Theory of Gravitation*, p. 25.

56. Yourgrau and Mandelstam, *Variational Principles in Dynamics and Quantum Theory*, p. 93. [Hilbert thought the conservation law for the energy-momentum tensor could replace Maxwell's equations. Mathematically, this is partly true—it is possible (though awkward) to calculate the time evolution of the four-vector potential from the four differential conservation equations obeyed by the energy-momentum tensor. However, physically this makes no sense, because it can't be done without knowing the explicit form of the energy-momentum tensor, and that explicit form is determined by the field equations, and only by the field equations. Thus, Hilbert's scheme leads nowhere unless the field equations are known—and if they are, then the scheme is pointless.

The attempt to treat electromagnetism as an aspect of gravitation was resurrected in a different form by G. Y. Rainich (1925) and by J. A. Wheeler and C. Misner (1957) whose analysis established that Hilbert's attempt was incomplete. It is indeed possible to extract the electromagnetic equations from the gravitational field equations, but to do so requires several supplementary equations for the Riemann curvature tensor, because not every conceivable geometry can be attributed to the presence of an electromagnetic field.

Corry et al., op. cit., also assert that Hilbert's equations are not covariant (and this canard has been repeated uncritically by Isaacson in a recent biography). The fact is that Hilbert's variational equations are covariant, but he supplements them, correctly, by extra, noncovariant, coordinate conditions that are needed to make the solution unique, as is well known to anybody who has ever tried to construct a solution of the Einstein equations. And, in fact, Einstein's paper in *Annalen* does not even state the covariant form of the field equation, but proceeds directly to a noncovariant form, based on a noncovariant coordinate condition. Einstein and Hilbert adopted a very special choice of coordinates to simplify the equation, and indeed it is possible to simplify the equation in such a way, but today relativists prefer not to prejudice such a simplification. They prefer to adjust the coordinates for the best fit to the physical system, because that makes it easier to interpret the meaning of the solution of the equations. Coordinates are merely an accounting method, and the choice of accounting method is merely a matter of convenience.]

57. Einstein, in Schilpp, *Albert Einstein, Philosopher-Scientist*, vol. I, p. 31.
58. Eddington, *Space, Time and Gravitation*, p. v.
59. S. Chandrasekhar, *Am. J. Phys.* **47**, 212 (1079).
60. Elsenstaedt, *The Curious History of Relativity*, p. 218; quoted from J. Crelinsten, "The Reception of General Relativity among American Astronomers: 1910–1930," dissertation, University of Montreal, 1981.
61. Weyl, *Space-Time-Matter*, p. 227.
62. J. Crelinsten, *Historical Studies in the Physical Sciences* 14, 1 (1984).
63. Chandrasekhar, op. cit.
64. Tolman, *Relativity, Thermodynamics, and Cosmology*, pp. 254–57.
65. Clark, *Einstein, The Life and Times*, p. 336.
66. Einstein uses the title "*Allgemeine Relativitätstheorie*" for his theory; this is often mistranslated as *General Theory of Relativity*, but what is "general" is the relativity, not the theory. Eddington was one of the few English writers who did not make this mistake; he translated the title correctly as *General Relativity Theory*.
67. CPE6, p. 280.
68. [For instance, we might expect that the inertial effects we observe correlate with the direction of the mass concentration at the Galactic center, but neither Einstein's theory nor observation indicate any such directional dependence. As best we know, inertia is isotropic. Furthermore, the solution of Einstein's equation demands boundary conditions in the space surrounding a given mass, and the local inertial properties are largely determined by these boundary conditions, not by Einstein's equation per se.]
69. CPE6, p. 288.
70. Ehrenfest, *Collected Scientific Papers*, p. 328.
71. H. C. Ohanian, *Am. J. Phys.* **45**, 903 (1977). This examines several ways of detecting tidal effects in a freely falling reference frame.

72. French, *Einstein, A Centenary Volume*, p. 115.

73. www.esa.int/esapub/br/br209.pdf, accessed December 7, 2007.

74. Kopff, *Die Einsteinsche Relativitätstheorie.*

75. The very first textbook, *Space-Time-Matter*, was published in 1918 by Hermann Weyl, then a professor at the Zurich Polytechnic and later at Princeton University. Weyl gave a somewhat clearer presentation of the implications of Mach's notion than Einstein, but he uncritically accepted Einstein's Principle of Equivalence.

76. Eddington, *The Mathematical Theory or Relativity*, pp. 40, 41.

77. Synge, *Relativity, The General Theory*, p. ix.

78. Fock, *The Theory of Space, Time, and Gravitation*, p. 8.

79. CPE6, p. 291.

80. [For cognoscenti: Einstein's overtly stated requirement merely involved *covariance* of the equations; but the requirement he actually used implicitly was *invariance*. The distinction was made clear by J. Anderson in his *Principles of Relativity Physics*, but unfortunately the distinction is still often ignored in relativity textbooks. In essence, invariance requires that the equations be covariant *and* that this covariance be achieved by appropriate insertions of the metric tensor (and its derivatives) into the equations, *without* insertion of any other "auxiliary" quantities.]

81. CPE6, p. 292.

82. The older documents of the Bezirksgericht Zurich, such as Einstein's divorce documents, are now stored at the Irchel campus of the university, in the suburbs of Zurich.

83. His formidable and devoted secretary, Helen Dukas, who stood guard over all the Einstein papers after his death, told Abraham Pais, "In 1923, the entire 121, 572 Kronor was indeed transmitted to her [Mileva]." Pais naively accepted this statement (Pais, *'Subtle is the Lord . . .'*, p. 503). But Dukas is here either lying outright to protect the reputation of her late employer or confusing the transfer of the money out of Germany with transfer to Mileva. Details of Einstein's handling of the money are revealed in Schwarzenbach, *Das Verschmähte Genie*, pp. 161, 162.

84. Fölsing, op. cit., p. 819.

85. Schwarzenbach, op. cit., p. 188. Quoted from the weekly *Wir Brückenbauer*, November 19, 1965.

86. CPE8A, p. 205.

87. CPE5, p. 565.

88. Ibid., p. 597.

Chapter 10: "The world is a madhouse"

1. S. Chandrasekhar, *Am. J. Phys.* **47**, 212 (1979). Also quoted in the *Life of Arthur Stanley Eddington.*

2. Pais, *'Subtle is the Lord . . .'*, p. 305.

3. Ibid., p. 305

4. Chandrasekhar, op. cit.

5. Newton, *Opticks*, Query 1.

6. Pais, op. cit., p. 200.

7. In Eddington's calculation it makes no difference whether the "light particle" is treated

by Newtonian mechanics or by special-relativistic mechanics (with a gravitational force proportional to the energy).

8. Chandrasekhar, op. cit.
9. Eddington, *Space, Time and Gravitation*, p. 115.
10. Ibid., p. 116.
11. J. Crelinsten, *Historical Studies in the Physical Sciences* **14**, 1 (1984).
12. Chandrasekhar, op. cit.
13. CPE8A, p. 225.
14. Ibid., p. 231.
15. Ibid., p. 293.
16. CPE7, p. 79.
17. Harrison, *Cosmology*, p. 280. Also Misner, Thorne, and Wheeler, *Gravitation*, p. 756, where a summary of cosmology is given.
18. The assertion is actually wrong. A static distribution of stars of the kind described by Newton is not in equilibrium; it will collapse inward. The static distribution merely corresponds to the instant of maximum expansion of a universe that has halted its expansion and is about to recollapse.
19. Pauli, *Theory of Relativity*, p. 180.
20. [This was shown by E. A. Milne and W. H. McCrea in 1934; see Bondi, *Cosmology*, Chapter IX, and Misner, Thorne, and Wheeler, op. cit, p. 759.]
21. CPE6, p. 543.
22. [Einstein was wrong in this. In any universe of uniform curvature, the mass distribution completely determines the geometry. Boundary conditions play no role in any of these universes, because the spacetime metric is determined to within a scale factor by the conditions of homogeneity and isotropy (according to the Robertson-Walker construction), and this scale factor is completely determined by the field equations. If Mach's principle is taken to mean that the mass distribution completely determines the geometry (which is debatable), then all these universes satisfy Mach's principle.]
23. CPE8A, p. 386.
24. Harrison, op. cit., p. 209.
25. Pais, op. cit., p. 288.
26. In response to Friedmann's results, Einstein contrived a proof that seemed to show that Friedmann was wrong [Zeitschr. für Physik, **11**, 326 (1922)]. Friedmann pointed out a calculation error in this proof, and Einstein conceded this [ibid., **16**, 228 (1923)], but he insisted that Friedmann's results lacked physical significance.
27. Gamow, *My World Line*, p. 150.
28. S. Weinberg, *Physics Today*, November 2005, p. 31.
29. Renn, *Albert Einstein, Ingenieur des Universums*.
30. CPE8A, p. 228.
31. CPE7, p. 138.
32. Pais, op. cit., p. 287.
33. Actually the factor is 10^{120}.
34. Overbye, *Einstein in Love*, p. 357.
35. Source unknown.
36. Ring formation by stars was first contemplated in a paper by O. Chwolson in 1924, and

later in a paper by Einstein. But this early work considered only ring formation by stars acting as deflectors, not galaxies. For stars acting as deflectors, the rings are too small to be observed, although we can see a blip in the image intensity when one star passes close to another.

37. Fölsing, op. cit., p. 496.
38. Ibid., p. 497.
39. CPE9, p. 213.
40. Chandrasekhar, op. cit.
41. *The Times,* November 7, 1919.
42. *The New York Times,* November 9, 1919.
43. CPE7, p. 214.
44. Ibid.
45. 800,000 dead, 1,700,000 wounded.
46. *Illustrirte Zeitung,* December 14, 1919.
47. Levenson, *Einstein in Berlin,* p. 228.
48. Calaprice, *The New Quotable Einstein,* p. 301.
49. Pais, *Einstein Lived Here,* p. 185.
50. *Kladderadatsch,* December 3, 1922.
51. Fölsing, op. cit. p. 513.
52. *L'echo de Paris,* April 8, 1922.
53. *Contact,* Spring 1921.
54. www.reelclassics.com/Movies/Casablanca/astimegoesby-lyrics.htm, accessed March 16, 2008.
55. Mostly from Calaprice, op. cit.
56. Pais, op cit., p. vi. See the Preface for an explanation of my translation of *raffiniert* in Einstein's German original. The German word is decidedly uncomplimentary, and we even have an example from Einstein's own hand that he understood the word to be uncomplimentary. In a letter to Elsa, he describes Mileva as "raffiniert und verlogen." The attempt to translate this as "subtle and mendacious" does not scan, whereas "cunning and mendacious" does.
57. Pais, op cit., p. vi.
58. Calaprice. op. cit., p. 229.
59. Fölsing, op. cit, p. 516.
60. Levenson, op. cit., p. 222.
61. Born, *The Born-Einstein Letters,* p. 38.
62. It might be doubted that Einstein was serious about this prohibition of the publication. He deliberately refrained from enforcing the prohibition by legal means, and he remained on good terms with Moszkowski, continuing to treat him as a friend.
63. Levenson, op. cit., p. 223.
64. CPE9, pp. 262, 263.
65. Fölsing, op. cit., p. 577.
66. The Princeton lectures of 1921 constitute the first four chapters of this book. Later editions of this book include in extra chapter on the expanding universe and an extra chapter on an attempt at a unified theory.
67. Hermann, op. cit., p. 309.

68. Pais, op cit., p. 310.
69. Fölsing, op. cit, p. 622.
70. Ibid., p. 480.
71. Ibid., p. 611.
72. Pais, op cit., p. 503.
73. Fölsing, op. cit., p. 614.
74. Ibid., p. 616.
75. CPE6, p. 212.

Chapter II: "Does God play dice?"

1. CPE7, p. 115.
2. Ibid., p. 345.
3. Ibid., p. 347.
4. Ibid., p. 345.
5. CPE10, p. 404.
6. Ibid., p. 419.
7. CPE7, p. 112.
8. Neffe, *Einstein, Eine Biographie*, p. 298.
9. Fölsing, *Albert Einstein, Eine Biographie*, p. 599.
10. Schönbeck, *Albert Einstein und Philipp Lenard*, p. 1.
11. Gehrcke, *Kritik der Relativitätstheorie*, p. 10.
12. Schönbeck, op. cit., p. 31.
13. Ibid., p. 33.
14. CPE7, p. 119. [This is actually a rather subtle example. Readers acquainted with physics might find it amusing to analyze how mechanical energy produced by the engine is expended to maintain the uniform motion of the countryside in the locomotive's reference frame.]
15. CPE7, p. 312.
16. Gehrcke, *Massensuggestion der Relativitätstheorie*.
17. A. Kleinert, in Renn, ed., *Albert Einstein—Ingenieur des Universums, Hundert Autoren für Einstein*, p. 226.
18. M. Wazeck, www.bpb.de/publikationen/IQ6OEE.html, accessed December 7, 2007.
19. Letter in *Scientific American*, February 1925.
20. Pais, *'Subtle is the Lord . . .'*, p. 171.
21. Galison, *Einstein's Clocks, Poincaré's Maps*, p. 297.
22. Pais, op. cit., p. 170.
23. Fölsing, op. cit., p. 229.
24. Zukav, *The Dancing Wu Li Masters*, p. 226.
25. Calaprice, *The New Quotable Einstein*, p. 194.
26. [The contraction of the electric fields of a moving charge is immediately obvious from the Liénard-Wiechert potentials, and it was known since the late 1890s.]
27. Lorentz could also have explained the time dilation by considering a periodic dynamic system, such as a point charge orbiting around another point charge. He had the necessary theoretical tools—he had obtained the correct expressions for the longitudinal

and transverse relativistic masses, that is, he knew the relativistic equation of motion. But he did not consider this problem.

28. [It is possible to examine the contraction of the longitudinal dimension of the orbit of an electron in the "old" Bohr-Sommerfeld quantum theory (although, as far as I know, this was never attempted). It would have been of limited use, because extrapolation from this to the contraction of a rigid rod is questionable; a full quantum treatment of the rod is needed.]

29. W.F.G. Swann, *Rev. Mod. Phys.* **13**, 197 (1941). [Technically: the length of the ground state of the solid body in motion at a given speed is shorter than the length of a similar body at rest by the expected length-contraction factor.]

30. J. S. Bell, *Progress in Scientific Culture* 1, No. 2, summer 1976; reprinted in Bell, *Speakable and Unspeakable in Quantum Mechanics*. [Bell did his calculation by classical mechanics, by direct numerical integration. Instead of performing such a numerical calculation it is easier to appeal to a general result of perturbation theory, according to which a slow perturbation (low acceleration) leaves the adiabatic invariants (or the quantum numbers) of the state unchanged. Hence it suffices to compare the lengths of identical orbital states at rest and in motion (as proposed by Swann).]

31. Many textbooks serve up the so-called light clock as an example of a more or less dynamical explanation of time dilation. This consists of two mirrors held at a fixed distance between which a light pulse "ticks" back and forth. But this is an incomplete example, because if the light clock is oriented longitudinally (instead of transversely), then its operation involves a length contraction of the distance between the mirrors, which, in this example, remains unexplained. Furthermore, it is a very limited, special example; it does not explain why the frequencies of atoms or of mechanical clocks suffer time dilation.

Because physical explanations of time dilation are not readily available in the physics literature, popularizers of relativity sometimes feel tempted to make up sophomoric explanations of their own, often with fallacious arguments. For instance, in the "popular" book *Einstein's Cosmos* by M. Kaku, the author imagines a police officer who chases after a light pulse with a very, very fast police car, so in the reference frame of the street, the police officer is speeding just behind the light pulse, traveling almost "neck and neck" with the light pulse. In the reference frame of the police car, the light pulse is still speeding away from the police officer at exactly the speed of light. To explain the relativistic time dilation, the author then says, "The only way to reconcile these two pictures is to have the brain of the officer slow down . . ." (p. 64). But this is obviously wrong, because it would not "reconcile the pictures" for a police car that is moving in the *opposite* direction, away from the light pulse. [In fact, in the relativistic addition law for velocities parallel to the direction of motion of the reference frames, the time-dilation factor plays no direct role (it cancels against the length-contraction factor), and the constant speed of light emerges entirely from the relativity of synchronization.]

The same mistake was made by the organizers of the exhibit *Einstein—Ingenieur des Universums* presented in Berlin in 2005 in honor of the "Einstein Year 2005." A film loop in this exhibit attributed the constant speed of light to the time dilation, as in Kaku's book. And this mistake was reproduced in the PBS NOVA program *Einstein's Big Idea*, shown in October 2005.

A different mistake was made by the mathematician J. Bronowski, the author of the TV series *The Ascent of Man* presented by the BBC in 1973 (later published as a book, under the same title). Bronowski tried to explain time dilation by saying that when a traveler races away from a stationary clock, say, a clock on a church tower, the light signals that reach his eye from the clock face take more and more time to catch up, minute by minute, because they have to travel longer and longer distances; hence, if the traveler looks over his shoulder, and waits for the light signals to reach his eye, he will see the minute hand taking more than 1 minute to advance from one tickmark on the clock face to the next, which means the clock's time seems dilated. Bronowski is indeed correct in saying that the minute hand will seem to take more than 1 minute to advance from one tickmark to the next, but this is not a time-dilation effect, but merely a *Doppler shift* (if Bronowski had thought about a different traveler racing *toward* the clock, his argument would have led him to the conclusion that the clock's time is *contracted*, which would have given him a clue about his mistake).

32. Three of his papers of 1907 use *Relativitätsprinzip* in the title.
33. Lorentz uses this terminology in his first publication on the FitzGerald-Lorentz contraction, in a book entitled *Versuch einer Theorie der elektrischen und optischen Erscheinungen in bewegten Körpern* (1895).
34. Recent examples are Milvich, *The Fall of Einstein's Relativity* (2003) and Sachs, *Relativity in Our Time: From Physics to Human Relations* (1994); also M. Sachs, *Physics Today* **24**, 23 (1971).
35. Whittaker, *History of the Theories of Aether and Electricity*, vol. II, p. 40.
36. Born, *The Born-Einstein Letters*, p. 199.
37. Pais, op. cit., p. 168.
38. M. Born, *Helv. Phys. Acta, Suppl.* IV, 244 (1956).
39. Pais, op. cit., p. 172.
40. Pauli, *Theory of Relativity*, p. 5.
41. Segré, *Faust in Copenhagen*, p. 72.
42. Ibid., p. 262.
43. Ibid., p. 73.
44. Westfall, *Never at Rest*, p. 143.
45. Fölsing, op. cit., p. 623.
46. Bodanis, $E = mc^2$, p. 218.
47. Fölsing, op. cit., p. 700.
48. Ibid., p. 448.
49. Lohmeier and Schell, *Einstein, Anschütz, und der Kieler Kreiselkompass*, p. 118, letter from Anschütz to Einstein.
50. Fölsing, pp. 557, 558.
51. Ibid., p. 678.
52. French, *Einstein, A Centenary Volume*, p. 43.
53. Isaacson, *Einstein*, p. 370.
54. Some of Einstein's defenders claim that "maybe he needed the money" or that maybe we should allow Einstein the grandiloquent response of Walt Whitman: "Do I contradict myself? Very well, then I contradict myself." But this does not excuse Einstein. Whether it was need for cash, or greed for cash, or maybe just a love to tinker, Ein-

stein's flirtations with the military-industrial complex raise serious questions about his pacifist convictions.

55. Calaprice, op. cit., p. 170.

56. Pais, op. cit., p. 432.

57. Ibid., p. 433.

58. CPE2, p. 585.

59. Fölsing, op. cit., p. 551.

60. Pais, op. cit., p. 436.

61. Fölsing, op. cit., p. 664.

62. Born, Lecture, *Helv. Phys. Acta, Suppl.* IV, 244 (1956).

63. Fölsing, op. cit., p. 644.

64. Ibid., p. 661.

65. Ibid.

66. Born, *The Born-Einstein Letters*, p. 91.

67. Ibid., p. 82.

68. CPE6, p. 364 and also *Physik. Z.* **18**, 121 (1918).

69. Bohr, in Schilpp, *Albert Einstein, Philosopher-Scientist*, vol. I, p. 218.

70. Pais, op. cit., p. 445.

71. Fölsing, op. cit., p. 670.

72. Pais, op. cit, pp. 446, 447.

Chapter 12: "The graveyard of disappointed hopes"

1. M. Born, *Helv. Phys. Acta, Suppl.* IV, 244 (1956).

2. Smolin, *The Trouble with Physics*, p. 49.

3. [The two obvious defects in the equations that Einstein attempted to use in place of the Maxwell equations are that they are third-order differential equations (whereas the Maxwell equations are first-order) and that the new equations do not allow for electric charge, that is, they are equations without sources.]

4. Isaacson, *Einstein*, p. 509.

5. French, *Einstein, A Centenary Volume*, p. 46.

6. Bird and Sherwin, *American Prometheus: The Triumph and Tragedy of J. Robert Oppenheimer*, p. 46.

7. Stern, *The Oppenheimer Case*, p. 338.

8. Whittaker, *A History of the Theories of Aether and Electricity*, vol. I, p. 240.

9. CPE8B, p. 670.

10. Ibid., p.727.

11. Pauli, *Theory of Relativity*, p. 230.

12. CPE9, pp. 38, 39.

13. Pais, *'Subtle is the Lord . . .'*, p. 332.

14. Eddington, *The Mathematical Theory of Relativity*, p. 219. [Eddington proposed that the metric tensor be proportional to a tensor $R'_{\mu\nu}$ which is a combination of the Ricci tensor augmented by various terms involving Christoffel symbols and derivatives of the four potential. But this lacked physical motivation, and the metric tensor constructed in this way does not, in general, have the right signature.]

15. Pais, op. cit., p. 343.

16. French, op. cit., p. 47.

17. [In Eddington's parallel transport, two infinitesimal parallel-transported vectors form a closed parallelogram (but all its sides have different lengths!). In Einstein's parallel transport (with asymmetric Christoffel symbols), the two transported vectors do not form a closed parallelogram. In the Riemannian case, the vectors form a closed parallelogram, and the lengths of the sides are pairwise equal.]

18. Pais, op. cit., p. 343.

19. Ibid., p. 344.

20. Ibid.

21. Fölsing, *Albert Einstein, Eine Biographie*, p. 684.

22. The French mathematician Elie Cartan later reported that he had discovered this geometry years earlier, in 1922, *and told Einstein about it.* Fölsing, op. cit., p. 907n.

23. Pais, op. cit., p. 346.

24. Ibid.

25. Ibid.

26. *The New York Times*, February 3, 1929

27. Pais, op. cit., p. 343.

28. Ibid., p. 347.

29. Fölsing, op. cit., p. 732

30. Gamow, *Thirty Years That Shook Physics*, p. 193.

31. Isaacson, op. cit., p. 421.

32. Gamow, op. cit., p. 210.

33. Fölsing, op. cit., p. 758.

34. Ibid., p. 739.

35. Ibid., p. 743.

36. Ibid.

37. Schwarzenbach, *Das Verschmähte Genie*, p. 143.

38. Ibid., p. 154.

39. Fölsing, op. cit., p. 747.

40. Ibid., p. 749.

41. Ibid., p. 752.

42. Sime, *Lise Meitner*, p. 37.

43. Hermann, *Max Planck*, p. 86.

44. Friedrich, *Blood & Iron*, p. 397.

45. Hermann, op. cit., p. 113.

46. Brian, *Einstein, A Life*, p. 251.

47. Turnbull, *The Letters of F. Scott Fitzgerald*, Letter, June 3, 1920, to John Grier Hibben.

48. http://etcweb.princeton.edu/CampusWWW/Companion/physics_department.html, accessed March 24, 2008.

49. www.ias.edu/spfeatures/john_von_neumann/von-neumann-s-legacy, accessed December 15, 2007.

50. *The New York Times*, May 4, 1935.

51 . Pais, op. cit., p. 494; Isaacson, op. cit., p. 450.

52. D. Kennefick, *Physics Today*, September 2005, p. 43.

53. Ibid.

54. Isaacson, op. cit., p. 251.

55. This had already been established by Eddington in the 1920s, but it was forgotten and then rediscovered in the 1960s by the mathematician Martin Kruskal.

56. Fölsing, op. cit., p. 822.

57. E. Tretkoff, *APS News*, December 2005, p. 2.

58. Pais, op. cit., p. 350.

59. [Some of the equations that supposedly describe the electromagnetic fields in his unified theory are third-order differential equations, whereas all the Maxwell equations are first-order differential equations (see M. A. Tonnelat, *Einstein's Unified Field Theory*, pp. 37, 71). It is a real puzzle why Einstein imagined that these third-order equations were the correct description of electromagnetic fields, or why some people should have taken this trash seriously.]

60. Years later, John Wheeler showed how a distortion of the geometry of spacetime can mimic an electric charge. He imagined a wormhole with electric field lines that enter on one side of the wormhole and emerge from the other. An external observer, who cannot see what happens to these field lines inside the horizon, will perceive one end of the wormhole as a positive charge, and the other as a negative charge. But Einstein could not have had this picture in mind, because black holes and wormholes were unknown in his days. Later research showed that not only was the electrodynamic part of Einstein's unified theory incomplete, but the gravitational part was in conflict with the Principle of Equivalence. [When an exact solution for the electric and gravitational fields surrounding an electric charge was obtained by Kursonoglu in 1949, it turned out that—according to this solution—electric energy does not have the same gravitational interactions as other energy. This can be seen from the contribution that the electric charge makes to the g_{00} component of the metric tensor. In general relativity it is e^2/r^2, as expected from the energy density in the electric field; but in the unified theory it is e^4/r^4. See Tonnelat, *Einstein's Unified Field Theory*.]

61. Isaacson, op. cit., p. 514.

62. Einstein, *The Meaning of Relativity*, p. 163.

63. Born, *The Born-Einstein Letters*, p. 182.

64. Isaacson, op. cit., p. 538.

65. Ibid., p. 515.

66. Fölsing, op. cit., p. 792.

67. Michelmore, *Einstein, A Life*, p. 261.

68. Dyson, in Calaprice, *The New Quotable Einstein*, p. ix.

69. Isaacson, op. cit., p. 517.

70. Dyson, in Calaprice, op. cit., p. x.

Postmortem

1. Not counting books and not counting papers that are reviews, translations, republications, or corrections of earlier work.

2. Joyce, *Ulysses*, p. 188.

3. Koestler, *The Sleepwalkers*, p. 332.

4. Calaprice, *The New Quotable Einstein*, p. 18.
5. Einstein, in Schilpp, *Albert Einstein-Philosopher Scientist*, vol. I, p. 8.
6. Koestler, op. cit., p. 519.
7. Ibid., p. 325.
8. There is no indication in his collected papers that he responded to any of these criticisms at all.
9. Einstein published his final, revised edition of *Relativity, The Special and the General Theory* in 1952
10. Pais, *'Subtle is the Lord . . .'*, p. 131.
11. Details about this approach to Einstein's equations can be found in H. C. Ohanian and R. Ruffini, *Gravitation and Spacetime*, Sections 3.2 and 7.2.
12. Woit, Not Even Wrong, p. 155.
13. CPE8A, p. 88.
14. I disagree with J. Norton's view that mistakes in the search for general relativity were mainly in the first category. (J. Norton, in Howard and Stachel, eds., *Einstein and the History of General Relativity*, p. 151.) Einstein's missteps arose from a poor grasp of the meaning of the mathematical structures in spacetime, not from faults in his physical hypotheses. Hilbert's ability to compete with Einstein in the search for the field equations makes this perfectly clear.
15. Crawford, *The Annual of Bernard Shaw Studies*, vol. 15, pp. 233, 234.
16. Baudelaire, *Oeuvres complètes*, p. 552.

Bibliography

Ackroyd, P. *Isaac Newton.* London: Vintage Books, 2007.

Aczel, A. D. *God's Equation.* New York: Delta, 1999.

Aharoni, J. *The Special Theory of Relativity.* New York: Dover, 1985.

Anderson, J. L. *Principles of Relativity Physics.* New York: Academic Press, 1967.

H. Arzeliès. *Rayonnement et dynamique du corpuscule chargé fortement accéléré.* Paris: Gauthiers-Villars, 1966.

Audoin, C., and B. Guinot. *The Measurement of Time.* Cambridge, UK: Cambridge University Press, 2001.

Baudelaire, C. *Oeuvres complètes.* Edited by M. A. Ruff. Paris: Editions du Seuil, 1968.

Bell, J. S. *Speakable and Unspeakable in Quantum Mechanics.* Cambridge, UK: Cambridge University Press, 1987.

Bergmann, P. *The Riddle of Gravitation.* New York: Scribner's, 1968.

Biagioli, M. *Galileo, Courtier.* Chicago: University of Chicago Press, 1993.

Bird, K., and M. J. Sherwin. *American Prometheus: The Triumph and Tragedy of J. Robert Oppenheimer.* New York: Knopf, 2005.

Bjerknes, C. J. *Albert Einstein, the Incorrigible Plagiarist.* Downer's Grove, IL: XTX Inc., 2002.

———. *Anticipations of Einstein.* Downer's Grove, IL: XTX Inc., 2003.

Bodanis, D. *E = mc², A Biography of the World's Most Famous Equation.* New York: Berkeley, 2000.

Bohm, D. *The Special Theory of Relativity.* London: Rutledge, 1996.

Bondi, H. *Cosmology.* Cambridge, UK: Cambridge University Press, 1961.

Boorstin, D. J. *The Discoverers.* New York: Random House, 1983.

Born, M. *The Born-Einstein Letters.* New York: Walker and Co., 1971.

———. *Einstein's Theory of Relativity.* New York: Dover, 1962.

———. *Experiment and Theory in Physics.* Cambridge, UK: Cambridge University Press, Cambridge, 1943.

———. *Physics in My Generation.* New York: Springer Verlag, 1969.

Brachner, A., G. Hartl, and C. Sichau. *Abenteuer der Erkenntnis, Albert Einstein und die Physik des 20. Jahrhunderts.* Munich: Deutsches Museum, 2005.

Brian, D. *Einstein, A Life.* Hoboken, NJ: Wiley, 1996.

Bronowski, J. *The Ascent of Man.* Boston: Little, Brown, and Co., 1973.

Brougham, H. P., and E. J. Routh. *Analytical View of Sir Isaac Newton's Principia.* New York: Johnson Reprint Corp., 1972.

Cahill, T. *The Best American Travel Writing 2006.* New York: Houghton Mifflin, 2006.

Calaprice, A. *The New Quotable Einstein.* Princeton: Princeton University Press, 2005.

Campbell, L., and W. Garnett. *The Life of James Clerk Maxwell.* New York: Johnson Reprint Corp., 1969.

Castiglione, B. *The Book of the Courtier.* New York: Anchor Books, 1959.

Cercignani, C. *Ludwig Boltzmann, The Man Who Trusted Atoms.* Oxford: Oxford University Press, 1998.

Clark, R. W. *Einstein, The Life and Times.* New York: Avon Books, 1972.

Cohen, I. B. *The Birth of a New Physics.* New York: W. W. Norton & Co., 1985.

Cohen, I. B., and G. E. Smith, eds. *The Cambridge Companion to Newton.* Cambridge, UK: Cambridge University Press, 2002.

Cohen, J. M., and M. J. Cohen, *A Dictionary of Modern Quotations.* Hammondsworth, England: Penguin Books, 1971.

Cook, M. *Faces of Science.* New York: W. W. Norton & Co., 2005.

Crawford, F. D., ed. *The Annual of Bernard Shaw Studies.* University Park, PA: Pennsylvania State University Press, 1995.

De Haas-Lorentz, G. L. *H. A. Lorentz, Impressions of His Life and Work.* Amsterdam: North-Holland, 1957.

Dick, S. J. *Sky and Ocean Joined, The U.S. Naval Observatory.* Cambridge, UK: Cambridge University Press, 2003.

Drake, S. *Galileo at Work.* Chicago: University of Chicago Press, 1978.

Dukas, H., and B. Hoffmann. *Albert Einstein, The Human Side.* Princeton: Princeton University Press, 1979.

Dürrenmatt, F. *Albert Einstein.* Zurich: Diogenes Verlag, 1979.

———. *Die Physiker.* Zurich: P. Schifferli, 1962.

Dyson, F. *The Scientist as Rebel.* New York: New York Review of Books, 2006.

Eddington, A. *The Mathematical Theory of Relativity.* Cambridge, UK: Cambridge University Press, 1963.

———. *Space, Time, and Gravitation.* Cambridge: Cambridge University Press, 1968.

Ehrenfest, P. *Collected Scientific Papers.* Amsterdam: North-Holland, 1959.

Einstein, A. *Einstein's 1912 Manuscript on the Special Theory of Relativity.* New York: George Braziller, 1996.

———. *The Meaning of Relativity.* Princeton: Princeton University Press, 1955.

——. *Out of My Later Years*. Lanham, MD: Littlefield, Adams & Co., 1967.

——. *Relativity, The Special and the General Theory*. New York: Three Rivers Press, 1961.

Einstein, A., and L. Infeld. *The Evolution of Physics*. New York: Simon and Schuster, 1938.

Eisenstaedt, J. *The Curious History of Relativity*. Princeton: Princeton University Press, 2006.

Everitt, C.W.F. *James Clerk Maxwell*. New York: Scribner's, 1974.

Fahie, J. J. *Galileo, His Life and Work*. London: John Murray, 1903.

Fantoli, A. *Galileo, for Copernicanism and for the Church*, 2nd ed. Vatican City: Vatican Observatory Publications, 1996.

Feldhay, R. *Galileo and the Church: Political Inquisition or Critical Dialogue?* Cambridge, UK: Cambridge University Press, 1995.

Fermi, L. *Atoms in the Family*. Chicago: University of Chicago Press, 1954.

Festa, E. *Galileo, La lotta per la scienza*. Roma-Bari: Editori Laterza, 2007.

Feuer, L. S. *Einstein and the Generations of Science*. New York: Basic Books, 1974.

Feynman, R., R. B. Leighton, and M. Sands. *The Feynman Lectures in Physics*. Reading, MA: Addison-Wesley, 1963.

Finocchiaro, M. A. *The Galileo Affair*. Berkeley: University of California Press, 1989.

Fock, V. *The Theory of Space, Time, and Gravitation*. New York: Pergamon Press, 1964.

Fölsing, A. *Albert Einstein, Eine Biographie*. Frankfurt: Suhrkamp, 1993.

French, A. P. *Einstein, A Centenary Volume*. Cambridge, MA: Harvard University Press, 1979.

Friedrich, O. *Blood & Iron*. New York: Harper Perennial, 1996.

Fritsch, H. *An Equation that Changed the World*. Chicago: University of Chicago Press, 1994.

Galilei, G. *Dialogue Concerning the Two Chief World Systems*. Translated by S. Drake. New York: Modern Library, 2001.

——. *Dialogue on the Great World Systems*. Translated by T. Salusbury, edited by G. de Santillana. Chicago: University of Chicago Press, 1953.

——. *Two New Sciences*. Translated by S. Drake. Madison, WI: University of Wisconsin Press, 1974.

Galison, P. *Einstein's Clocks, Poincaré's Maps*. New York: W. W. Norton & Co., 2003.

Gamow, G. *My World Line; an informal biography*. New York: Viking, 1970.

——. *Thirty Years That Shook Physics*. Garden City, NY: Anchor Books, 1966.

Gehrcke, E. *Kritik der Relativitätstheorie*. Berlin: H. Meusser, 1924.

——. *Massensuggestion der Relativitätstheorie*. Berlin: H. Meusser, 1924.

Gleick, J. *Isaac Newton*. New York: Vintage Books, 2004.

Greene, B. *The Elegant Universe*. New York: W. W. Norton & Co., 1999.

——. *The Fabric of the Cosmos*. New York: Vintage Books, 2004.

Gribbin, J. *The Fellowship*. Woodstock: Overlook Press, 2005.

——. *The Scientists*. New York: Random House, 2004.

Guicciardini, N. *Reading the Principia*. Cambridge, UK: Cambridge University Press, 1999.

Harman, P. M., and E. E. Shapiro, eds. *The Investigation of Difficult Things*. Cambridge, UK: Cambridge University Press, 1992.

Harrison, E. R. *Cosmology*. Cambridge, UK: Cambridge University Press, 1981.

Heisenberg, W. *Physics and Beyond, Encounters and Conversations*. New York: Harper & Row, 1971.

——. *Physics and Philosophy*. New York: Harper & Row, 1962.

Hentschel, A. M., and G. Grasshoff. *Albert Einstein, "Jene glücklichen Berner Jahre,"* Bern: Stämpfli Verlag, 2005.

Herivel, J. *The Background to Newton's Principia.* Oxford: Clarendon Press, 1965.

Hermann, A. *Einstein.* Hamburg: Piper Verlag, 2004.

———. *Max Planck.* Hamburg: Rowohlt, 1973.

Hershman, D. J., and J. Lieb. *The Key to Genius.* Amherst, NY: Prometheus Books, 1988.

———. *Manic Depression and Creativity.* Amherst, NY: Prometheus Books, 1998.

Highfield, P., and P. Carter. *The Private Lives of Albert Einstein.* New York: St. Martin's Press, 1993.

Hoffman, K. *Otto Hahn: Achievement and Responsibility.* New York: Springer-Verlag, 2001.

Hoffmann, B. *Albert Einstein, Creator and Rebel.* New York: Viking, 1972.

Horgan, J. *The End of Science.* New York: Broadway Books, 1997.

Howard, D., and J. Stachel, eds. *Einstein and the History of General Relativity.* Boston: Birkhäuser, 1989.

Isaacson, W. *Einstein: His Life and Universe.* New York: Simon and Schuster, 2007.

Jackson, J. D. *Classical Electrodynamics,* 2nd ed. New York: Wiley, 1975.

Jammer, M. *Concepts of Mass in Classical and Modern Physics.* Mineola, NY: Dover Publications, 1997.

———. *The Conceptual Foundations of Quantum Mechanics.* New York: McGraw-Hill, 1966.

Jost, R. *Das Märchen vom Elfenbeinernen Turm.* Berlin: Springer-Verlag, 1995.

Joyce, J. *Ulysses.* New York: Random House, 1946.

Kaku, M. *Einstein's Cosmos.* New York: W. W. Norton & Co., 2004.

Klawans, H. *Newton's Madness.* New York: Harper & Row, 1990.

Koestler, A. *The Sleepwalkers.* New York: Grosset and Dunlap, 1963.

———. *The Watershed .* Garden City, NY: Anchor Books, 1960.

Kopff, A. *Die Einsteinsche Relativitätstheorie.* Leipzig: Gressner & Schramm, 1920.

Kox, A. J., and J. Eisenstaedt, eds. *The Universe of General Relativity.* Boston: Birkäuser, 2005.

Kuhn, T. S. *The Copernican Revolution.* New York: Vintage, 1957.

Lenard, P. *Über Relativitätsprinzip, Äther, und Gravitation.* Leipzig: Hirzel, 1921.

Levenson, T. *Einstein in Berlin.* New York: Bantam Books, 2003.

Levi-Civita, T. *The Absolute Differential Calculus.* New York: Dover Publications, 1977.

Litz, A. W., and C. MacGowan. *The Collected Poems of William Carlos Williams.* New York: New Directions Publishing, 1991.

Livingston, D. M. *The Master of Light.* New York: Scribner's, 1973.

Lohmeier, D., and B. Schell, eds. *Einstein, Anschütz und der Kieler Kreiselkompass.* Kiel: Raytheon Marine, 2005.

Lombroso, C. *The Man of Genius,* 2nd ed. New York: Scribner's, 1905.

Lorentz, H. A. *The Theory of Electrons.* New York: Dover Publications, 1952.

Lorentz, H. A., A. Einstein, H. Minkowski, and H. Weyl. *The Principle of Relativity.* New York: Methuen, 1923.

Machamer, P., ed. *The Cambridge Companion to Galileo.* Cambridge, UK: Cambridge University Press, 1998.

McMullin, E., ed. *The Church and Galileo.* Notre Dame, IN: University of Notre Dame Press, 2003.

Mason, S. F. *A History of the Sciences.* New York: Collier Books, 1962.

Mehra, J. *Einstein, Hilbert, and the Theory of Gravitation.* Dordrecht: Reidel, 1974.

Michelmore, P. *Einstein, Profile of the Man.* New York: Dodd, Mead, 1962.

Miller, A. I. *Albert Einstein's Special Theory of Relativity.* Reading, MA: Addison-Wesley, 1981.

———. *Einstein, Picasso.* New York: Basic Books, 2001.

Milvich, B. *The Fall of Einstein's Relativity.* Basalt, CO: Basalt Printing and Publishing, 2003.

Misner, C. W., K. S. Thorne, and J. A. Wheeler. *Gravitation.* San Francisco: Freeman, 1973.

Monk, R. *Bertrand Russell, The Ghost of Madness.* New York: Free Press, 2001.

More, L. T. *Isaac Newton, A Biography.* New York: Scribner's, 1934.

Neffe, J. *Einstein, Eine Biographie.* Hamburg: Rowohlt, 2005.

Newton, I. *Mathematical Principles of Natural Philosophy and His System of the World.* Translated by A. Motte (revised by F. Cajori). Berkeley: University of California Press, 1962.

———. *Opticks.* New York: Dover Publications, 1952.

———. *The Principia.* Translated by I. B. Cohen and A. Whitman. Berkeley: University of California Press, 1999.

Nichols, P. *A Voyage for Madmen.* New York: Harper Collins, 2001.

Norton, J. D. *The Historical Foundations of Einstein's General Theory of Relativity,* Dissertation, University of New South Wales, 1981.

Ohanian, H. C. *Classical Electrodynamics.* Hingham, MA: Infinity Science Press, 2007.

Ohanian, H. C., and R. Ruffini. *Gravitation and Spacetime.* New York: W. W. Norton & Co., 1994.

Oppenheimer, J. R. *The Flying Trapeze: Three Crises for Physicists.* New York: Harper & Row, 1969.

Overbye, D. *Einstein in Love.* New York: Viking Penguin, 2000.

Pais, A. *Einstein Lived Here.* Oxford: Clarendon Press, 1994.

———. *Niels Bohr's Times.* Oxford: Clarendon Press, 1991.

———. *'Subtle is the Lord . . .'.* Oxford: Clarendon Press, 1982.

Pauli, W. *Theory of Relativity.* London: Pergamon Press, 1958.

Poincaré, H. *Science and Hypothesis.* New York: Dover Publications, 1952.

———. *Science and Method.* New York: Dover Publications.

Ranke, L. von. *History of the Popes.* New York: Colonial Press, 1901.

Ratzinger, J., Cardinal. *A Turning Point for Europe? The Church in the Modern World—Assessment and Forecast.* San Francisco: Ignatius Press, 1994.

Reichenbach, H. *The Philosophy of Space and Time.* New York: Dover Publications, 1957.

Redondi, P. *Galileo Heretic.* Princeton: Princeton University Press, 1987.

Renn, J., ed. *Albert Einstein—Ingenieur des Universums, Einstein's Leben und Werk im Kontext.* Weinheim: Wiley-VCH Verlag, 2005.

———. *Albert Einstein—Ingenieur des Universums, Hundert Autoren für Einstein.* Weinheim: Wiley-VCH Verlag, 2005.

Richelson, J. T. *Spying on the Bomb.* New York: W. W. Norton & Co., 2006.

Rigden, J. S. *Einstein 1905, The Standard of Greatness.* Cambridge, MA: Harvard University Press, 2005.

Rindler, W. *Essential Relativity,* 2nd ed. New York: Springer-Verlag, 1977.

Robinson, A., ed. *Einstein, A Hundred Years of Relativity.* New York: Abrams, 2005.

Rosenkranz, Z. *Albert Einstein, privat und ganz persönlich.* Bern: Historisches Museum, 2004.

Rowland, W. *Galileo's Mistake.* New York: Arcade Publishing, 2003.

Ryden, B. *Introduction to Cosmology.* San Francisco: Addison-Wesley, 2003.

Sachs, M. *Einstein vs. Bohr.* La Salle, IL: Open Court, 1988.

————. *Relativity in our Time: From Physics to Human Relations.* London: Taylor and Francis, 1994.

Santillana, G. de. *The Crime of Galileo.* Chicago: University of Chicago Press, 1955.

Seelig, C. *Einstein, A Documentary Biography.* London: Staples Press, 1956.

Segrè, E. *From X-rays to Quarks.* San Francisco: W. H. Freeman and Co., 1980.

Segré, G. *Faust in Copenhagen.* New York: Viking, 2007.

Seife, C. *Alpha and Omega.* New York: Penguin Book, 2003.

Schilpp, P. A. *Albert Einstein: Philosopher-Scientist.* New York: Harper & Row, 1959.

Schönbeck, C. *Albert Einstein und Philipp Lenard.* Berlin: Springer-Verlag, 2000.

Schwarzenbach, A. *Das Verschmähte Genie.* Munich: Deutsche Verlagsanstalt, 2005.

Schwinger, J. *Einstein's Legacy.* New York: Dover Publications, 1986.

Sime, R. L. *Lise Meitner.* Berkeley: University of California Press, 1996.

Simonton, D. K. *Origins of Genius.* New York: Oxford University Press, 1999.

Simonyi, K. *Kulturgeschichte der Physik.* Budapest: Akadémiai Kiadó, 1990.

Smart, J. J. C., ed. *Problems of Space and Time.* New York: Macmillan, 1964.

Smolin, L. *The Trouble with Physics.* New York: Houghton Mifflin, 2006.

Sobel, D. *Galileo's Daughter.* New York: Walker and Co., 1999.

————. *Longitude.* New York: Walker and Co., 1995.

Sobel, D., and W. J. Andrewes. *The Illustrated Longitude.* New York: Walker and Co., 1998.

Sommerfeld, A. *Electrodynamics.* New York: Academic Press, 1964.

Stern, P. M. *The Oppenheimer Case.* New York: Harper & Row, 1969.

Storr, A. *Churchill's Black Dog, Kafka's Mice, and Other Phenomena of the Human Mind.* New York: Grove Press, 1988.

Sulloway, F. J. *Born to Rebel.* New York: Pantheon Books, 1996.

Synge, J. L. *Relativity, The General Theory.* Amsterdam: North-Holland Publishing, 1971.

————. *Talking About Relativity.* Amsterdam: North-Holland Publishing, 1970.

Tauber, G., ed. *Albert Einstein's Theory of General Relativity.* New York: Crown Publishers, 1979.

Thorne, K. S. *Black Holes & Time Warps.* New York: W. W. Norton & Co., 1994.

Tolman, R. *Relativity, Thermodynamics, and Cosmology.* Oxford: Clarendon Press, 1934.

Tomalin, N., and R. Hall. *The Strange Last Voyage of Donald Crowhurst.* New York: Stein and Day, 1970.

Tonnelat, M. A. *Einstein's Unified Field Theory.* New York: Gordon and Breach, 1966.

Tuchman, B. *The Guns of August.* New York: Dell Publishing, 1962.

Turnbull, A., ed. *The Letters of F. Scott Fitzgerald.* New York: Scribner's, 1963.

Valentin, A. *The Drama of Albert Einstein.* Garden City, NY: Doubleday, 1954.

Verne, J. *From the Earth to the Moon.* New York: Dover Publications, 1960.

Waller, J. *Einstein's Luck.* Oxford: Oxford University Press, 2002.

Weart, S. R., and M. Phillips, eds. *History of Physics.* New York: American Institute of Physics, 1985.

Weaver, J. H. *The World of Physics.* New York: Simon and Schuster, 1987.

Weinberg, S. *Facing Up, Science and Its Cultural Adversaries.* Cambridge, MA: Harvard University Press, 2001.

Weissmann, G. *Galileo's Gout.* New York: Bellevue Literary Press, 2007.

Westfall, R. S. *Never at Rest.* Cambridge, UK: Cambridge University Press, 1980.

Weyl, H. *Space-Time-Matter*. New York: Dover Publications, 1952.

Wheeler, J. A. *A Journey into Gravity and Spacetime*. New York: Scientific American Library, 1990.

Wheeler, J. A., and K. Ford. *Geons, Black Holes & Quantum Foam*. New York: W. W. Norton & Co., 1998.

White, M. *Isaac Newton, the Last Sorcerer*. Reading, MA: Perseus Books, 1997.

White, M., and J. Gribbin, *Einstein, A Life in Science*. New York: Dutton, 1994.

Whittaker, E. T. *A History of the Theories of Aether and Electricity*. New York: Harper & Brothers, 1960.

Wickert, J. *Albert Einstein*. Reinbek bei Hamburg: Rowohlt Verlag, 1972.

Woit, P. *Not Even Wrong*. New York: Basic Books, 2006.

Woolf, H., ed. *Some Strangeness in the Proportion*. Reading, MA: Addison-Wesley, 1980.

Yourgrau, W., and S. Mandelstam. *Variational Principles in Dynamics and Quantum Theory*, 3rd ed. Philadelphia: Saunders Co., 1968.

Zukav, G. *The Dancing Wu Li Masters*. New York: Morrow and Co., 1979.

Index

Page numbers in *italics* refer to illustrations;
page numbers beginning with 339 refer to endnotes.